MESHLESS
METHODS
AND THEIR
NUMERICAL
PROPERTIES

MESHLESS METHODS
AND THEIR
NUMERICAL PROPERTIES

HUA LI

SHANTANU S. MULAY

CRC Press
Taylor & Francis Group
Boca Raton   London   New York

CRC Press is an imprint of the
Taylor & Francis Group, an **informa** business

CRC Press
Taylor & Francis Group
6000 Broken Sound Parkway NW, Suite 300
Boca Raton, FL 33487-2742

First issued in paperback 2017

No claim to original U.S. Government works
Version Date: 20121106

ISBN 13: 978-1-138-07231-2 (pbk)
ISBN 13: 978-1-4665-1746-2 (hbk)

Library of Congress Cataloging-in-Publication Data

Li, Hua, 1956-
   Meshless methods and their numerical properties / Hua Li, Shantanu S. Mulay.
     p. cm.
   Includes bibliographical references and index.
   ISBN 978-1-4665-1746-2 (hardback)
   1. Engineering mathematics. 2. Meshfree methods (Numerical analysis) I. Mulay, Shantanu S. II. Title.

TA335.L57 2013
518'.2--dc23
                                                   2012032569

Visit the Taylor & Francis Web site at
http://www.taylorandfrancis.com

and the CRC Press Web site at
http://www.crcpress.com

Dedicated first and foremost to my motherland,
and to Duer, Anne, and my parents

Hua Li

Dedicated to my parents, and Janaki, Kaustubh,
Supriti, and little angel Aadhip

Shantanu S. Mulay

Contents

Preface

For a long time, the finite element method (FEM) has been a standard tool for numerically solving a wide range of engineering problems. Today's real-world problems are becoming so highly complex that FEM alone is inadequate to solve them. Some of the limitations are listed below.

- It is not easy for FEM to generate a good quality mesh that should be correct according to the geometry and the specific requirements of a physical phenomenon. This problem will become more serious when we solve large deformation problems such as crack propagation, astrophysics phenomena, and extrusion.
- It is difficult for FEM to treat discontinuities properly as this process depends on mesh quality. Therefore, computational results may be incorrect due to high discontinuity if mesh is distorted.
- Adaptive meshing in FEM may cause degradation of accuracy in complex programs. It is impractical to solve systems of equations based on billions of elements.
- FEM is required to handle a lot of geometry degeneracy cases to generate correct and good quality mesh for complex geometry. This makes programs highly complex and slow while running.
- Mesh generation and mesh refinement in FEM are computationally expensive.

Due to these limitations of FEM, it is necessary to develop other numerical techniques that should be able to solve complex problems without generating mesh. A technique should also correctly handle governing equations by conserving essential parameters. As a result, the meshless or meshfree method is proposed as one such numerical technique to overcome these limitations.

Meshless methods have been developed in the past decade, and significant progress has been achieved recently for numerical computations of wide ranging engineering problems. These meshless methods do not require mesh for discretisation of problem domains, and they construct the approximate functions only via a set of nodes where no element is required for approximation of functions. They overcome the limitations of FEM. Several examples of the advantages of the meshless methods include

Computational cost is reduced significantly since no mesh is required.

Higher computational accuracy is achieved easily by simply adding nodes, especially for cases where more refinement is required.

High-order shape functions are constructed easily.

Compared with FEM, the meshless methods can easily handle large deformation and strongly nonlinear problems, since the connectivity among the nodes is generated as a portion of computation and it can change with time.

To date, about 15 books have been published on meshless methods. However, the authors focused primarily on methods that they developed. It is thus really necessary to publish a handbook type volume that provides the complete mathematical formulations for each of the most important and classic meshless methods that are well known and widely accepted and cover recent developments. It is also necessary to demonstrate a rigorous mathematical treatment of the numerical properties of meshless methods that will give sufficient confidence to users about the capabilities of particular meshless methods. This is especially important for readers who are interested in the individual meshless methods and seek full background information about all the most important and classic methods. This information will also be useful to an individual researcher who wants to embark on a journey of meshless method development.

A comprehensive introduction of the most important and classic meshless methods through complete mathematical formulations is thus warranted to provide overall insight into the meshless methods, theoretical understanding of the difference between FEM and meshless methods, and explanations of the detailed numerical computational characteristics of the methods. However, as noted, there is a lack of comprehensive publications on the formulations of the most important and classic methods. This monograph is thus written to systematically document the most important and classic meshless methods and the analyses of numerical properties.

In this book, the introduction for each of the most important and classic meshless methods is provided along with the complete mathematical formulations. In total, it presents 19 meshless methods, including the authors' recent contribution, in detail with full mathematical formulations and performance studies for the methods developed by the authors showing numerical properties such as convergence, consistency, stability, and adaptive analyses systematically.

Several engineering applications of the meshless methods are also included, for example, the CAD designing of MEMS devices, the nonlinear fluid structure analysis of near-bed submarine pipelines, and two-dimensional multiphysics simulations of pH-sensitive hydrogels.

This is the first monograph of its kind in which a comprehensive and systematic introduction of the most important and classic meshless methods is provided by complete mathematical formulations with full development information, although this is not the first book about meshless methods. It also covers the recent development of the meshless methods, mainly contributed by the authors. The methods are fully formulated mathematically and their numerical properties, such as convergence, consistency, stability, and adaptivity are studied in detail. Further, the benchmark results for engineering applications of the methods are also documented. Finally, this monograph is written in as simple a manner as possible so that it is informative and easy reading for researchers and can also serve as a rich reference source, for example, as a handbook for a graduate student who intends to work in the area of numerical computational techniques.

This monograph is intended to meet the needs of scientists and engineers in the broad areas of computational science and engineering. It will be especially useful for them as a reference book, and also if they wish to conduct further studies to extend their work to modeling and simulation of practical engineering problems. Another important primary audience is postgraduate students in the areas of computational theory, numerical methods, and discrete mathematics, especially those involved in developing new high-performance numerical methods. Possible secondary audiences include undergraduate students taking advanced numerical analysis courses covering discrete numerical analysis and methods. The chapters on the formulation of the selected classic meshless methods will be especially useful to these students. Correspondingly, course lecturers will also find this book a good reference source.

This book provides both casual and interested readers with insight into the special features and intricacies of meshless methods. It will also be invaluable to design engineers using CAD software for modeling and simulation of a wide range of engineering problems, serving as a useful reference containing benchmark formulations to compare and verify other numerical methods.

The authors would like to thank Professor Tom Hou of Caltech for his guidance in the stability analysis of the RDQ method, and Professors Khin-Yong Lam, Gui-Rong Liu, and J. N. Reddy for their constant support and encouragement. The authors would also like to thank J.Q. Cheng and Q.X. Wang for their invaluable contributions to this research.

Hua Li and Shantanu S. Mulay
School of Mechanical & Aerospace Engineering
Nanyang Technological University
Singapore

Authors

Dr. Hua Li earned BSc and MEng degrees in engineering mechanics from the Wuhan University of Technology, in the Peoples Republic of China in 1982 and 1987, respectively. He obtained his PhD in mechanical engineering from the National University of Singapore in 1999. From 2000 to 2001, Dr. Li was a postdoctoral associate at the Beckman Institute for Advanced Science and Technology, University of Illinois at Urbana-Champaign in the United States. In late 2005, he served as an invited visiting scientist in the Department of Chemical and Biomolecular Engineering of Johns Hopkins University. From 2001 to 2006, he was a research scientist at the A*STAR Institute of High Performance Computing.

Dr. Li is currently an assistant professor at the School of Mechanical and Aerospace Engineering at Nanyang Technological University. His research interests include the modeling and simulation of MEMS, focusing on the use of smart hydrogels in BioMEMS applications; the development of advanced numerical methodologies; and the dynamics of high-speed rotating shell structures.

He is the sole author of a monograph entitled *Smart Hydrogel Modeling* published by Springer. He has co-authored a book entitled *Rotating Shell Dynamics* published by Elsevier and book chapters on MEMS simulation and hydrogel drug delivery system modeling. He has authored or co-authored over 110 articles published in top international peer-reviewed journals. His research has been extensively funded by agencies and industries and he acted as the principal investigator of a computational BioMEMS project awarded under A*STAR's Strategic Research Programme in MEMS.

Dr. Shantanu S. Mulay earned a BEng in mechanical engineering from the Maharastra Institute of Technology of the University of Pune, India in 2001 and a PhD in mechanical engineering from Nanyang Technological University (NTU) in Singapore in 2011.

Since August 2010, Dr. Mulay has worked as a postdoctoral associate with Professor Rohan Abeyaratne of the Massachusetts Institute of Technology as part of the Singapore–MIT Alliance for Research and Technology (SMART).

Before joining NTU, Dr. Mulay worked in product enhancement of DMU (CATIA workbench) and the development of NISA (FEM product), where he gained exposure to a variety of areas such as the development of CAD translators, computational geometry, and handling user interfaces of FEM products. During his PhD program, Dr. Mulay worked extensively in the field of computational mechanics and developed a meshless random differential quadrature (RDQ) method. His work on the RDQ method has culminated in several chapters of this book.

Chapter 1

Introduction

1.1 BACKGROUND

One of the most important advances in the field of numerical methods was the development of the finite element method (FEM) in the late 1950s. In FEM, a continuum is divided into discrete elements that are connected together by a topological map and usually are called meshes. FEM is a robust and thoroughly developed method for computational science and engineering. However, FEM has several shortcomings, for example:

- Difficulty in meshing and re-meshing
- Low accuracy, especially for computing stresses
- Difficulties when dealing with certain class of problems
  - Simulating the dynamic crack growth with arbitrary paths that usually do not coincide with the original element mesh lines
  - Handling large deformation that leads to an extremely skewed mesh
  - Simulating the breakage of structures or components with large numbers of fragments
  - Solving dynamic contacts with moving boundaries
  - Solving multi-physics problems

A close examination of these problems associated with FEM will reveal their root causes, namely the use of elements or meshes. As long as the elements or meshes are used with FEM, the problems mentioned above will not have easy solutions. Therefore, the idea of getting rid of the elements and meshes is naturally proposed, such that the meshless or meshfree methods have been developed continuously for a long time.

Recently, the meshless methods have achieved remarkable progress, and many meshless methods have been developed. The studies of meshless methods have already become one of the hottest topics in the field of numerical simulation methodology of science and engineering problems. In brief, the meshless methods possess many advantages (Li and Liu, 2002). For example,

- They can save significant computation time because mesh generation is not required.
- They can easily solve large deformation and strong nonlinear problems because the connectivity among the nodes is generated as part of the computation and can be modified over time.
- High accuracy can be achieved easily, e.g., in the areas where more refinement is required, and nodes can be added quite easily.
- They can easily handle the damage of the components, such as fracture, which is very useful to simulate material breakage.
- They can easily solve the problems with requirement of multi-domains and multi-physics.

In general, the meshless methods can be grouped into two categories based on using or not using integration (Liu, 2003) or based on computational modelling (Li and Liu, 2002). The first category involves methods that do not require integration and are based on the strong forms of partial differential equations (PDEs), The second category includes meshless methods based on the weak forms of PDEs. In addition, a meshless method based on the combination of the strong form and weak form has also been developed and is known as the meshless weak–strong (MWS) form method.

In the meshless strong form methods, usually the PDEs are discretised at nodes by the collocation technique. This group includes the smooth particle hydrodynamics (SPH) method (Lucy, 1977; Gingold and Monaghan, 1977), the generalized finite difference method (Liszka and Orkisz, 1980; Liszka, 1984), the finite point method (Oñate et al., 1996a), the hp-meshless cloud method (Liszka et al., 1996), and the collocation method (Zhang et al., 2001). The meshless strong form method has several advantages: (1) it is simple to implement; (2) it is computationally efficient; and (3) it is a truly meshless method without any requirement to use a mesh for field variable approximation or integration.

Because of the above advantages, the meshless strong form methods have been successfully used in computational mechanics, especially for the problems of fluid mechanics. However, the meshless strong form methods also have obvious shortcomings. For example, they are often numerically unstable and less accurate, especially for problems governed by PDEs with Neumann boundary conditions.

The second category consists of meshless methods based on the weak forms of PDEs, including the use of the global weak form and the local weak form. The examples of meshless methods based on the global weak form, include the diffuse element method (Nayroles et al., 1992), the element-free Galerkin method (EFGM; Belytschko et al., 1994a and b; Belytschko et al., 1995a and b), the reproducing kernel particle method (RKPM; Liu et al., 1995; Liu et al., 1996; Liu and Jun, 1998), and the point interpolation method (PIM; Liu and Gu, 2001). In order to avoid global integration, the local weak forms are used for development of local meshless techniques such as the meshless local Petrov–Galerkin method (MLPG; Atluri, 2004; Atluri et al., 2006a and b). In addition, the meshless methods in combination with other numerical methods have also been studied (Gu and Liu, 2001).

The meshless methods based on the weak form have very attractive merits. First, they exhibit very good stability and excellent accuracy. The reason is probably that the weak form can smear the computational error over the integral domain and control the error level. The second benefit is that the Neumann boundary conditions can be naturally satisfied by the weak form, such that the stress boundary conditions are often called the natural boundary conditions in these methods.

The meshless weak form methods have been successfully employed in problems of solid mechanics. In particular, however, the meshless weak form methods mentioned above are "meshless" only in terms of the interpolation of the field variables, as compared with the usual FEM. Most of them have to use background cells or meshes, global or local, to integrate a weak form of the governing PDEs over the entire problem domain or the local integration domain. The numerical integration makes these methods computationally expensive, and the background mesh for integration makes them not truly meshless.

In addition, several other types of the meshless methods have also been proposed, such as the boundary node method (BNM; Mukherjee and Mukherjee, 1997), the variation of local point interpolation method (Li et al., 2004d), and others. The coupling methods between two meshless methods and between meshless and conventional methods, for example, the EFGM/FEM (Hegen, 1996) have also been developed.

The objective of this book is to introduce the basic concepts of the meshless method through the complete mathematical formulations of the classic methods that are well known and widely accepted. This book also aims to detail the recent development of the meshless methods contributed by the present authors and provide several examples of the engineering applications of meshless methods, for example, in the CAD design of MEMS devices and the nonlinear fluid structure analysis of near-bed submarine pipelines. It should be pointed out that several meshless methods have been developed to solve the specific engineering problems. This book will involve the most important, general, and classic meshless methods only, without covering all of them. For other meshless methods, readers can refer to the relevant papers and books.

1.2 ABOUT THIS MONOGRAPH

Several meshless methods have been developed to date. Based on the available literature, it is still difficult for a beginner to study this subject and for an advanced user to choose a specific meshless method for an application. The primary reason is that every meshless method has its own advantages and disadvantages from the formulation and application points of view. Some of the differences are subtle. Thus, the main objective of this monograph is to plug this gap by providing the complete mathematical formulations of several widely accepted classical and contemporary meshless methods. Along with the established meshless methods, several meshless methods developed by the authors are also presented. Their numerical analyses such as convergence, consistency, stability, and adaptivity, are discussed in detail. In this way a broad framework is developed that can guide future development of meshless methods.

Chapter 2

Formulation of classical meshless methods

2.1 INTRODUCTION

For a long time, the finite element method (FEM) has been a standard tool for numerically solving different engineering problems, especially those from the mechanical engineering. Today's real world problems such as astrophysics and diffusion phenomena are highly complex, such that FEM alone is inadequate to solve them. We now describe the several limitations of FEM.

One of the limitations of FEM is the difficulty of generating a good quality mesh that should be correct according to geometry and the specific requirements of a physical phenomenon. It is essential for a mesh to capture the order of continuity required by the underlying geometry. For example, it is not good for a linear triangular element to discretize the circular boundary that has a second order continuity; otherwise it will introduce an error. However, this problem becomes serious when a time-dependent large deformation is involved, such as crack propagation, astrophysics phenomenon, or extrusion is involved. Because the geometry is re-meshed after every time increment in solving these problems, it is possible for the newly generated mesh to contain distorted elements that may cause numerical problems during further computation.

Discontinuities in the domain such as cracks are difficult to treat by the FEM, resulting in poor quality of the mesh. When the mesh is distorted, the results are not always correct due to the presence of high discontinuities in the approximation of the field variables.

In general, the adaptive meshing in the FEM is known to improve the accuracy of the solution when the geometry is adequately simple and the distribution of field variables is sufficiently smooth and continuous. However, when the geometry becomes highly complex with the distribution of a field variable having a local high gradient, the robust adaptive algorithm may not fail but it may compromise the accuracy of the solution by generating a few distorted elements. Thus, the adaptive meshing in the

FEM may cause the degradation of accuracy and a complex program that may lead to the billions of elements that makes the solution of the system of equations very difficult.

In mathematics, a degenerated geometry is the limiting case, in which a particular class of object changes its nature so as to belong to a simpler class. For example, a circle with a radius equal to zero unit degenerates into a point, and a line with the length equal to zero unit degenerates into a point. Apart from these simpler cases, there are several other complex cases of degenerated geometry. If the degenerated geometry is not handled correctly during the generation of mesh, it may easily break the code of the FEM solver. Therefore, a lot of geometry degeneracy cases need to be handled while generating a good quality mesh for the complex geometry. This makes the implementation program of the FEM highly complex and slow during the execution.

Due to these limitations of the FEM, it is necessary to have a method that can solve complex problems without generating any mesh. The method should also correctly handle the governing differential equations with the appropriate boundary conditions.

2.2 FUNDAMENTALS OF MESHLESS METHODS

Various interpolation functions can be used in meshless methods that approximate the values of the field variables at the field nodes. The following are the important properties of a shape function when it is used in meshless methods.

1. It is required to satisfy the partition of unity condition as

$$\sum_{i=1}^{n} N_i(x) = 1 \tag{2.1}$$

where $N_i(x)$ is the value of the shape function at the ith field node.

2. It is required to satisfy the reproduction of linear field variable as

$$\sum_{i=1}^{n} N_i(x)\, x_i = x \tag{2.2}$$

which is the essential condition to pass a patch test in the FEM and a preferable condition for meshless methods.

3. It should possess the Kronecker delta function property, which is also a preferable condition, as

$$N_i(x_j) = \begin{cases} 1, & i = j \\ 0, & i \neq j \end{cases} , \quad \text{such that } i, j = 1, 2, \ldots, n \tag{2.3}$$

If the shape functions satisfy the delta function property, it is easy to impose the essential boundary conditions (BCs). However, even if it does not satisfy the delta function property, the essential BC can still be imposed with some special arrangements.

2.3 COMMON STEPS OF MESHLESS METHOD

The Figure 2.1 flowchart shows a general sequence of the working procedure of a meshless method and it is explained below step by step.

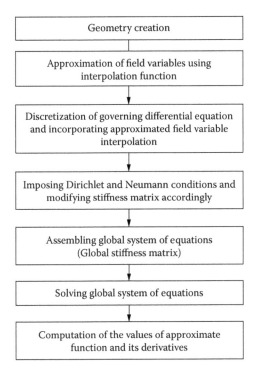

Figure 2.1 General flowchart of a meshless method.

2.3.1 Geometry creation

The geometrical model of the structure is created first. In order to obtain the coordinates in the computational domain, the randomly scattered or uniformly distributed field nodes are created in the domain and boundaries. The values of the field variables are computed at these field nodes. The density of the nodes depends on the accuracy requirement and the distribution of the field variables.

2.3.2 Approximation of field variable

A local domain is defined around each field node, where the surrounding nodes falling in this local domain are used for an interpolation of the field variable at the note of interest, using a suitable interpolation function as

$$f^h(x_I, y_I) = \sum_{i=1}^{n} N_i(x_I, y_I)\, u_i \tag{2.4}$$

where the approximate function value $f^h(x,y)_I$ at the Ith field node is interpolated by the surrounding field nodes $i = 1, 2, ..., n$, and $N_i(x_I, y_I)$ and u_i are the values of shape function and nodal parameter at the ith field node, respectively. The local interpolation domain may be created in any shape, such as a circle or rectangle, as shown in Figure 2.2.

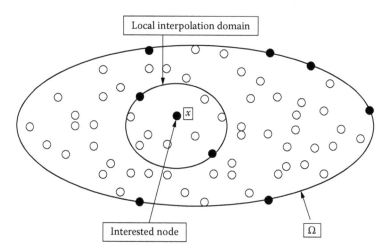

Ω **Whole computational domain**

Figure 2.2 Local domain for interpolation.

2.3.3 Discretisation of governing differential equation

In general, there are two approaches to handle the governing differential equation, namely the strong and weak forms.

2.3.3.1 Strong form approach

The derivative terms from the governing differential equations are discretised directly by this approach, and the system of equations is obtained in terms of the values of approximate function at the field nodes. It is often difficult to impose the essential BC exactly by the strong form approach, as the governing equation is satisfied at all the nodes in the internal domain and the boundary condition is satisfied at every boundary node. This makes the strong form approach very rigid, and sometimes is not workable during execution, but it is very easy to implement and may give accurate results if proper system equations are formed.

2.3.3.2 Weak form approach

The governing differential equation is converted to a corresponding weak form, and then solved by the numerical integration technique. The governing equations and boundary conditions are satisfied averagely over a domain instead of at individual nodes. This makes the weak form approach stable, but time consuming to implement because numerical integration is involved. In general, the numerical integration is performed by the Gauss quadrature with the Gauss points, and then the results are interpolated at the field nodes. This makes a further approximation of the values of function at the field nodes.

Whether to use a strong or weak form approach depends on how the meshless method is formulated. In general, if the problem is static, a system of algebraic equations is obtained after discretizing the governing equations. If the problem is transient, the temporal discretisation (explicit or implicit) is required by a suitable numerical method when the meshless method is employed.

2.3.4 Assembly of system of equations

A linear algebraic equation that is local in nature is obtained at every field node by the meshless method based on the strong or weak form approach. All the local equations are assembled together to form the global stiffness matrix and force vector ($K\,U = F$). At this step, the essential boundary conditions are imposed if the strong form-based meshless method is employed. If the interpolation function used in the meshless method does not possess the delta function property, the imposition of essential boundary conditions may be difficult. However, various techniques of imposing the essential boundary conditions are discussed in the work of Krongauz and Belytschko (1996).

2.3.5 Solving assembled system of equations

The final system of equations can be solved by either the direct or iterative methods.

2.3.5.1 Direct method

There are several direct methods to solve the linear system of equations, such as the Gauss elimination, LU decomposition, LDL^T decomposition, and Cholesky factorization.

2.3.5.2 Iterative method

There are also several iterative methods available, such as the Jacobi, Gauss-Seidel, conjugate gradient, and Krylov subspace methods. These methods involve repetitively solving a system of equations until the desired accuracy is achieved. In other words, the computational error in the solution is minimized by checking the residual value at the end of every iteration.

For static problems of displacement type, the displacements at every field node are obtained, and the strain and stress are computed by different relations. For free vibration or buckling problems, eigenvalues and eigenvectors are computed by the solver of the eigenvalues equation, such as Jacobi, QR, subspace iteration or Lanczos method. For dynamic problems, the modal superposition or direct integration method can be used by the implicit or explicit approaches in which time histories of the displacement, velocity, and acceleration are obtained.

In summary, different drawbacks of the classical FEM are discussed in this subsection, and we can see how they can be overcome by meshless methods. The fundamentals and common steps involved in meshless methods are studied. In the next subsection, a literature survey of various meshless methods is performed.

2.4 CLASSICAL MESHLESS METHODS

Although meshless methods have been a focus of research for the past 40 years, extensive studies were carried out continuously in the last decade when their potential to solve the large deformation and moving boundary problems was discovered. They are broadly categorized into two types of groups based on how the governing differential equation is solved, namely the strong (collocation-based) and weak forms (Galerkin approach-based).

Some of the earlier developed methods based on the weak form approach include smooth particle hydrodynamics (SPH; Lucy, 1977; Gingold and Monaghan, 1977), the diffuse element method (DEM; Nayroles et al., 1992), the element-free Galerkin method (EFGM; Belytschko et al., 1994a and b; Belytschko et al., 1995a and b), the natural element method (NEM; Braun and Sambridge, 1995), the reproducing kernel particle method (RKPM; Liu et al., 1995; Liu et al., 1996; Liu and Jun, 1998), the partition of unity-based FEM (PUFEM; Melenk and Babuška, 1996), the meshless local Petrov–Galerkin (MLPG) approach (Atluri and Zhu, 1998), the local boundary integral equation (LBIE) method (Zhu et al., 1998; Atluri et al., 2000), the point interpolation method (PIM; Liu and Gu, 2001), and the local Kriging method (Li et al., 2004c).

Most of these methods use the least square (LS) approximation or the reproducing kernel particle interpolation functions for the approximation of function, and they are combined with the Galerkin or variational weak form of the governing differential equation. Earlier developed methods based on the strong form approach include the finite point method (FPM; Oñate et al., 1996a and b), Hermite cloud method (Li et al., 2003), and the gradient smoothing method (GSM; Liu et al., 2008b).

All these methods were successfully applied to solve different types of engineering problems (Sukumar et al., 1998; Atluri et al., 1999a and b; Sukumar et al., 2001; Idelsohn et al., 2003; Li et al., 2004c; Li et al., 2005c; Liu et al., 2006; Gilhooley et al., 2008). In general, the meshless methods based on the weak form are considered numerically more stable than those based on the strong form. However, the strong form methods can capture the local high gradients well, and they are easy to implement and economical to compute.

Of all the meshless choices, several well established and classical methods are selected, and the corresponding formulations and working principles are discussed in detail.

2.4.1 Smooth particle hydrodynamics

The SPH method developed by Lucy (1977) and Gingold and Monaghan (1977) is considered an approximation for an integral determined as per the Monte Carlo procedure (Gingold and Monaghan, 1977). As well known, the Monte Carlo method can give a reasonable solution of multiple integrals with fewer field points than the finite difference method (FDM), such that it is feasible to reduce the computational work by employing the statistical smoothing methods like the Monte Carlo. Thus, at the heart of SPH lies the Monte Carlo method.

Lucy (1977) formulated and demonstrated the applicability of SPH for the two-dimensional (2-D) and three-dimensional (3-D) gas dynamical problems of astronomical interest, such as the fission problem for optically thick protostars. Let us consider that the function is approximated by an integral approximation as

$$\eta(r) = \int_V w(r - r')\, \xi(r')\, \rho(r')\, dV' \tag{2.5}$$

where $\rho \geq 0$, and $w(r - r')$ are window functions. As per the standard Monte Carlo theory, if a set of J points is randomly distributed in a volume V, and thus the probability of a point found in the volume element dV' at the distance r' is proportional to $\rho(r')\, dV'$, one can have

$$\tilde{\eta}(r) = \frac{1}{J} \sum_j w(r - r_j)\, \xi(r_j) \tag{2.6}$$

Here Equation (2.6) converges to $\eta(r)$ as $J \rightarrow \infty$. If we assume

$$\int_V w(r - r')\, dV' = 1, \text{ and } w = 0 \text{ for } |r - r_j| > \sigma \tag{2.7}$$

$\eta(r) \rightarrow \xi(r)\, \rho(r)$ as $\sigma \rightarrow 0$. This is followed by

$$\tilde{\eta}(r) \rightarrow \xi(r)\, \rho(r) \text{ as } J \rightarrow \infty \text{ and } \sigma \rightarrow 0 \tag{2.8}$$

In the work of Lucy (1977) concerning the simulation of the fission of optically thick protostars, the terms of spatial derivative from the fission governing equations are represented by the SPH method, as given in Equation (2.5). The problem is thus solved by applying the Monte Carlo theory to the discrete spatial representation to obtain the continuous spatial representation. If the window function w is approximated as a continuous function in the space, the mass density $\rho(r)$ is obtained from Equations (2.6) and (2.8) by setting $\xi(r) = 1$ as

$$\rho(r) = \frac{1}{J} \sum_j w(r - r_j) \tag{2.9}$$

Similarly, a function approximating the entropy density $s = \rho\, S$ is obtained by setting $\xi(r)$ equal to the specific density as $\xi(r) = S(r)$. Then

$$s(r) = \frac{1}{J} \sum_j S_j\, w(r - r_j) \tag{2.10}$$

Equations (2.9) and (2.10) will converge as $J \to \infty$ and $\sigma \to 0$. The window function w is chosen as a function of $|r - r_j|$ with the continuous second order derivative. For example, a function satisfying these conditions is given as

$$w = \begin{cases} \dfrac{105}{16\,\pi\,\sigma^3}\,(1+3\,z)\,(1-z)^3, & \text{for } z \le 1 \\[2ex] 0, & \text{for } z > 1 \end{cases} \tag{2.11}$$

where $z = (r/\sigma)$. Here Equation (2.11) is a bell-shaped function. The window function is required to have several properties, such as the normalization property $\int_\Omega w(x - x', h)\, dx' = 1$, namely the summation of window function over a local domain should be equal to 1; $\lim_{h \to 0} w(x - x', h) = \delta(x - x')$, namely the window function should approach zero when the smoothing length h approaches the zero. The window function should be compact, namely $w(x - x', h) = 0$ when $|x - x'| > k\,h$, where k is a constant related to the smoothing function, which defines the effective area of smoothing function. This effective area is nothing but the local support domain of point x.

The SPH method achieved classical status for the several particle-based methods developed later and it has several advantages. First, it is employed to obtain the continuous spatial representation from the spatially distributed discrete particles. Second, it uses the current spatial distribution of particles; thus the interconnectivity history of particles is not required. Third, the SPH method can be used in applications where it is necessary to capture long distance effects (nonlocal theories) at a specific location of interest (such as a crack tip) within the computational domain. As SPH is an approximate method, it is prone to some disadvantages if not applied correctly in a specific problem. First, it is sensitive to the number of particles used in the spatial approximation. Second, the accuracy of an approximation depends on the order of accuracy of the window function employed. Although the SPH method was initially proposed to simulate gas dynamical problems of astronomical interest, it can be applied in broad areas where the field variable distribution is continuously evolving in space with time.

2.4.2 Diffuse element method

The classical FEM has two main drawbacks. The first is that the derivatives are discontinuous across the element boundaries, while the approximated solution is continuous within the element. The second is that mesh generation with sufficiently good quality is a difficult task, especially for the

complex 3-D domains. Nayroles et al. (1992) proposed the DEM in order
to overcome these problems, in which the local polynomial-based interpolation in the FEM is replaced by a local weighted least squares (WLS) fitting that is valid only over the region around the concerned node $\bar{x}$.

The FEM uses the piecewise approximation of the unknown functions
over the element e as

$$u^e(x) = \sum_{j=1}^{m} p_j(x)\, a_j^e \tag{2.12}$$

where the vector p consists of m independent monomials, and a^e is a vector of m parameters that are constant over the element e. If the element e
totally has the n^e nodes, $u^e(x)$ takes the values of u_i, namely

$$\{u_i\} = [P_n]\{a^e\} \tag{2.13}$$

The matrix $[P_n]$ is invertible for $n^e = m$, leading to the standard expression
of shape function as

$$\{u_i\} = \langle p_j(x)\rangle\, [P_n]^{-1}\{a^e\} = \langle N_i(x)\rangle\{u_i\} \tag{2.14}$$

Equation (2.14) can also be obtained by the LS approach, by minimizing
the following expression with respect to a^e as

$$J = \sum_{i=1}^{n^e}\left[(u_i - u^e(x_i))\right]^2 = \sum_{i=1}^{n^e}\left[(u_i - \langle p_j(x_i)\rangle\{a^e\})\right]^2 \tag{2.15}$$

The minimization of Equation (2.15) gives an equation similar to that given
in Equation (2.14), if $n^e = m$. Equation (2.15) is used in the DEM with an
addition of a weighting function (WF) as

$$J = \sum_{i=1}^{n} w_i^e\left[(u_i - u^e(x_i))\right]^2 \tag{2.16}$$

where n is the number of total nodes in the local interpolation of the concerned field node $u^e(x)$. The total n^e nodes in an element are separated from
the total m monomial terms in an interpolation due to the use of DEM in

the FEM. The local nature of the function approximation is preserved as the WF w_i^e vanishes beyond certain distance. The stationarity of J with respect to a^e leads to

$$u^b(x) = \sum_{i=1}^{n} N_i(x) \, u_i, \quad \text{where} \ N_i(x) = \sum_{i=1}^{m} P_{xj}(x) \left[A^{-1}(x) \, B(x) \right]_{ji} \quad (2.17)$$

where

$$\left[A(x) \right] = \sum_{i=1}^{n} w_i^e(x) \, P_{ij} \sum_{k=1}^{m} P_{ik}, \ \left[B(x) \right] = \sum_{i=1}^{n} w_i^e(x) \, P_{ij} \quad (2.18)$$

The necessary condition to get the nonsingular $A^{-1}(x)$ matrix is the existence of m nodes at least in the local interpolation domain of $u^e(x)$. The approximate derivative of the function is computed by differentiating $p(x)$ with respect to x, by considering a^e as a constant, and given as

$$\left(\frac{\partial u}{\partial x} \right)_x = \left\langle \left(\frac{\partial p}{\partial x} \right)_x \right\rangle \{a^x\} \quad (2.19)$$

$$\therefore \left(\frac{\partial u}{\partial x} \right)_x = \left\langle \left(\frac{\partial N}{\partial x} \right)_x \right\rangle \{u_i\}, \text{where} \left\langle \left(\frac{\partial N}{\partial x} \right)_x \right\rangle = \left\langle \frac{\partial p}{\partial x} \right\rangle [A^x]^{-1} [B^x] \quad (2.20)$$

If w_i^e is continuous with respect to x, the shape functions are also continuous with respect to x, and the derivatives of the function up to the mth order exist. It is noted that, if the WF is constant over the domain, the diffuse approximation is similar to LS fitting of a polynomial function. If the WF is constant only over a fixed subdomain, the diffuse approximation may become identical with the finite element approximation. The FEM shape functions given in Equation (2.14) are replaced by the diffuse approximation shape functions given in Equation (2.17), while implementing the diffuse approximation in the FEM. The elements are not used for an approximation of the function, but they may still be used for the numerical integration over the domain.

There are several advantages offered by the DEM. First, some of the derivatives of function approximation are discontinuous in FEM at element boundaries. This is overcome in DEM by employing local weighted least squares polynomial fitting. Second, the DEM provides local but continuous distribution of function and its gradients. It is also observed by solving several mechanics problems that the gradients of the function computed by the DEM are better than those of the FEM (Nayroles et al., 1992). Third, it is easy to introduce DEM in the existing FEM programs.

There are few disadvantages of DEM as well due to the formulation. First, the weighting function should be continuous enough, as the feasibility of the function and its gradients is highly dependent on the choice of weighting function. Thus, the weighting function needs to be studied carefully before actually using it in the DEM program. Second, the function and its gradients are still approximated by an aggregation technique. There are several possible applications of the DEM. The existing FEM codes can be modified easily by the DEM, possibly to capture the local high gradients in the distribution of field variables. New plate elements can be created with DEM by building approximate functions subjected to the constraints. The DEM is especially attractive in the simulations of crack or damage propagation.

2.4.3 Element-free Galerkin method

This method was developed by Belytschko et al. (1994a and b, 1995a and b, 1996), by the MLS approximation to construct the trial and test functions for the variational principle in weak form. The several key points of this method are highlighted as

- The EFGM does not exhibit volumetric locking, even for the linear basis functions.
- The rates of convergence of the EFGM are significantly better than those of the FEM.
- The EFGM is quite effective for the problems of linear elastic fracture.

The notable differences between the EFGM and MLPG method are

- The essential boundary condition is imposed by the penalty method in the MLPG technique, while it is achieved by Lagrange multipliers in the EFGM.
- The variational weak form is used in the EFGM where the trial and test functions are approximated by the MLS approximation. The Petrov–Galerkin weak form is used in the MLPG method where the trial functions are approximated by the MLS approximation, and the test functions are approximated by the Gaussian weight function used in the MLS approximation.

As the weight function $w_i(x)$ plays an important role in the computational accuracy of the final solution obtained by the EFGM, it is chosen in such a way that it should give higher values for the nearer nodes and progressively lower values away from the concerned node. Therefore, the weight functions in the form of $w_i(x) = w_i(d^{2k})$ are studied, where $d = |x - x_i|$ is the distance between the two nodes x and x_i. The constant k is determined such that the first m derivatives of $w_i(d^{2k})$ with respect to d are continuous. In order to study the effects of different weighting functions in the EFGM, two weighting functions are considered, namely the exponential and conical functions. The mechanical equilibrium equation over the domain Ω is given as

$$\nabla \bullet \sigma + B = 0 \, (u = \bar{u} \text{ on } \Gamma_u \text{ and } \sigma \bullet n = \bar{t} \text{ on } \Gamma_t) \tag{2.21}$$

The variational weak form of Equation (2.21) is given as

$$\int_\Omega \delta(\nabla.v) : \sigma \, d\Omega - \int_\Omega \delta v.B \, d\Omega = \int_{\Gamma_t} \delta v.\bar{t} \, d\Gamma + \int_{\Gamma_u} \delta \lambda.(u - \bar{u}) \, d\Gamma + \int_{\Gamma_u} \delta v.\lambda \, d\Gamma \tag{2.22}$$

where $u \in H^1$ and $\lambda \in H^0$ are the trial and Lagrange multiplier functions, respectively, and δv and $\delta \lambda$ are the test functions for u and λ, respectively. The H^1 and H^0 are the Sobolev spaces of the degrees one and zero, respectively. The discrete form of Equation (2.22) is obtained by substituting the approximate solution u and the test function δv by Equation (2.17). The λ is replaced as

$$\lambda(x) = N_I(s) \, \lambda_I \text{ and } \delta\lambda(x) = N_I(s) \, \delta\lambda_I \tag{2.23}$$

where $N_I(s)$ and s are the Lagrange interpolant and the arc length along the boundary, respectively. In order to solve Equation (2.22), a cell structure is used and it is independent of the field nodes. Finally, the discrete form of Equation (2.22) is solved by $n_Q \times n_Q$ Gauss quadrature using the rule $n_Q = \sqrt{m} + 2$, where m is the number of total nodes in a cell.

EFGM offers several advantages. First, it eliminates the creation of elements in the classical FEM. Second, the postprocessing of strains and stresses is required in the FEM to obtain smooth field plots, whereas no such postprocessing in required in the EFGM as these fields are already smooth. Third, the performance of the EFGM seems to be independent of nodal point arrangement. Also, an incompressible material can be treated by EFGM without any modifications.

The method has several disadvantages as well. First, the accuracy of solution may be affected by irregularly placed nodes. Second, the performance of EFGM depends on the choice of weighting function. Third, the construction of underlying zones for the purposes of numerical integral quadrature can be an awkward and difficult task. The use of Lagrange multipliers to impose the Dirichlet boundary condition also complicates the solution process. Several test problems of mechanics such as bending, fracture, crack propagation, an infinite plate with a central hole, and steady-state heat conduction are solved successfully by the EFGM (Belytschko et al., 1994a and b, 1996), and very high rates of the convergence are achieved by the large number of quadrature points in the computational domain. If the EFGM is considered as a continuous form, it is identical to the free Lagrange method if the correct weight function is defined (Belytschko et al., 1994a). Thus, the EFGM is attractive for problems of crack propagation and fracture growth. It can also be well applied for problems in heat and elasticity.

2.4.4 Natural element method

The NEM proposed by Braun and Sambridge (1995) is a general Galerkin form of the weighted residual method, in which the natural neighbour coordinates are used as the geometrical trial functions. It is required to understand several concepts before proceeding further.

Any set of arbitrarily distributed nodes in a plane has a unique set of natural neighbours around each node. Any two nodes are said to be natural neighbours if their Voronoi cells (Braun and Sambridge, 1995; Sukumar et al., 1998, 2001) have a common boundary. The Voronoi cell about each node is a part of the plane closest to that node. For a given nodal distribution, the natural neighbours are uniquely determined.

Delaunay triangulation is obtained by connecting all pairs of natural neighbours together. Consider a test point x in the Voronoi diagram of the set N, as shown in Figure 2.3. If the point x is considered as a node along with the set of N nodes, the natural neighbours of x are those nodes that form an edge of a triangle with the point x in the new tessellation (triangulation). The natural neighbours of x are found to be the 1 to 5 nodes, as shown in Figure 2.3. The perpendicular bisectors from the point x to its natural neighbours are drawn, and the Voronoi cell T_x, namely the closed polygon of a, b, c, d, e and a, is obtained.

In the Sibson interpolation, the natural neighbour coordinates are used as the interpolating functions in the natural neighbour interpolation (Sukumar et al., 1998, 2001), and the area is taken as a measure of the natural neighbour coordinates for the 2-D domain. The natural neighbour coordinate $\phi_I(x)$ of the point x with respect to its natural neighbour I, where $\forall I: I \in [1 \text{ to } 5]$, is defined as the ratio of the overlap area $A_I(x)$ of the Voronoi cell T_I to the total area $A(x)$ of the Voronoi cell T_x around the point x as given by

$$\phi_I(x) = \frac{A_I(x)}{A(x)} \tag{2.24}$$

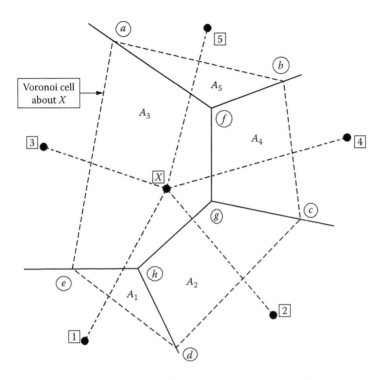

Figure 2.3 Computation of natural neighbour coordinates in the NEM.

where $A(x) = \sum_{I=1}^{5} A_I(x)$. The areas A_1 to A_5, as shown in Figure 2.3, are called the second order cells, and the area covered by the closed polygon, a, b, c, d, e and a, is called the first-order Voronoi cell. The ratio of the areas $\varphi_I(x)$ is taken as the shape function $N_I(x)$ in the NEM. For example, $N_3(x)$ is computed as

$$N_3(x) = \frac{A_3(x)}{A(x)} \tag{2.25}$$

where $A_3(x)$ is given by the polygon a, f, g, h, e and a. When the point x coincides with the node x_I, the shape function $N_I(x) = 1$ and the other shape functions are equal to zero. As a result, the shape functions constructed by the natural neighbour coordinates possess the property of the Dirac delta function $N_I(x_J) = \delta_{IJ}$ and the property of partition of unity $\sum_{I=1}^{5} N_I(x) = 1$

The shape functions in the NEM are computed by the natural neighbour coordinates as explained earlier, and the terms of approximate function from the weak form of the governing equation, as given in Equation (2.22), are substituted by the trial function $f^b(x) = \sum_i N_i(x - x_i) f_i$. The important difference between the NEM and FEM is that the function in the FEM is approximated within the element and it is discontinuous over the element boundaries, while the function in the NEM is approximated by the natural neighbour coordinates that are continuous and differentiable everywhere, except at the interested node itself. In order to compute the spatial derivatives in the NEM, a generalized and versatile approach was developed (Braun and Sambridge, 1995) that works with the n dimensions.

NEM presents several advantages over the classical FEM. First, the natural neighbour coordinates are used as trial functions; therefore no discontinuities are present across the element boundaries. Second, these trial functions are also orthogonal and isoparametric, just like the classical FEM, but they place no constraints on the positions of nodes.

There are several computational aspects to be considered, if the NEM is implemented. In the classical FEM, the weak form equation in the integral form is solved by the Gauss quadrature and sometimes gives accurate results, as the trial function is approximated by the polynomials. But, the trial functions are not simple polynomial expressions in the NEM and a trial and error approach is required when the integral equation is solved by the Gauss quadrature. Scalar quantities are normally evaluated and stored at the nodes of numerical mesh (material points), whereas tensors must be evaluated at the integration points inside the elements.

A problem arises in NEM for calculations involving tensor quantities (Cauchy stress) that need to be stored from one time step to the next one. As the triangulation is updated at each time step in NEM, the geometry of the element changes and a new set of integration points has to be defined. Thus, tensors known at the old integration points have to be interpolated or mapped onto the new ones. In classical FEM, node numbers are assigned to minimize the bandwidth of the stiffness matrix K in the linear system of equations $KU = F$. The node connectivity is more complex in NEM, such that it may not be always possible to optimize the node numbers that will minimize the bandwidth of stiffness matrix K. Furthermore, the NEM method involves several concepts from computational geometry, and thus the users must be knowledgeable about them before the NEM can be used. Typically, the NEM can be employed in problems involving the fluid structure interactions and solid mechanics (Sukumar et al., 1998).

2.4.5 Reproducing kernel particle method

The RKPM method is a discretised form of the reproducing kernel method (RKM) developed by Liu et al. (1995, 1996) using the wavelet theory and the convolution theorem. They studied the effects of discretisation in

Fourier analysis. The role of interpolation functions is as a low pass filter that reduces the amplitudes of signals with frequencies higher than the cutoff frequency. A dilation parameter, as found in the wavelet theory, is introduced in the interpolation function to extend the convolution theory to the broader field.

The convolution of two functions in the function domain (x) is equivalent to the multiplication of two functions in the Fourier transform domain (ξ). The conventional convolution is defined as

$$u^R(x) = u(x) * \phi(x) = \int_{-\infty}^{+\infty} u(y)\ \phi(x-y)\ dy \tag{2.26}$$

where $*$ is the symbol of convolution. The approximate function $u^R(x)$ in the domain Ω is given by the RKM as

$$u^R(x) = \int_{\Omega} u(\tilde{x})\ \overline{\phi}_a(x-\tilde{x})\ d\tilde{x} \tag{2.27}$$

The given function $u(x)$ is reproduced up to any desired order by choosing a suitable window function $\overline{\phi}_a$. The expansion of $u(x)$ about x by Taylor series is given as

$$u(\tilde{x}) = u(x) - (x-\tilde{x})\ u'(x) + \frac{1}{2!}\ (x-\tilde{x})^2\ u''(x) + \ldots + \frac{(-1)^n}{n!}\ (x-\tilde{x})^n\ u^{(n)}(x) \tag{2.28}$$

Substituting Equation (2.28) into (2.27) gives

$$\left.\begin{array}{l} u^R(x) = u(x) \int_{\Omega} \overline{\phi}_a(x-\tilde{x})\ d\tilde{x} - u'(x) \int_{\Omega} (x-\tilde{x})\overline{\phi}_a(x-\tilde{x})\ d\tilde{x} + \dfrac{u''(x)}{2!} \\[4mm] \int_{\Omega} (x-\tilde{x})^2\ \overline{\phi}_a(x-\tilde{x})\ d\tilde{x} + \ldots + \dfrac{(-1)^n\ u^{(n)}(x)}{n!} \int_{\Omega} (x-\tilde{x})^n\ \overline{\phi}_a(x-\tilde{x})\ d\tilde{x} \end{array}\right\} \tag{2.29}$$

The integral terms from Equation (2.29) are used to define the moment as

$$\overline{m}_k(a,\ x) = \int_{\Omega} (x-\tilde{x})^k\ \overline{\phi}_a(x-\tilde{x})\ d\tilde{x},\ k = 0,\ 1,\ \ldots,\ n \tag{2.30}$$

Rearranging Equation (2.29) by (2.30) leads to

$$u^R(x) = u(x)\,\bar{m}_0(a,\,x) - u'(x)\,\bar{m}_1(a,\,x) + \frac{u''(x)}{2!}\,\bar{m}_2(a,\,x) +$$

$$...+ \frac{(-1)^n u^{(n)}(x)}{n!}\,\bar{m}_n(a,\,x) \qquad (2.31)$$

In order to reproduce the function up to the nth order, the reproducing condition needs to be satisfied as

$$\bar{m}_k(a,\,x) = \delta_{k0}, \qquad k = 0,\,1,\,2,\,...,\,n \qquad (2.32)$$

The window function $\bar{\phi}_a(x - \tilde{x})$ is given as

$$\bar{\phi}_a(x - \tilde{x}) = P^T(x - \tilde{x})\,b(a,\,x)\,\phi_a(x - \tilde{x}) \qquad (2.33)$$

where

$$P(x - \tilde{x}) = \begin{bmatrix} 1 \\ (x - \tilde{x}) \\ \vdots \\ (x - \tilde{x})^n \end{bmatrix} \text{ and } b(a,\,x) = \begin{bmatrix} b_0(a,\,x) \\ b_1(a,\,x) \\ \vdots \\ b_n(a,\,x) \end{bmatrix} \qquad (2.34)$$

The $\bar{\phi}_a$ is an arbitrarily chosen window function that does not necessarily satisfy the reproducing conditions. The product of vectors $P^T(x - \tilde{x})$ and $b(a,\,x)$ is called the correction function $C_a(x;\,x - \tilde{x})$ to the window function, and the unknown coefficients $b(a,\,x)$ are computed by the reproducing condition. Substituting Equation (2.33) into (2.30) gives

$$\bar{m}_k(a,\,x) = \int_\Omega (x - \bar{x})^k \left[\sum_{k=0}^n (x - \tilde{x})^k\,b_k(a,\,x)\,\phi_a(x - \tilde{x}) \right] d\tilde{x} \qquad (2.35)$$

$$\therefore \bar{m}_k(a,\,x) = b_0(a,\,x)\,m_k(a,\,x) + b_1(a,\,x)\,m_{k+1}(a,\,x) + ... + b_n(a,\,x)\,m_{k+n}(a,\,x) \qquad (2.36)$$

where $m_k(a, x)$ is the kth moment of the original window function $\phi_a(x - \tilde{x})$. Equation (2.36) can be written in the matrix form as

$$
\begin{bmatrix}
\bar{m}_0(a, x) \\
\bar{m}_1(a, x) \\
\vdots \\
\bar{m}_n(a, x)
\end{bmatrix}
=
\begin{bmatrix}
m_0(a, x) & m_1(a, x) & \cdots & m_n(a, x) \\
m_1(a, x) & m_2(a, x) & \cdots & m_{n+1}(a, x) \\
\vdots & & \ddots & \vdots \\
m_n(a, x) & m_{n+1}(a, x) & \cdots & m_{2\,n}(a, x)
\end{bmatrix}
\begin{bmatrix}
b_0(a, x) \\
b_1(a, x) \\
\vdots \\
b_n(a, x)
\end{bmatrix}
$$

$$(2.37)$$

$$
\therefore \bar{m}(a, x) = M(a, x)\, b(a, x) \tag{2.38}
$$

where M is called the moment matrix and written as

$$
M(a, x) = \int_{\Omega} P(x - \tilde{x})\, \phi_a(x - \tilde{x})\, P^T(x - \tilde{x})\, d\tilde{x} \tag{2.39}
$$

Equation (2.32), giving the reproducing condition, is written in the vector form as

$$
\begin{bmatrix}
\bar{m}_0(a, x) \\
\bar{m}_1(a, x) \\
\vdots \\
\bar{m}_n(a, x)
\end{bmatrix}
=
\begin{bmatrix}
1 \\
0 \\
\vdots \\
0
\end{bmatrix}
= P(0) \tag{2.40}
$$

Combining Equation (2.40) with Equation (2.38) results in

$$
M(a, x)\, b(a, x) = P(0) \tag{2.41}
$$

The unknown coefficients $b(a, x)$ are obtained as

$$
b(a, x) = M^{-1}(a, x)\, P(0) \tag{2.42}
$$

Substituting Equation (2.42) into Equation (2.33) gives the window function $\bar{\phi}_a(x - \tilde{x})$, which leads to the evaluation of approximate function by the RKM, if it is further substituted in Equation (2.27). The discrete form of Equation (2.27) is the RKPM function (Liu et al., 1996), and given as

$$
u^R(x) = \sum_{I=1}^{NP} u(x_I)\, \bar{\phi}_a(x - x_I)\, \Delta v_I \tag{2.43}
$$

where NP represents the total particles in the discrete system and Δv_I is the nodal volume assigned to each particle. The window function is normalized by the dilation parameter a to reduce the noise in the solution as

$$\bar{\phi}_a(x - x_I) = \frac{1}{a} \bar{\phi}\left(\frac{x - x_I}{a}\right) \tag{2.44}$$

The discrete moment term is given as

$$\tilde{m}_k(a, x) = \sum_{I=1}^{NP} (x - x_I)^k \, \bar{\phi}_a(x - x_I) \, \Delta v_I \tag{2.45}$$

Equation (2.33) gives the originally chosen window function $\phi_a(x - \tilde{x})$. The RKPM shape function is given as

$$N_I(x) = C_a(x; x - x_I) \, \phi_a(x - x_I) \, \Delta v_I \tag{2.46}$$

and the approximate function is then given as

$$u^R(x) = \sum_{I=1}^{NP} N_I(x) \, u_I \tag{2.47}$$

Because of the addition of the correction function term in the RKPM, the field variables are almost accurately reproduced if the required order of the field variable distribution is included in the vector of polynomial basis. As the RKPM is inherited from the SPH method, it also inherited the SPH characteristics such as the smoothing of a function value over a local domain by the approximation of an integral function.

Aluru (2000) and Aluru and Li (2001) discussed several forms of the RKPM such as the fixed, moving, and multiple fixed forms. The main difference between them is the way the kernel or window function is handled. The kernel is fixed at the interested node in case of the fixed RKPM, where the approximate function is given as

$$f^b(x, y) = \int_{\Omega} C(x, y, u, v) \, K(x_k - u, y_k - v) \, f(u, v) \, du \, dv \tag{2.48}$$

where $C(x, y, u, v)$ is unknown correction function, $f^b(x, y)$ is an approximation of the function $f(x, y)$ at a node (x, y), and $K(x_k - u, y_k - v)$ is the kernel function fixed at the node (x_k, y_k). The order of approximate function $f^h(x,y)$ is determined by the order of the monomial basis functions used in the correction function $C(x,y,u,v) = P^l(u,v)\, c(x,y)$, where $P^T(u,v) = \{b_1(u,v),\ b_2(u,v),\ ...,\ b_m(u,v)\}$ is the column vector of the mth order monomial and $c(x,y)$ is the mth order unknown row vector. The moment matrix in the fixed RKPM is constant for a local domain due to the fixed kernel (Aluru and Li, 2001).

The approximate function in the moving kernel method is given as

$$f^h(x,y) = \int_\Omega C(x,y,u,v)\, K(x-u,y-v)\, f(u,v)\ du\ dv \qquad (2.49)$$

As seen from Equation (2.49), the moment matrix in the moving kernel technique is the function of x and y, and no longer constant. The approximate function by the multiple fixed kernel method is given as

$$f^h(x,\ y) = \int_\Omega C(x,\ y,\ s,\ t)\, K_{s,\,t}(s-x,\ t-y)\, f(s,\ t)\ ds\ dt \qquad (2.50)$$

It is observed from Equation (2.50) that the kernel is fixed at the node (s, t), which is in the local interpolation domain of the central node (x, y). The kernel is fixed at several nodes and therefore called the multiple fixed KPM.

Aluru and Li (2001) developed the finite cloud method by combining the fixed RKPM with the collocation method, and performed the convergence analysis. The approximate derivatives in the finite cloud method (2001) were computed by the diffuse derivatives approach (Nayroles et al., 1992). Aluru (2000) also developed a point collocation method based on the RKPM and performed the convergence analysis. He showed that the assignment of nodal volumes in the particle-based function approximation is not an issue with the point collocation method, but it is an issue in the Galerkin approach-based RKPM. Several variations of the RKPM can be adapted in the weak or strong form-based formulation while solving the problems of mechanics, elasticity, heat transfer, and microelectromechanical systems.

2.4.6 Partition of unity finite element method

The classical FEM often presents some difficulties during the solution of problems with rough or local oscillatory behaviour. In order to overcome this, the partition of unity finite element method (PUFEM) was proposed by Melenk and Babuška (1996) and Babuška and Melenk (1997). The prominent features of PUFEM include

- The ability to include *a priori* knowledge about the local behaviour of the field variable in the finite element space
- The ability to construct more suitable finite element spaces for the higher order governing equations
- Generalization of the classical *h, p,* and *hp* versions of the FEM.

In the classical FEM, the space of function approximation consists of piecewise polynomials that possess good properties of local approximation, and are conforming and continuous across the boundaries of the element. As a result, a function with good property of local approximation may lead to the good quality of solution by the FEM. There are many functions other than polynomials that have the property of local approximation. If the FEM space is constructed with these functions, the solutions of the differential equation are locally approximated very well. Therefore, the PUFEM offers the means to construct a conforming space from any given system of local approximation space without sacrificing the approximation properties.

Let $\Omega \subset R^n$ be an open set, $\{\Omega_i\}$ be an open cover of Ω satisfying a pointwise overlap condition, and $\{\phi_i\}$ be a Lipschitz partition of unity subordinate to the cover $\{\Omega_i\}$ satisfying

$$\text{supp } \phi_i \subset \text{ closure } \Omega_i \qquad \forall i \tag{2.51}$$

$$\sum_i \phi_i = 1 \text{ on } \Omega \tag{2.52}$$

$$\left\| \phi_i \right\|_{L^\infty(R^n)} \leq C_\infty \tag{2.53}$$

$$\left\| \nabla \phi_i \right\|_{L^\infty(R^n)} \leq \frac{C_G}{\text{diam } \Omega_i} \tag{2.54}$$

where C_∞ and C_G are constants. The $\{\phi_i\}$ is called a (M, C_∞, C_G) partition of unity subordinate to the cover $\{\Omega_i\}$. The partition of unity $\{\phi_i\}$ is said to be of the m degree, such that $m \in N_0$ if $\{\phi_i\} \subset C^m(R^n)$. The covering sets $\{\Omega_i\}$ are called patches.

The PUFEM method is nothing but generalized h and p versions if the spaces of local approximation are chosen as the spaces of polynomials. Several examples are discussed (Melenk and Babuška, 1996) for the construction and various choices of the local approximation spaces with better approximation properties than those of the polynomial of the degree p.

However, a proper choice of the partition of unity $\{\phi_i\}$ is required to put the given sets of local approximation spaces together to get a conforming global space. A general choice of the partition of unity is given here. Let $\{\Omega_i\}$ be the collection of overlapping patches that cover Ω, and $\{\psi_i\}$ be the functions that are supported by the patches $\{\Omega_i\}$. Using the normalization, we have

$$\phi_i = \frac{\psi_i}{\sum_j \psi_j} \tag{2.55}$$

Equation (2.55) gives the partition of unity subordinate to the cover $\{\Omega_i\}$, in which only those j are considered that satisfy $\Omega_i \cap \Omega_j \neq 0$. The function ϕ_i inherits the smoothness of ψ_i, such that this normalization technique gives one possible construction of the finite element spaces with a higher regularity.

The main advantage of the PUFEM is that it allows the use of nonpolynomial shape functions in the classical FEM formulation without sacrificing the conformity, i.e., interelement continuity. It also presents a few drawbacks. First, the choice of the basis of PUFEM space is tricky. A second drawback is the implementation of essential boundary conditions and the third is the integration of the elements of the stiffness matrix. The potential field of the application of PUFEM is where the classical polynomial-based FEM fails, such as problems involving rough or highly oscillatory behaviour of a field variable distribution.

2.4.7 Finite point method

In the FPM proposed by Oñate et al. (1996a and b), the WLS technique is used for interpolation of the data and the point collocation method is used to solve the governing differential equations. Several methods for point data interpolation such as the LS, WLS, MLS, and RKPM approximations were studied, and the corresponding shape functions were compared with the FEM (Oñate et al., 1996a). The methods of point data interpolation are sensitive to the number of points added in the interpolation. The FPM is finally devised by the WLS interpolation with the Gaussian weighting function and the point collocation method for the evaluation of governing equations.

In the point data interpolation methods, if Ω_k represents the interpolation domain for a function $u(x)$, and $j = 1, 2, ..., n$ interpolation points used such that $x_j \in \Omega_k$, the unknown function $u(x)$ is given as

$$u(x) = \sum_{i=1}^{m} p_i(x)\, \alpha_i \tag{2.56}$$

where $\alpha_i = [\alpha_1, \alpha_2, ..., \alpha_m]^T$ is a vector of unknown coefficients and $p_i(x)$ contains the monomials or basis functions. Equation (2.56) represents the model for the approximation of function. The function $u(x)$ is sampled at the n points such that $x_j \in \Omega_k$, then the approximate function is given as

$$u^b = \begin{Bmatrix} u_1^b \\ u_2^b \\ \vdots \\ u_n^b \end{Bmatrix} = \begin{Bmatrix} P_1^T \\ P_2^T \\ \vdots \\ P_n^T \end{Bmatrix} \alpha = C \, \alpha \tag{2.57}$$

For the FEM, $m = n$, C is thus a square matrix, and the unknown coefficients are obtained by

$$\alpha = C^{-1} \, u^b \tag{2.58}$$

For the LS method, $m \neq n$, C is thus not a square matrix, and the function approximation cannot fit all values u^b. This problem can be solved by minimizing the square of difference between the approximate function and its model equation in Equation (2.56) as

$$J = \sum_{j=1}^{n} \left[(u_j - p_j^T \, \alpha) \right]^2 \tag{2.59}$$

Minimization of Equation (2.59) results in

$$\alpha = \bar{C}^{-1} \, u^b \text{ with } \bar{C}^{-1} = A^{-1} \, B \tag{2.60}$$

where

$$[A(x)] = \sum_{j=1}^{n} P(x_j) \, P^T(x_j), \text{ and } [B(x)] = [p(x_1), \, p(x_2), \, ..., \, p(x_n)] \tag{2.61}$$

The LS method can be improved in a local region by weighting the squared distances in Equation (2.59), as shown in Equation (2.16), and it thus is called the WLS method. The weighting function w_i^e is chosen such that it takes the value of unity in the vicinity of the interested node and vanishes beyond the local region Ω_k. In the WLS method with $m \neq n$, the vector α of unknown coefficients should thus be determined by the minimization as given in Equations (2.17) and (2.18).

The weighting function w_i^e is translated over the domain in the MLS method such that it takes the maximum value at the point k with the coordinates x_k where the unknown function $u(x)$ is evaluated. The following function is thus minimized at the point k as

$$J = \sum_{j=1}^{n} \phi_k(x_j - x_k) \left[(u_j^h - p_j^T \, \alpha \right]^2 \qquad (2.62)$$

where ϕ_k changes the shape and span as per the location x_k, which is an arbitrary coordinate position. In the case of a constant grid spacing, ϕ_k becomes

$$\phi_k(x_j - x_k) = \phi(x_j - x) \qquad (2.63)$$

We can see from Equation (2.62) that the selection of location x_k is difficult as it presents an infinite number of possibilities. Therefore, it is convenient to specify the weighting function at the finite number of chosen points only, similar to WLS with the type of $\phi_i(x_j - x_i)$. A good computational procedure with this definition is described as

$$\phi_k(x_j - x_k) = \phi_j(x_k - x_j) \qquad (2.64)$$

where x_k is an arbitrary location $(x_k = x)$, and j is a fixed point in the domain. With this substitution, the functional to be minimized is rewritten from Equation (2.62) as

$$J = \sum_{j=1}^{n} \phi_j(x - x_j) \left[(u_j^h - p_j^T \, \alpha(x) \right]^2 \qquad (2.65)$$

This equation gives

$$\left. \begin{aligned} [A(x)] &= \sum_{j=1}^{n} \phi_j(x - x_j) \, P(x_j) \, P^T(x_j), \\ B(x) &= \left[\phi_1(x - x_j) \, p(x_1), \ \phi_2(x - x_j) \, p(x_2), \ \ldots, \ \phi_n(x - x_j) \, p(x_n) \right] \end{aligned} \right\} \qquad (2.66)$$

It is noted in Equation (2.65) that the unknown parameter α_i is not constant and varies with the position x.

In the FPM (Oñate et al., 1996a), the governing equations are discretised by the point collocation method, and the LS approximation is used by simply choosing $w_i = \delta_i$, namely the Dirac delta function, resulting in the governing and boundary equations as

$$[A(\hat{u})]_i - b_i = 0 \quad \text{in } \Omega \tag{2.67}$$

$$B(\hat{u})]_i - t_i = 0 \quad \text{in } \Gamma_t, \text{and } \hat{u}_i - u_p = 0 \quad \text{in } \Gamma_u \tag{2.68}$$

where Ω, Γ_t and Γ_u are the internal domain and Neumann and Dirichlet boundaries, respectively. Equations (2.67) and (2.68) are simplified by substituting the shape functions as

$$K\,U^h = f \tag{2.69}$$

The solution of Equation (2.69) give the values of approximate solution u_i.

There are several advantages offered by the FPM as compared with the classical FEM and FDM. As the total number of interpolation points n is not equal to the total polynomial order m, i.e., $n \geq m$, the FPM offers a freedom to choose n and m independently. It also gives us the freedom to choose the weighting function and its shape to achieve the desired effect of weighting in the function approximation. The FPM also offers considerable advantage over standard FDM in handling the nonstructured distribution of points.

The FPM has a few disadvantages as well due to the use of WLS interpolation. The local value of the unknown function is the same as the nodal parameter value in classical FEM, i.e., $\hat{u}(x_i) = u_i^h$, where as in WLS-based FPM $\hat{u}(x_i) \neq u_i^h$. Therefore, one more level of interpolation is required to compute the actual values of approximate functions. In WLS interpolation $n \geq m$, the interpolation matrix is rectangular and not square. Thus it must be made square by the least square approximation technique. As the FPM is a strong form-based meshless method, it is suitable in fluid flow problems such as convection–diffusion equations.

2.4.8 Meshless local Petrov–Galerkin method

This method was proposed by Atluri and Zhu (1998), Atluri and Shen (2002), Atluri (2004), and Atluri (2005). A local interpolation domain is created around each field node and the function is approximated at each field node by the MLS technique with the local interpolation domain. The governing equation is converted to the weak form, but the trial and test functions are chosen from the different spaces as with the Petrov–Galerkin

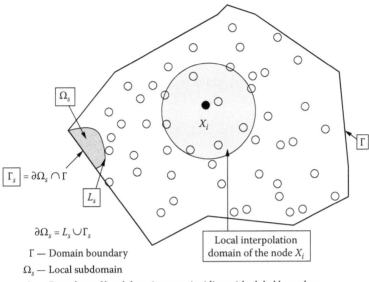

Ω_s

X_i

$\boxed{\Gamma}$

$\boxed{\Gamma_s} = \partial\Omega_s \cap \Gamma$

$\boxed{L_s}$

$\partial\Omega_s = L_s \cup \Gamma_s$

Γ — Domain boundary

Ω_s — Local subdomain

Local interpolation domain of the node X_i

L_s — Boundary of local domain not coinciding with global boundary

Γ_s — Boundary of local domain coinciding with global boundary

Figure 2.4 **Different domains in MLPG method.**

method. The trial function *u* is approximated by the MLS technique, and the test function *v* is approximated by the weighting function used in the MLS approximation of the trial function. The local symmetric weak form (LSWF) is developed by the Petrov–Galerkin approach and the final equations are solved by the Gauss quadrature method. The essential BC is imposed by the penalty method. Thus, the primary objective of this method is to eliminate the use of a mesh while still using the weak form-based formulation.

The essential BC is imposed by the penalty method. The formulation of MLS approximation is discussed first. Consider a computational domain, as shown in Figure 2.4. In general, the global domain Ω is subdivided into several subdomains Ω_s with the local boundary $\partial\Omega_s = L_s \cup \Gamma_s$, where L_s and Γ_s are the parts of the boundary of subdomain Ω_s. L_s does not coincide with the global boundary Γ, while Γ_s coincides with Γ.

Let $p_j(x)$ be a vector of monomials in the *n*-dimensional space such that $\forall\, n \in [1, 2, 3]$. Therefore, the monomials for 1-D space are given as

$$P^T(x) = [1,\ x],\quad m = 2\ \text{for a linear basis} \tag{2.70a}$$

$$P^T(x) = [1,\ x,\ x^2],\quad m = 3\ \text{for a quadratic basis} \tag{2.70b}$$

The monomials for 2-D space are given as

$$P^T(x) = [1,\ x,\ y,\ x^2,\ xy,\ y^2], \quad m = 6 \ \text{ for a quadratic basis} \qquad (2.70c)$$

Consider a subdomain Ω_x around the node at x with the n field nodes in the interpolation domain such that the MLS approximation $u^b(x)$ is given as

$$u^b(x) = \sum_{j=1}^{m} P_{xj}(x)\, a_j(x) \qquad (2.71)$$

where the unknown coefficients $a_j(x)$ are determined by minimizing the weighted discrete L_2 norm as

$$E = \sum_{i=1}^{n} w(x - x_i) \left[\sum_{j=1}^{m} P_{ij}(x)\, a_j(x) - u_i \right]^2 \qquad (2.72)$$

where $w(x - x_i) = W_i(x)$ is the weighting function. Equation (2.72) is minimized with respect to a_j by

$$\frac{\partial E}{\partial a_j} = 2 \sum_{i=1}^{n} w_i(x) \left[\sum_{j=1}^{m} P_{ij}(x)\, a_j(x) - u_i \right] P_{ij} = 0 \qquad (2.73)$$

After Equation (2.73) is simplified, we have

$$A(x)\, a(x) = B(x)\, U \qquad (2.74)$$

where

$$[A(x)] = \sum_{i=1}^{n} w_i(x)\, P_{ij} \sum_{k=1}^{m} P_{ik}, [B(x)] = \sum_{i=1}^{n} w_i(x)\, P_{ij} \qquad (2.75)$$

Inverting $A(x)$ from Equation (2.74) results in

$$a(x) = A^{-1}(x)\, B(x)\, U \qquad (2.76)$$

Substituting Equation (2.76) into Equation (2.71) and then simplifying the result, we have

$$u^b(x) = \sum_{i=1}^{n} N_i(x)\, u_i, \text{ where } N_i(x) = \sum_{j=1}^{m} P_{xj}(x)\,[A^{-1}(x)\,B(x)]_{ji} \qquad (2.77)$$

Equation (2.77) is the final form for the MLS approximation. It should be noted from Equation (2.77) that $N_i(x_j) \neq 1$, namely the MLS approximation does not possess the property of the delta function. If A^{-1} exists, the rank of matrix $A(x)$ should be at least m. The Gaussian function is used as the weighting function, as given below

$$w_i(x) = \begin{cases} \dfrac{e^{[-(d_i/c_i)^{2k}]} - e^{[-(r_i/c_i)^{2k}]}}{1 - e^{[-(r_i/c_i)^{2k}]}}, & 0 \le d_i \le r_i \\[2mm] 0, & d_i \ge r_i \end{cases} \qquad (2.78)$$

where $d_i = |x - x_i|$ is the distance between the node located at x, its interpolation node at x_i, c_i is the constant controlling the shape of weight function $w_i(x)$, and r_i is the size of support for the weight function $w_i(x)$ that determines the support of node at x_i.

After the MLS approximations are developed at the field nodes, the LSWF is developed over a local subdomain Ω_s. The shape of the local subdomain is taken as a sphere in a 3-D domain and a circle in a 2-D domain. The general Poisson equation over the internal domain Ω with boundary $\Gamma = \Gamma_u \cup \Gamma_q$ is given as

$$\nabla^2 f(x) = \bar{g}(x) \quad x \in \Omega, \text{ such that } f = \bar{f} \text{ on } \Gamma_u \text{ and } \frac{\partial f}{\partial \bar{n}} = \bar{q} \text{ on } \Gamma_q \quad (2.79)$$

The generalized weak form of Equation (2.79) is given as

$$\int_{\Omega_s} \left(\nabla^2 f - \bar{g}\right) v\, d\Omega - \alpha \int_{\Gamma_{su}} (f - \bar{f})\, v\, d\Gamma = 0 \qquad (2.80)$$

where α is a penalty factor, f and v are the trial and test functions, respectively, and Γ_{su} is a part of the boundary $\partial\Omega_s$ of the subdomain Ω_s over which the Dirichlet boundary condition is specified. Note from Figure 2.4 that if the subdomain Ω_s is completely inside the global domain Ω, the Γ_{su} term vanishes automatically. As the MLS approximation does not possess the delta function property, the Dirichlet boundary conditions are imposed by the penalty method with $\alpha \gg 1$. Equation (2.80) is simplified according to Figure 2.4, by applying integration by parts and the divergence theorem as

$$\int_{\Omega_s} (f_{,i}\ v_{,i} + \bar{g}v)\ d\Omega + \alpha \int_{\Gamma_{su}} (f - \bar{f})\ v\ d\Gamma = \int_{L_s} q\ v\ d\Gamma + \int_{\Gamma_{su}} q\ v\ d\Gamma + \int_{\Gamma_{sq}} \bar{q}\ v\ d\Gamma \ (2.81)$$

where Γ_{sq} is a part of $\partial\Omega_s$, in which the Neumann boundary condition $\bar{q}$ is specified. If the subdomain Ω_s is completely inside the domain Ω, $L_s = \partial\Omega_s$ such that the integral terms over Γ_{sq} and Γ_{su} boundaries vanish. The test function v is chosen so that it vanishes along L_s. This is accomplished using the Gaussian weight function from the MLS approximation as the test function v. The radius r_i of the support in the Gaussian weight function is replaced by the radius r_o of the subdomain Ω_s such that v vanishes at r_o. The LSWF given in Equation (2.81) is then rearranged as

$$\int_{\Omega_s} f_{,i}\ v_{,i}\ d\Omega + \alpha \int_{\Gamma_{su}} f\ v\ d\Gamma - \int_{\Gamma_{su}} q\ v\ d\Gamma = \int_{\Gamma_{sq}} \bar{q}\ v\ d\Gamma - \int_{\Omega_s} \bar{g}\ v\ d\Omega + \alpha \int_{\Gamma_{su}} \bar{f}\ v\ d\Gamma$$

$$(2.82)$$

Equation (2.82) is solved at each node to get one algebraic equation in the form of nodal values. The terms f and v are replaced by Equations (2.77) and (2.78), respectively, to obtain the discrete form of Equation (2.82).

The convergence rates and the accuracy of the solution are considerably higher by the MLPG method as compared with the FEM. No smoothing technique is required to compute the derivatives. A few drawbacks are associated with the MLPG method. The essential BC has to be imposed by the penalty method, as it is not easy to directly impose it by the MLS approximation.

The MLPG method is employed to solve several problems of engineering mechanics (Atluri and Zhu, 1998; Atluri et al., 1999a; Atluri, 2004; Atluri, 2005) and achieve higher rates of convergence, as compared with the FEM. Atluri (2004, 2005) and Atluri and Shen (2002) discussed the MLPG method in detail for applications in fluid and solid mechanics, respectively. Atluri and Shen (2002) discussed different global and local trial and test functions and provided a broad framework, under which different MLPG forms can be derived by combining them with different local trial and test functions.

Several MLPG mixed schemes are also proposed, such as the finite difference method through MLPG (Atluri et al., 2006b), the finite volume method through MLPG (Atluri et al., 2004), and the MLPG mixed collocation method (Atluri et al., 2006a). These MLPG mixed schemes and the MLPG method are used to solve various engineering problems such as large deformation (Han et al., 2005), 3-D contact problems (Han et al.,

2006), heat conduction (Wu et al., 2007), crack analysis in the 2-D and 3-D domains (Sladek et al., 2007b), thermo-piezoelectricity (Sladek et al., 2007a), and shell deformation (Jarak et al., 2007). Cai and Zhu (2008) modified the MLPG method to overcome the drawbacks of the Shepard partition of unity (PU) approximation by employing novel PU-based Shepard and LS interpolations.

Liu et al. (2008a) developed a meshless method for the analysis of structural dynamic problems by constructing trial functions with the natural neighbor concept and combining them with the general MLPG method. Several researchers also modified the MLPG method to solve specific engineering problems such as the Navier-Stokes and energy equations (Arefmanesh et al., 2008), limit analysis of plastic collapse (Chen et al., 2008), microelectromechanical systems (Dane and Sankar, 2008), topology optimization (Li and Atluri, 2008; Zheng et al., 2009), elastoplastic fracture analysis (Long et al., 2008), steady state and transient heat conduction in a 3-D solid (Sladek et al., 2008c), boundary and initial value problems in piezoelectric and magnetoelectric elastic solids (Sladek et al., 2008a), thermal bending of Reissner-Mindlin plates (Sladek et al., 2008b), and 3-D potential problems (Pini et al., 2008). Based on these applications, the MLPG method works well to solve many types of engineering problems.

2.4.9 Local boundary integral equation method

If the Galerkin approach-based FEM (GFEM) is compared with the boundary element method (BEM), the GFEM is more popular due to the local nature of the basis functions. This means that the GFEM basis functions are nonzero over a specific element and zero outside of it, while the BEM reduces the dimension of the problem by one, with the terms of trial function and its derivatives present under the integral sign applied over the global boundary of the domain only.

The solution of a problem by the GFEM leads to banded, sparse, and symmetric matrices; the BEM yields full and unsymmetrical matrices. The EFGM was developed by Belytschko et al. (1994a and b, 1995a and b, 1996) to overcome the difficulty of mesh generation in the FEM for complex 3-D domains. However, the EFGM still involves integrals over the auxiliary elements and creates difficulties in imposing the essential boundary condition due to the use of MLS approximation. To combine the advantages of the GFEM, BEM, and EFGM, Zhu et al. (1998) and Atluri et al. (2000) proposed the LBIE method. It involves boundary integration only over a local boundary around the concerned node and uses the MLS approximation to interpolate the function values. The final stiffness matrix given by the LBIE method is sparse and banded, and in spite of the use of MLS approximation, the essential boundary conditions are imposed exactly.

A companion solution concept (Zhu et al., 1998) is introduced in the LBIE method. The LBIE for the solution of trial function inside the domain Ω of the given problem involves the trial function in the integral terms, only over the local boundary $\partial\Omega_s$ around the concerned node. This is in contrast with the BEM, where the trial function and its gradient over the global boundary Γ of the domain Ω are involved under the integral terms. In the LBIE, however, only if a source point is positioned on Γ, the integrals over the local boundary $\partial\Omega_s$ involve both the trial function and its gradient. The continuity requirement in the approximation of the trial function can be greatly relaxed in the formulation of LBIE. No derivative of the shape functions is required while constructing the stiffness matrix of the system, at least for the nodes in the internal domain of computation. The LBIE approach is explained as follows.

Let us consider a Poisson equation applied over a domain Ω enclosed by Γ where $\Gamma = \Gamma_u \cup \Gamma_t$, and Γ_u and Γ_t are the boundaries involving the essential and natural boundary conditions, respectively, namely

$$\nabla^2 u(x) = p(x) \qquad x \in \Omega \tag{2.83}$$

$$u = \bar{u} \text{ on } \Gamma_u, \text{ and } \frac{\partial u}{\partial \bar{n}} = \bar{q} \text{ on } \Gamma_t \tag{2.84}$$

where $\bar{u}$ and $\bar{q}$ are the prescribed values of the potential and normal flux vector, respectively, on the boundaries Γ_u and Γ_t, respectively, and $\bar{n}$ is the unit outward direction normal to Γ. A weak form of Equation (2.83) is constructed as

$$\int_{\Omega} [\nabla^2 u(x) - p(x)] \, u^* \, d\Omega = 0 \tag{2.85}$$

where u^* and u are the test and trial functions, respectively. The chosen test function u^* satisfies the equation

$$\nabla^2 u^*(x, y) + \delta(x, y) = 0 \tag{2.86}$$

where $\delta(x, y)$ is the Dirac delta function. Applying integration by parts twice to Equation (2.85) leads to

$$u(y) = \int_{\Gamma} u^*(x, y) \, \frac{\partial u(x)}{\partial n} \, d\Gamma - \int_{\Gamma} u(x) \, \frac{\partial u^*(x, y)}{\partial n} \, d\Gamma - \int_{\Omega} u^*(x, y) \, p(x) \, d\Omega \tag{2.87}$$

where x and y are the generic and source points, respectively. Although the source point y is located within the domain Ω, it is known from the potential theory that Equation (2.87) holds over the entire domain Ω and the boundaries. As a result, Equation (2.87) is called the global boundary integral equation (GBIE), and it is used to compute the value of an unknown variable u at the source point y. If no entire domain Ω is involved, only a subdomain Ω_s containing the source point y is considered such that $\Omega_s \in \Omega$, then Equation (2.87) should also hold over Ω_s; therefore

$$u(y) = \int_{\partial\Omega_s} u^*(x, y) \frac{\partial u(x)}{\partial n} \, d\Gamma - \int_{\partial\Omega_s} u(x) \frac{\partial u^*(x, y)}{\partial n} \, d\Gamma - \int_{\Omega_s} u^*(x, y) \, p(x) \, d\Omega \quad (2.88)$$

where $\partial\Omega_s$ is the boundary of Ω_s. Equation (2.88) evaluates the unknown function u at the source point y by performing the integration over the closed boundary surrounding the concerned point, and over the subdomain entirely enclosed within the closed boundary. Equation (2.87) of GBIE contains the terms of either the potential $\bar{u}$ or normal flux $\bar{q}$ defined at every point on Γ, which makes it a well posed problem. No boundary conditions are known *a priori* while solving Equation (2.88). Therefore, in order to get rid of the term $u(x)$ of gradient over the subdomain $\partial\Omega_s$, the concept of companion solution associated with the fundamental solution is introduced. This is defined as the solution of Dirichlet problem over the subdomain Ω_s as

$$\nabla^2 u' = 0 \quad \text{on } \Omega_s \text{ and } u' = u^*(x, y) \quad \text{on } \partial\Omega_s \quad (2.89)$$

The fundamental solution u^* is regular everywhere except the source point y. Using $u^{**} = u^* - u'$ in Equation (2.85) and performing the integration by parts twice result in

$$\int_{\Omega_s} -u(x) \, \nabla^2 u^*(x, y) \, d\Omega = \int_{\partial\Omega_s} -u^{**}(x, y) \frac{\partial u(x)}{\partial n} \, d\Gamma - \int_{\partial\Omega_s} u(x) \frac{\partial u^{**}(x, y)}{\partial n} \, d\Gamma -$$

$$\int_{\Omega_s} u^{**}(x, y) \, p(x) \, d\Omega \quad (2.90)$$

Note that $-\nabla^2 u^{**} = -\nabla^2 u^* + \nabla^2 u' = \delta(x, y)$ in Ω_s and $u^{**} = 0$ along $\partial\Omega_s$. Equation (2.90) is further modified by the definition of companion solution as

$$u(y) = -\int_{\partial\Omega_s} u(x) \frac{\partial u^{**}(x, y)}{\partial n} \, d\Gamma - \int_{\Omega_s} u^*(x, y) \, p(x) \, d\Omega \quad (2.91)$$

Only the unknown variable u appears in the integral form of local boundary. Equation (2.91) is called the LBIE and is applicable to any shape of the subdomain Ω_s. Among the several advantages of the LBIE method, the essential BC is easily imposed. Second, no special integration scheme is needed to evaluate volume and boundary integrals. Third, the stiffness matrix is banded. The local domain creation in LBIE may be a tedious task. The LBIE method is successfully employed to solve several mechanics problems (Atluri et al., 2000).

2.4.10 Point interpolation method

The PIM is an interpolation technique proposed by Liu and Gu (2001). The terms of approximate function from the weak form governing equation as in Equation (2.82) are replaced by the expression of shape functions computed by the PIM. Consider a function $u(x)$ defined over a domain Ω and discretised by randomly distributed field nodes. The value of function $u(x)$ at each field node is interpolated by the surrounding field nodes via the polynomial as

$$u^h(x) = P^T(x)\, A(x) \tag{2.92}$$

where the function $u(x)$ is interpolated by the field nodes x_i ($i = 1, 2, \ldots, n$), $P^T(x)$ and $A(x)$ are the vectors of monomials and unknown coefficients, respectively, as given in Equation (2.70) and

$$A^T(x) = [a_1,\ a_2,\ \ldots,\ a_m] \tag{2.93}$$

The coefficients $A(x)$ in Equation (2.92) are computed by satisfying Equation (2.92) at all the n field nodes surrounding the concerned node at x. The approximate function at the ith node is given as

$$u_i = P^T(x_i)\, A \tag{2.94}$$

where u_i is the nodal value of u at the node $x = x_i$. Equation (2.94) is imposed at all the n field nodes, and then the result is given in the matrix form as

$$\mathbf{u}^e = P_Q\, A \tag{2.95}$$

where

$$\mathbf{u}^e = [u_1,\ u_2,\ \ldots,\ u_n]^T,\ \text{and } P_Q^T = [P(x_1),\ P(x_2),\ \ldots,\ P(x_n)] \tag{2.96}$$

The vector of unknown coefficients from Equation (2.95) is given as

$$A = P_Q^{-1} u^e \qquad (2.97)$$

Substituting Equation (2.97) into Equation (2.92) results in

$$u^h(x) = \sum_{i=1}^{n} P^T(x) \, P_Q^{-1} \, u^e \qquad (2.98)$$

The shape functions by the PIM are then given as

$$N(x) = P^T(x) \, P_Q^{-1} \qquad (2.99)$$

It is noted that $N_i(x = x_i) = 1$, $\sum_{i=1}^{n} N_i(x) = 1$, and $m = n$ to get the square matrix P_Q in Equation (2.95), namely the number of total monomial terms in $P^T(x)$ should be equal to the number of total interpolation nodes in the local domain of the field node located at x. Based on the number of total interpolation nodes, the appropriate monomial terms are selected from Pascal's triangle. The essential boundary conditions are easily imposed, as the PIM shape functions satisfy the Delta function property.

When the PIM is used, the trial and test functions from Equation (2.82) are replaced by the shape functions obtained by the PIM interpolation (Liu and Gu, 2001; Liu, 2003), and the integrals are solved by the Gauss quadrature with a separate cell structure that is independent of the actual field nodes. In the EFGM, the interpolation is based on the arbitrarily distributed field nodes but $m \neq n$. This means that the total monomial terms in $P^T(x)$ are not equal to the total nodes in the interpolation domain, while in the PIM $m = n$.

Certain characteristics and drawbacks of the PIM should be considered before implementation. The inverse of P_Q in Equation (2.97) may not exist if the required numbers of the independent values of x and y coordinates are not present (Liu and Gu, 2001). In this situation, either x or y nodal coordinates of the field nodes in the local interpolation domain of the concerned node may be perturbed slightly to get the nonsingular matrix P_Q.

Consider the 1-D interpolation computed at the interested node x by 10 field nodes with the second order of the field variable distribution. The vector $P^T(x)$ consists of the monomials up to the ninth order, as per the PIM method, namely $P^T(x) = [1, x, x^2, ..., x^9]$, and $m = 10$. As a result, the second order distribution of the field variable is approximated by the ninth order polynomial.

According to Runge's phenomenon (Dahlquist and Björck, 2003), this may cause oscillations along the boundaries of domain for the uniformly distributed field nodes. Therefore, the field nodes should be arranged in a sufficiently random manner to ensure that the PIM will yield a good approximation of field variables. The definition of an influence domain around a central concerned node is another difficult task. Since the PIM is an interpolation technique, it can be combined with any of the existing meshless methods for a specific application. Several test problems of the mechanics are solved by the PIM (Liu and Gu, 2001; Liu, 2003), and good convergence rates are reported.

2.4.11 Gradient smoothing method

In general, the meshless methods based on the strong form need a special procedure to numerically compute the derivatives of functions at the field nodes. A novel GSM method was proposed by Liu et al. (2008b), in which the techniques of gradient smoothing and directional derivatives are adopted to compute the first and second order approximate derivatives at the node of interest by assigning weights to the nodes surrounding the interested node. The governing equations are discretised in the strong form by the standard collocation method. The GSM is developed to be applied easily to complex geometries discretised by random nodes.

The problem domain Ω is initially discretised by the triangular cells, as shown in Figure 2.5, with total M field nodes. For the ith field node, the smoothing domain Ω_i is constructed by joining the centroids and mid-edge points of the triangular cells around the ith field node, as shown in

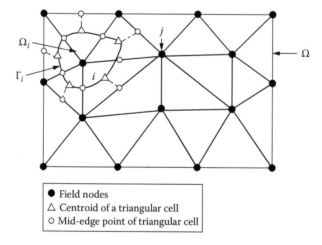

● Field nodes
△ Centroid of a triangular cell
○ Mid-edge point of triangular cell

Figure 2.5 GSM working principle.

Figure 2.5. No overlapping is present between any two smoothing cells. A smooth operation to the gradient of function is done to compute the approximate derivatives as

$$\nabla^h \ u(x_i) = \int_{\Omega_i} \nabla^h u(x) \ \Phi(x - x_i) \ d\Omega_i \tag{2.100}$$

where Φ is a smoothing function. Performing an integration by parts in Equation (2.100) results in

$$\nabla^h \ u(x_i) = \int_{\Gamma_i} u^h(x) \ n(x) \ \Phi(x - x_i) \ d\Gamma - \int_{\Omega_i} u(x) \ \nabla\Phi(x - x_i) \ d\Omega_i \tag{2.101}$$

In fact, any smoothing function can be used in the GSM. Here a weighted Shepard function is used as

$$\Phi(x - x_i) = \frac{\phi(x - x_i)}{\displaystyle\sum_{j=1}^{M} \phi(x - x_j) \ A_j} \tag{2.102}$$

where $A_i = \int_{\Omega_i} d\Omega$ is the area of total smoothing domain Ω_i. The function ϕ can be used as a piecewise constant function as

$$\phi(x - x_i) = \begin{cases} 1 & x \in \Omega_i \\ 0 & x \notin \Omega_i \end{cases} \tag{2.103}$$

Therefore, the smoothing function Φ becomes

$$\Phi(x - x_i) = \begin{cases} 1/A_i & x \in \Omega_i \\ 0 & x \notin \Omega_i \end{cases} \tag{2.104}$$

Substituting Equation (2.104) into (2.105), the smoothed gradient of the field variable is computed as

$$\nabla^h \ u(x_i) = \int_{\Gamma_i} u^h(x) \ n(x) \ \Phi(x - x_i) \ d\Gamma$$

$$\therefore \nabla^h \ u(x_i) = \frac{1}{A_i} \int_{\Gamma_i} u^h(x) \ n(x) \ d\Gamma \tag{2.105}$$

The second term in Equation (2.105) is equal to zero as the smoothing function Φ is constant. As a result, the area integral becomes the line integral in Equation (2.107) along the portions of boundary Γ_i as

$$\nabla^b u_i = \frac{1}{A_i} \sum_{j=1}^{m_i} \left(L_{ij}^{(L)} \ n_{ij}^{(L)} \ u_{ij}^{(L)} + L_{ij}^{(R)} \ n_{ij}^{(R)} \ u_{ij}^{(R)} \right) \tag{2.106}$$

where m_i is the number of total cells covering the ith field node, $L_{ij}^{(L)}$ and $L_{ij}^{(R)}$ are the lengths of the smoothing boundary along the left and right sides of the edge i–j, respectively, $N_{ij}^{(L)}$ and $N_{ij}^{(R)}$ are the outward unit normals over the left and right sides of the smoothing boundary with respect to the edge $i - j$, respectively, and similarly $u_{ij}^{(L)}$ and $u_{ij}^{(R)}$ are the values of the field variable. The second order approximate gradient of the field variable is evaluated by following the procedure in Equations (2.100) to (2.106).

$$\nabla^2 u(x_i) = \frac{1}{A_i} \int_{\Gamma_i} \nabla^b u(x) \ n(x) \ d\Gamma \tag{2.107}$$

Several schemes were developed to compute the values of $u_{ij}^{(L)}$ and $u_{ij}^{(R)}$. The convergence analysis of the GSM is performed in detail through the problems with the uniform and irregular computational domains (Liu et al., 2008b). In brief, the GSM was developed for the approximation of derivatives and coupled with the simple collocation method to get a system of algebraic equations with the values of field variables as the unknowns.

For small systems, the GSM is more efficient than the FEM and gives more accurate results for stresses. For large systems, the FEM is faster than GSM, but the results obtained by the GSM are more accurate than results from the FEM. The GSM is also easily applied to 3-D problems but presents several drawbacks. First, the construction of a smoothing domain by pre-existing mesh may be a difficult task. Second, the user must be well versed on the rules for the construction of difference schemes. The GSM is typically applied to elasticity and mechanics problems (Liu et al., 2008b).

2.4.12 Radial point interpolation-based finite difference method

The conventional FDM requires the uniform grid of points to discretize the derivative terms in the governing differential equation. To extend the applicability of FDM, Liu et al. (2006) developed the radial point interpolation-based finite difference method (RFDM) and applied it to mechanics problems. The radial point interpolation (RPI) is used to approximate the values of function at the field nodes, and the FDM is used to discretize the terms of derivative from the governing differential equation.

Two types of nodal patterns are employed to impose the FDM on the field nodes distributed randomly. The domain is first sprinkled with the field nodes distributed randomly, then a background uniform grid is created for the FDM. The nodes on the grid of FDM are interpolated by the field nodes distributed randomly with the RPI method. The governing equation is discretised by the FDM at the background uniform grid and the terms of function values at the FDM grid nodes are replaced by their corresponding interpolations in terms of the function values at the field nodes distributed randomly. As a result, the system of algebraic equations is obtained finally with the rows corresponding to the number of total background FDM nodes and the columns corresponding to the number of total field nodes distributed randomly.

The RBF is used for data fitting, based on arbitrarily distributed nodes. The RBF shape function is created at any interested node by the nodes in the local support domain of the interested node. Let the function $u(x)$ be defined over the domain Ω, and the interested node be located at the coordinate x_Q, where there are n nodes in the local support domain of the node located at x_Q. As a result, the value of $u(x)$ is approximated by the n field nodes using the RBF as

$$u^h(x, x_Q) = \sum_{i=1}^{n} R_i(x)\, a_i(x_Q) + \sum_{j=1}^{m} p_j(x)\, b_j(x_Q) = \mathrm{R}^T(x)\, a(x_Q) + \mathrm{p}^T(x)\, b(x_Q) \quad (2.108)$$

where n and m represent the total nodes in the support domain and the number of monomial terms in the polynomial basis, respectively. $a(x_Q)$ and $b(x_Q)$ are the unknown coefficients vectors of $\mathrm{R}^T(x)$ and $\mathrm{p}^T(x)$, respectively, and given as

$$a^T(x_Q) = \{a_1 \quad a_2 \quad \dots \quad a_n\}, \text{ and } b^T(x_Q) = \{b_1 \quad b_2 \quad \dots \quad b_n\} \quad (2.109)$$

The polynomial terms must satisfy the extra condition to ensure unique approximation

$$\sum_{i=1}^{n} p_j(x_i)\, a_i = \mathrm{P}_m^T\, a = 0, \quad j = 1,\, 2,\, \dots,\, m \quad (2.110)$$

Several types of RBF are available. One is the multi-quadrics (MQ) RBF adopted in the RFDM and given as

$$R_i(x, y) = [r_i^2 + (\alpha_c \ d_c)^2]^q \tag{2.111}$$

where r_i is the Euclidean distance between the interpolation point located at x and its neighbourhood node at x_i. Forcing Equation (2.108) to pass through all the nodes in the support domain leads to

$$U_s = R_Q \ a(x_Q) + P_m \ b(x_Q) \text{ with } P_m^T \ a(x_Q) = 0 \tag{2.112}$$

where

$$R_Q = \begin{bmatrix} R_1(r_1) & R_2(r_1) & \cdots & R_n(r_1) \\ R_1(r_2) & R_2(r_2) & \cdots & R_n(r_2) \\ \vdots & \vdots & \ddots & \vdots \\ R_1(r_n) & R_2(r_n) & \cdots & R_n(r_n) \end{bmatrix}_{n \times n} , P_m = \begin{bmatrix} 1 & x_1 & y_1 & \cdots & p_m(x_1) \\ 1 & x_2 & y_2 & \cdots & p_m(x_2) \\ \vdots & \vdots & \vdots & \ddots & \vdots \\ 1 & x_n & y_n & \cdots & p_m(x_n) \end{bmatrix}_{n \times m} \tag{2.113}$$

The term r_k in $R_i(r_k)$ from Equation (2.113) is defined as

$$r_k = \sqrt{(x_k - x_i)^2 + (y_k - y_i)^2} \tag{2.114}$$

Equation (2.112) in the matrix form is given as

$$\tilde{U}_s = \begin{bmatrix} U_s \\ 0 \end{bmatrix} = \begin{bmatrix} R_Q & P_m \\ P_m^T & 0 \end{bmatrix}_{(n+m) \times (n+m)} \left\{ \begin{array}{c} a(x_Q) \\ b(x_Q) \end{array} \right\}_{(n+m) \times 1} = G \ a_0 \tag{2.115}$$

where

$$a_0^T = \left\{ a_1 \ a_2 \ \cdots \ a_n \ b_1 \ b_2 \ \cdots \ b_m \right\}, \tilde{U}_s = \left\{ u_1 \ u_2 \ \cdots \ u_n \ 0 \ 0 \ \cdots \ 0 \right\} \tag{2.116}$$

$$G = \begin{bmatrix} R_Q & P_m \\ P_m^T & 0 \end{bmatrix} \tag{2.117}$$

Matrix G is symmetric. The solution of Equation (2.115) leads to

$$a_0 = G^{-1} \tilde{U}_S \tag{2.118}$$

Substituting Equation (2.118) into Equation (2.108) results in

$$u^b(x, x_Q) = \left\{ \quad R^T(x) \quad p^T(x) \quad \right\} G^{-1} \tilde{U}_S \tag{2.119}$$

$$\therefore u^b(x, x_Q) = \tilde{\Phi}^T(x) \tilde{U}_S, \text{ where } \tilde{\Phi}^T(x) = \left\{ \quad R^T(x) \quad p^T(x) \quad \right\} G^{-1} \tag{2.120}$$

Equation (2.120) can be rewritten as

$$u^b(x, x_Q) = \tilde{\Phi}^T(x) \tilde{U}_S = \sum_{i=1}^{n} \phi_i \, u_i \tag{2.121}$$

Equation (2.121) is the final expression of the RBF shape functions. The approximate derivatives of the field variables are computed by simply differentiating Equation (2.121) as

$$u^l_{,d}(x) = \tilde{\Phi}^T_{,d}(x) \tilde{U}_S \tag{2.122}$$

where, d denotes the derivative in either the x or y coordinate direction.

The working principle of RFDM is shown in Figure 2.6. If there are m FD grid points and n random field nodes and the governing differential equation contains the derivative term $\partial u/\partial x$, the derivative term is discretised at the FD grid point (x_i, y_j) by the classical FDM with the central difference scheme as

$$\left(\frac{\partial u}{\partial x} \right)_{i,j} = \frac{1}{2h} \left(u_{i+1,j} - u_{i-1,j} \right) \tag{2.123}$$

The term $u(x_i, y_j)$ from Equation (2.123) is replaced by Equation (2.121) as

$$u(x_i, y_j) = \sum_{k=1}^{n} \phi_k \, u_k \tag{2.124}$$

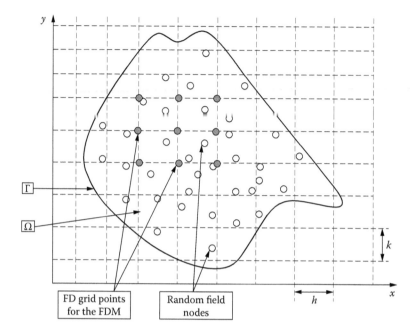

Figure 2.6 RFDM working principle.

Substituting Equation (2.124) into Equation (2.123) leads to a single algebraic equation as

$$\left(\frac{\partial u}{\partial x}\right)_{i,j} = \{k_i\}_{1\times n}\{u_k\}_{n\times 1} \tag{2.125}$$

Similarly, Equation (2.123) is applied at all the m FD grid points resulting in a matrix as

$$[K]_{m\times n}\ \{U\}_{n\times 1} = \{F\}_{m\times 1} \tag{2.126}$$

The stiffness matrix K in Equation (2.126) is rectangular with $m \geq n$. To solve Equation (2.126) uniquely, rank of the matrix K should be at least equal to n. The rectangular system of equations is solved by the LS approach such that $[K]_{n\times m}^T$ is multiplied on both sides of Equation (2.126), resulting in the square stiffness matrix as

$$K_{n\times m}^T\ K_{m\times n}\ U_{n\times 1} = K_{n\times m}^T\ F_{m\times 1} \tag{2.127}$$

$$\overline{K}_{n\times n}\ U_{n\times 1} = \overline{F}_{n\times 1} \tag{2.128}$$

Equation (2.128) is solved for the values of unknown field variables at the random field nodes.

The applicability of the classical FDM is extended over the irregular distribution of nodes by employing RPI method in the RFDM. The final stiffness matrix is made square by the least square technique. The main drawback of RFDM may be the basing of the relationship between the number of finite difference grid points and field nodes on trial and error. A user must ensure that sufficient finite difference grid points exist in the computational domain to achieve fairly accurate results. Several mechanics problems with regular and irregular domains were solved and convergence analyses were performed by the RFDM (Liu et al., 2006). The convergence rates obtained by the RFDM are reasonably good even for the irregular domains.

2.4.13 Generalized meshfree (GMF) approximation

This method is proposed by Park (2009), Park et al. (2011) and Wu et al. (2011). The main theme of GMF approximation is to unify all existing meshless approximations, and develop new approximations by different basis functions. This method provides great flexibility in selecting the basis functions to achieve convex and nonconvex function approximations. It even allows the mixture of convex and nonconvex approximations. This property of GMF approximation results in weak Kronecker delta function property at the boundaries, which allows the direct imposition of essential boundary conditions without any complicated mathematical treatment.

The basic approach for the 1-D case is stated as follows. Let the output data be $u(x_1), u(x_1), \ldots, u(x_n) \in R$. The approximation function $u^b(x): R^d \to R$ is constructed with the basis functions $\Gamma_i(x): R^d \to R$ as

$$u^b(x) = \sum_{i=1}^{n} C_i(x)\Gamma_i(x) = \sum_{i=1}^{n} u(x_i)\Psi_i(x) \tag{2.129}$$

where $C_i(x)$ are unknown coefficients, $u(x_i)$ are given data, and $\Psi_i(x)$ are the shape functions. Any shape function such as the Shepard function can be used.

$$\Psi_i(x) = \frac{\phi_a(x; X_i)}{\sum_{j=1}^{n} \phi_a(x; X_j)} \tag{2.130}$$

where $X_i = (x - x_i)$ is used. Substituting Equation (2.130) with (2.129), we get

$$C_i(x) = \frac{u(x_i)\phi_a(x; X_i)}{\sum_{j=1}^{n} \phi_a(x; X_j)} \text{ and } \Gamma_i(x) = 1 \tag{2.131}$$

Thus, Equation (2.129) is written as

$$u^h(x) = \sum_{i=1}^{n} C_i(x)\Gamma_i(x) = \sum_{i=1}^{n} \frac{u(x_i)\phi_a(x;X_i)}{\sum_{j=1}^{n}\phi_a(x;X_j)} \tag{2.132}$$

For linear reproducibility of a polynomial,

$$\sum_{i=1}^{n} \Psi_i(x)(x - x_i) = 0 \tag{2.133}$$

As Shepard function does not satisfy this linear constraint, the basis function is enriched similar to the RKPM. For a single constraint in Equation (2.133), one enrichment function is introduced and the basis function is expressed as

$$\Gamma_i(x,\lambda) = 1 + E[\lambda(x)] \tag{2.134}$$

where $E[\lambda(x)]$ is an enrichment function and $\lambda(x)$ is an unknown to be determined. The first term imposes the partition of unity of the shape function, and the second term satisfies any constraint on the shape function. There are several possible choices of the enrichment function based on the choice of $\lambda(x)$ (Park, 2009). For example, a constant choice of $\lambda(x)$ leads usually to a Shepard function. Similarly, for a linear choice of $\lambda(x)$ leads to

$$\tag{2.135}$$

$$u^h(x,\lambda) = \sum_{i=1}^{n} C_i(x,\lambda)\Gamma_i(x,\lambda) = \sum_{i=1}^{n} \frac{u(x_i)\phi_a(x;X_i)[1+\lambda(x)\sum_{k=1}^{m} e_k X_i^k]}{\sum_{j=1}^{n}\phi_a(x;X_j)[1+\lambda(x)\sum_{k=1}^{m} e_k X_j^k]}$$

where

$$\Gamma_i(x_i,\lambda) = 1 + \lambda(x)\sum_{k=1}^{m} e_k X_i^k \tag{2.136}$$

where e_k represents the coefficients of polynomial X_i^k. The shape function $\Psi_i(x)$ thus becomes

$$\tag{2.137}$$

$$\Psi_i(x,\lambda) = \frac{\phi_a(x;X_i)[1+\lambda(x)\sum_{k=1}^{m} e_k X_i^k]}{\sum_{j=1}^{n}\phi_a(x;X_j)[1+\lambda(x)\sum_{k=1}^{m} e_k X_j^k]}$$

Substituting Equation (2.137) in (2.133) leads to

$$(2.138)$$

$$\lambda(x) = \frac{-\sum_{j=1}^{n} \phi_a(x; X_j) X_j}{\sum_{j=1}^{n} \phi_a(x; X_j) X_j \ \sum_{k=1}^{m} e_k X_j^k}$$

In this manner, any number of the constraint equations can be imposed by identifying the corresponding enrichment functions $\lambda(x)$.

The GMF approximation formulation is based on the equations above. The first-order GMF approximation in 1-D is given as

$$(2.139)$$

$$\Psi_i(x, \lambda) = \frac{\psi_i}{\psi} = \frac{\phi_a(x; X_i) \Gamma_i(X_i, \lambda)}{\sum_{j=1}^{n} \phi_a(x; X_j) \Gamma_j(X_j, \lambda)}$$

subjected to

$$R(x) = \sum_{i=1}^{n} \Psi_i X_i = 0, \quad \text{for linear constraint} \qquad (2.140)$$

where $\psi = \sum_{i=1}^{n} \psi_i = \sum_{i=1}^{n} \phi_a(x; X_i) \Gamma_i(X_i, \lambda)$. The property of the partition of unity is automatically satisfied in the GMF approximation by the normalization in Equation (2.140). Therefore, the GMF approximation is completed when λ that satisfies Equation (2.140) is identified. The enriched basis function and weight function play important roles in controlling the smoothness and convexity in the GMF approximation. For detail formulations, interested readers can refer to Park (2009), Park et al. (2011), and Wu et al. (2011).

The stability analysis by Von Neumann method is also performed (Park et al., 2011) to identify stable discretisation schemes and study the effects of spatial nodal spacing on the numerical wave propagation speed within the computational domain. These results show that the GMF approximation method is suitable to solve time-dependent problems from elasticity. It is also suitable to solve the problems of fracture mechanics and crack propagation (Park, 2009).

2.4.14 Maximum entropy (ME) approximation method

The numerical integration of weak form-based meshfree methods is traditionally performed by background cells (triangular or quadrilateral elements). When the meshfree basis functions are nonpolynomial, and when the support of basis functions no longer coincides with the union of background cells used in the numerical integration, the patch test is not passed to machine precision. This suggests inaccuracies in the numerical integration of weak form integrals. It is a well known fact that the standard displacement-based Galerkin formulation exhibits severe stiffening or volumetric locking while modelling near-incompressible materials. This is because Poisson's ratio approaches the value of 1/2. These numerical problems have been eliminated in the ME approximation meshfree method proposed by Ortiz et al. (2010) and Ortiz et al. (2011).

The ME approximation method is based on the maximum entropy principle. Shannon (1948) introduced the concept of entropy in information theory. Suppose $x(x_1, x_2, \ldots, x_n)$ is a set of events of a complete system. The corresponding set of a probability of their occurrence is $P(p_1, p_2, \ldots, p_n)$, where n is the total number of events. The probability satisfies the following equations

$$p_i \geq 0, (i = 1, 2, \ldots, n) \text{ and } \sum_{i=1}^{n} p_i = 1 \tag{2.141}$$

Then the entropy of the system is defined as

$$H(\mathrm{p}) = H(p_1, p_2, \ldots, p_n) = -\sum_{i=1}^{n} p_i \log_b p_i \tag{2.142}$$

where b is the base of the logarithm used. The entropy $H(\mathrm{p})$ indicates the amount of uncertainty in a system. The entropy H is minimum, i.e., $H(\mathrm{p}) = 0$ when all $p_i = 0$, and H is maximum, i.e., $H(\mathrm{p}) = \log_b(n)$ when all $p_i = 1/n$. As per the ME principle proposed by Jaynes (1957), the probabilities $P(p_1, p_2, \ldots, p_n)$ are unknown. Only thing known is that the expected value of the function $f_h(x)$ is given as

$$\langle f_h(x) \rangle = \sum_{i=1}^{n} p_i f_h(x_i), \quad (r = 1, 2, \ldots, m) \tag{2.143}$$

where m is the total number of expected functions, and the symbol $\langle \ \rangle$ indicates the average value. As the number of unknowns m exceeds the number of constraint equations, the least biased values of p_i can be obtained when the entropy becomes maximum.

$$\text{Maximize} \left[H(\mathrm{p}) = -\sum_{i=1}^{n} p_i \log_b p_i \right] \tag{2.144}$$

In the context of the meshfree ME approximation method, the probability corresponds to basis or shape functions ϕ_a associated with the nodes located at x_a where $a = 1,2,...,n$. The connection between maximum entropy (max-ent) basis functions and the construction of polygonal interpolants was established by Sukumar (2004). The principle of maximum entropy is employed here to obtain linearly complete interpolants on polygonal domains (Sukumar, 2004). Let us consider an element with n nodes having coordinates x_i, where $(i = 1,2,...,n)$. Then the interpolant

$$u^b(x) = \sum_{a=1}^{n} \phi_a(x) u(x_a) \tag{2.145}$$

where $\phi_a(x)$ are the shape functions that satisfy both conditions in Equation (2.141). As per the ME principle, the values of $\phi_a(x)$ are obtained such that

$$\text{Maximize} \left[H(x,\phi_a,W_a) = -\sum_{a=1}^{n} \phi_a(x) \log_b \left(\frac{\phi_a(x)}{W_a(x)} \right) \right] \tag{2.146}$$

subject to

$$\phi_a(x) \geq 0, \quad \sum_{a=1}^{n} \phi_a(x) = 1, \quad \text{and} \quad \sum_{a=1}^{n} \phi_a(x)(x_a - x) = 1 \tag{2.147}$$

If $W_a[x;(x_a - x)] = 1$ for a whole domain, then it is a global ME approximation. If $W_a[x;(x_a - x)]$ is a kernel function with a fixed support size, the result is a local ME approximation. This ME approximation is finally solved by Lagrange multipliers to get a system of nonlinear equations (Ortiz et al., 2010).

The ME approximation meshfree method is suitable for compressible and nearly incompressible elasticity problems (Ortiz et al., 2010). A modified Gaussian integration scheme was proposed and alleviated numerical integration errors (Ortiz et al., 2010). The advantage of the ME approximation is that it is convex and possesses weak Kronecker delta function property at the boundaries. The disadvantage is that it is difficult to extend the method to higher order approximations since the first k moments of arbitrary fixed points do not belong to the moment space given by the convex hull of the nodal points (Park, 2009).

In summary, the working principles of several meshless methods based on the strong and weak forms are discussed in this subsection. The Galerkin-based weak form approach is extensively employed. Although most of the meshless methods do not require meshes or interconnectivity information among the field nodes, they still need an auxiliary set of nodes used either for numerical integration or for derivative discretisation.

2.5 SUMMARY

In this chapter, the fundamentals and the common steps involved in the meshless method are discussed. They are not much different from the classical FEM, except for the computation of shape function and the evaluation of the integral terms.

To achieve a deeper understanding of the working principles of meshless methods, several well established methods based on the strong or weak forms of the governing differential equations are discussed. It may be noted from the literature review that although extensive research work has been done on meshless methods based on the weak form approach, meshless methods based on the strong form approach must be explored further.

Chapter 3

Recent developments of meshless methods

3.1 INTRODUCTION

In this chapter, the five meshless methods are presented. Three of them are in the strong form: the Hermite-cloud method, random differential quadrature (RDQ) method, and the point weighted least-squares (PWLS) method. The other two are in the weak form: the local Kriging (LoKriging) method and the variation of local point interpolation method (vLPIM). All the five meshless methods presented here were contributed by the present authors.

3.2 HERMITE-CLOUD METHOD

The present Hermite-cloud method employs the Hermite interpolation theorem for the construction of interpolation functions and the point collocation technique for the discretisation of partial differential equations (PDEs) and boundary conditions. It is based on an extension of the well known RKPM developed by Liu et al. (1995, 1996) and Liu and Jun (1998). In the classical RKPM, via the selection of appropriate window functions, the approximate solutions of the unknown functions are constructed by the reproducing kernel method.

In general, the reproducing kernel method refers to a class of mathematical operators that can reproduce the function by integration transform over the defined domain. The more popular examples include the classical Fourier and Laplace transforms. The different forms of window functions result in the different corresponding reproducing kernel methods. One of the major contributions of classical RKPM is the development of a corrected kernel window function, consisting of a correction function and a so-called kernel function. Subsequently, other kernel functions were also developed, such as the moving kernel, fixed kernel and multiple fixed kernel functions (Aluru and Li, 2001).

The Hermite-cloud method represents a significant development of the meshless methods. It employs the Hermite theorem to construct the interpolation functions in combination and the point collocation technique for the discretisation of PDEs. The Hermite-cloud method is based on the idea of the classical RKPM but the fixed kernel is used instead as the kernel function. By the Hermite interpolation theorem, the shape functions are constructed first to correspond to both the unknown functions and the first-order derivatives of these unknown functions, respectively.

For a given set of PDEs with the Dirichlet and/or Neumann boundary conditions, the approximate solutions of both the unknown functions and their first-order derivatives are constructed in terms of unknown point values by Hermite interpolation functions. Certain differential-type auxiliary conditions are also constructed for the generation of a complete set of equations for the governing partial differential boundary value (PDBV) problem. By arbitrarily scattering a set of points in the computational domain including the edges and using the point collocation technique, the partial differential governing equations and the boundary conditions are discretised. The complete set of the discrete algebraic equations is obtained and then solved with respect to the unknown point values. Finally the PDBV problem is solved numerically.

3.2.1 Formulation of Hermite-cloud method

As well known, usually the reproducing kernel method can be referred to as a class of mathematical operators that can reproduce a function by integration transform over a defined domain, for example, in Fourier transform or Laplace transform. We can take a one-dimensional (1-D) real function $f(x)$ as an example in a domain Ω, and then we have a generic mathematical expression of the reproducing kernel method as

$$f(x) = \int_{\Omega} \Phi(x - u) f(u) \, du \qquad (3.1)$$

where $\Phi(x)$ is a real window function. Theoretically, an ideal window function is required to be orthogonal and its integration over the domain Ω is required to be unity in order to exactly reproduce $f(x)$. The selection of appropriate window functions differentiates the different reproducing kernel methods. However, it is often not easy to select appropriate function because the window function is required to satisfy both the above mentioned conditions to exactly reproduce the unknown function $f(x)$. The classical RKPM introduces a correction function $C(x,u)$ and a kernel function $K(x)$, and constructs the real window function $\Phi(x)$ as

$$\Phi(x - u) = C(x,u) K(x - u) \qquad (3.2)$$

By substituting this window function (3.2) into Equation (3.1), the approximation $\tilde{f}(x)$ of the unknown function $f(x)$ is then represented as

$$\tilde{f}(x) = \int_{\Omega} C(x,u) \, K(x - u) \, f(u) \, du \qquad (3.3)$$

For a two-dimensional unknown real function $f(x,y)$, the approximate expression $\tilde{f}(x,y)$ can be similarly written as

$$\tilde{f}(x,y) = \int_\Omega C(x,y,u,v)K(x-u,y-v)f(u,v)\,du\,dv \qquad (3.4)$$

As mentioned above, when the fixed kernel is employed as the present kernel function $K(x,y)$, Equation (3.4) is rewritten as

$$\tilde{f}(x,y) = \int_\Omega C(x,y,u,v)K(x_k-u,y_k-v)f(u,v)\,du\,dv \qquad (3.5)$$

where the point (x_k, y_k) is located at the center of the fixed kernel corresponding to the kernel function $K(x_k-u, y_k-v)$. Depending on the characteristics of PDBV problems, the kernel function may be constructed in different forms of weighted window functions, for example, the Gaussian function, the spline function, and the radial basis function. In the present Hermite-cloud method, a cubic spline function is employed to construct the kernel function as

$$K(x_k-u,y_k-v) = W^*((x_k-u)/\Delta x)W^*((y_k-v)/\Delta y)/(\Delta x\Delta y) \qquad (3.6)$$

where $W^*(z)$ is the cubic spline window function given as

$$W^*(z) = \begin{cases} 0 & |z| \ge 2 \\ (2-|z|)^3/6 & 1 \le |z| \le 2 \\ (2/3) - z^2(1-0.5\,|z|) & |z| \le 1 \end{cases} \qquad (3.7)$$

where $z = (x_k - u)/\Delta x$ or $z = (y_k - v)/\Delta y$. Here Δx and Δy denote the cloud sizes with the center node (x_k, y_k) in the x- and y-directions, respectively. These are adjusted according to the point coordinates and computational accuracy requirement due to the consistency conditions of the reproducing kernel technique.

A basis of vector space is defined by the set of linearly independent functions such that a new function can be expressed by the linear combination of these basis functions. Thus, the correction function $C(x,y,u,v)$ in Equation (3.5) is expressed as a product of a βth order column coefficient vector $C^*(x,y)$ and a βth order row basis function vector $B(u,v)$,

$$C(x,y,u,v) = B(u,v)C^*(x,y) = \{b_1(u,v),b_2(u,v),\ldots,b_\beta(u,v)\}\,\{c_1,c_2,\ldots,c_\beta\}^T \qquad (3.8)$$

In the basis function vector $B(u,v) \in R^{\beta}$ of Equation (3.8), $b_i(u,v)$ and $(i = 1, 2, ..., \beta)$ are linearly independent basis functions, where β denotes the degree of polynomials of the basis function. Usually the defined form of the basis function vector depends on the PDBV problem to be solved. For example, for a 1-D PDE system, it may be defined as $B(u) = \{1, u, u^2, ... u^{\beta-1}\}$. For a 2-D second order PDE system, a quadratic basis function $B(u, v)$ is defined as

$$B(u,v) = \{b_1(u,v), b_2(u,v), \ldots, b_{\beta}(u,v)\} = \{1, u, v, u^2, uv, v^2\} \quad (\beta = 6)$$

(3.9)

For the coefficient vector $C^{*T}(x,y) = \{c_1, c_2, ..., c_{\beta}\}$ of Equation (3.8), the coefficients $c_i (i = 1, 2, ..., \beta)$ are unknown, and they are determined by the consistency conditions as

$$b_i(x,y) = \int_{\Omega} C(x,y,u,v) K(x_k - u, y_k - v) b_i(u,v)\, du\, dv \quad (i = 1, 2, ..., \beta) \quad (3.10)$$

There are two distinct differences between the RKPM (Liu et al., 1996) and the presently proposed Hermite-cloud method. The first is that the Hermite-cloud method is based on the fixed reproducing kernel technique, while the RKPM is based on the moving reproducing kernel technique. The second is that the Hermite-cloud method directly solves the discretised strong form of the governing PDEs without the integration requirement, while the RKPM solves the weak form of the governing PDEs and thus requires weak form integration.

For the strong form discretisation, a set of points is scattered arbitrarily in the problem domain Ω and along its edges. By substituting Equation (3.8) into Equation (3.10), the present consistency conditions expressed by Equation (3.10) are rewritten in discrete form as

$$b_i(x,y) = \sum_{n=1}^{N_T} C(x,y,u_n,v_n) K(x_k - u_n, y_k - v_n) b_i(u_n,v_n) \Delta S_n$$

$$= \sum_{n=1}^{N_T} B(u_n,v_n) C^*(x,y) K(x_k - u_n, y_k - v_n) b_i(u_n,v_n) \Delta S_n \quad (i = 1, 2, ..., \beta)$$

(3.11)

where N_T is the total number of the arbitrarily scattered points covering both the interior computational domain Ω and the surrounding edges. The subscript n represents the nth scattered point. ΔS_n is defined as the cloud area of the nth point.

Based on the definition of basis-function vector $B(u,v)$, the linearly independent basis functions $b_i(u,v)$ ($i = 1, 2, ..., \beta$) are known. As a result, Equation (3.11) generates a set of linear algebraic equations with respect to the coefficients c_i ($i = 1, 2, ..., \beta$). By rewriting Equation (3.11) with respect to the coefficient vector $C^{*T}(x,y) = \{c_1, c_2, ..., c_\beta\}$ in matrix form, one can obtain

$$A(x_k, y_k)C^*(x,y) = B^T(x,y) \tag{3.12}$$

or

$$C^*(x,y)A^{-1}(x_k, y_k) = B^T(x,y) \tag{3.13}$$

It is found from Equation (3.12) that $A(x_k, y_k)$ is a symmetric constant matrix and it is associated with the center point (x_k, y_k) of the fixed cloud,

$$A_{ij}(x_k, y_k) = \sum_{n=1}^{N_T} b_i(u_n, v_n) K(x_k - u_n, y_k - v_n) b_j(u_n, v_n) \Delta S_n \quad (i, j = 1, 2, ..., \beta) \tag{3.14}$$

By substituting Equations (3.8) and (3.13) into (3.5), the approximate solution $\tilde{f} = (x,y)$ of the unknown real function $f(x,y)$ in Equation (3.5) is obtained by

$$\tilde{f}(x,y) = \int_\Omega B(u,v)C^*(x,y)K(x_k - u, y_k - v)f(u,v)\,du\,dv$$

$$= \int_\Omega B(u,v)A^{-1}(x_k, y_k)B^T(x,y)K(x_k - u, y_k - v)f(u,v)\,du\,dv \tag{3.15}$$

and it can be written in the discretised form as

$$\tilde{f}(x,y) = \sum_{n=1}^{N_T} (B(u_n, y_k)A^{-1}(x_k, v_n)B^T(x,y)K(x_k - u_n, y_k - v_n)\Delta S_n)$$

$$f_n = \sum_{n=1}^{N_T} N_n(x,y)f_n \tag{3.16}$$

where f_n is the unknown point value at the nth point, and $N_n(x,y)$ are defined as the shape functions that are simply polynomials in x and y. Therefore, any derivative of the shape functions can be derived directly by differentiation of the basis functions in the $B(x, y)$ vector. It is further evident that the present shape functions $N_n(x, y)$ satisfy the consistency conditions defined as Equation (3.10) or (3.11) for all the independent basis functions $b_i(x, y)$ $(i = 1, 2, ..., \beta)$. In particular, when $b_1(x, y) = 1.0$, $b_2(x, y) = x$ and $b_3(x, y) = y$ are taken in the discretised consistency conditions of Equation (3.11), one can have

$$1.0 = \sum_{n=1}^{N_T} B(u_n, v_n) C^*(x, y) K(x_k - u_n, y_k - v_n) \Delta S_n = \sum_{n=1}^{N_T} N_n(x, y) \quad (i = 1) \quad (3.17)$$

$$x = \sum_{n=1}^{N_T} B(u_n, v_n) C^*(x, y) K(x_k - u_n, y_k - v_n) x_n \Delta S_n = \sum_{n=1}^{N_T} N_n(x, y) x_n \quad (i = 2) \quad (3.18)$$

$$y = \sum_{n=1}^{N_T} B(u_n, v_n) C^*(x, y) K(x_k - u_n, y_k - v_n) y_n \Delta S_n = \sum_{n=1}^{N_T} N_n(x, y) y_n \quad (i = 3)$$

$$(3.19)$$

The first-order derivatives of the unknown real function $f(x,y)$ with respect to the variables x and y are given as

$$g_x(x, y) = \frac{\partial f(x, y)}{\partial x}, \quad g_y(x, y) = \frac{\partial f(x, y)}{\partial y} \qquad (3.20)$$

According to the Hermite interpolation theorem, if $g_x(x,y)$ and $g_y(x,y)$ are considered additional unknown functions, they may be also discretised by imposing Equation (3.16) in a similar manner. As a result, the corresponding approximate solutions can be written as

$$\tilde{g}_x(x, y) = \sum_{m=1}^{N_s} M_m(x, y) g_{xm} \qquad (3.21)$$

$$\tilde{g}_y(x, y) = \sum_{m=1}^{N_s} M_m(x, y) g_{ym} \qquad (3.22)$$

where $N_S(\leq N_T)$ is the total number of the randomly scattered points, and $M_m(x, y)$ are the corresponding shape functions for the unknown first-order differential functions $g_x(x, y)$ and $g_y(x, y)$. The shape functions $M_m(x, y)$ can be constructed in similar manner as $N_n(x,y)$, but the basis functions are $B(x, y) \in R^{\beta-1}$ and g_{xm} and g_{ym} are the corresponding unknown point values at the mth point.

Based on the Hermite interpolation theorem, a true meshless approxima-tion $\tilde{f}(x,y)$ of the unknown real function $f(x,y)$ is finally constructed in the following form,

$$
\tilde{f}(x,y) = \sum_{n=1}^{N_T} N_n(x,y)f_n + \sum_{m=1}^{N_S}\left(x - \sum_{n=1}^{N_T} N_n(x,y)x_n \right)M_m(x,y)g_{xm}
$$
$$
+ \sum_{m=1}^{N_S}\left(y - \sum_{n=1}^{N_T} N_n(x,y)y_n \right)M_m(x,y)g_{ym}
$$

(3.23)

The Hermite-based interpolation approximation has so far been constructed in Equation (3.23) with many computational advantages. Most notably, the computational accuracy at scattered discrete points in the domain is much refined not only for the approximate function solutions, but also for the first-order derivatives of these approximate functions. This can be easily understood from the subsequent formulation in Equations (3.36) to (3.39), where the final unknown vector of the set of the algebraic equations derived by the discretised partial differential governing equations is composed of all the three unknown point value vectors $\{f_n\}_{1 \times N_T}$, $\{g_{xm}\}_{1 \times N_S}$ and $\{g_{ym}\}_{1 \times N_S}$. Therefore, the Hermite-cloud method is able to directly compute the approximate solutions of both the unknown function and its corresponding first-order derivatives simultaneously. Apart from that, the Hermite-based interpolation approximation expressed by Equation (3.23) is also employed for the construction of necessary auxiliary conditions as follows.

As mentioned above, the functions $g_x(x,y)$ and $g_y(x,y)$ are introduced as the additional unknown functions such that the auxiliary conditions are required to generate a complete set of PDBV equations. Based on the mathematical definitions of $g_x(x,y)$ and $g_y(x,y)$, the auxiliary conditions are naturally developed by imposing the first-order partial differentials with respect to x and y on the constructed approximate solution $\tilde{f}(x,y)$ expressed by Equation (3.23) as,

$$
\tilde{f}_{,x}(x,y) = \sum_{n=1}^{N_T} N_{n,x}(x,y)f_n
$$
$$
+ \sum_{m=1}^{N_S}\left(1 - \sum_{n=1}^{N_T}(N_{n,x}(x,y)x_n) \right)M_m(x,y)g_{xm} - \sum_{m=1}^{N_S}\left(\sum_{n=1}^{N_T}(N_{n,x}(x,y)y_n) \right)M_m(x,y)g_{ym}
$$
$$
+ \sum_{m=1}^{N_S}\left(x - \sum_{n=1}^{N_T}(N_n(x,y)x_n) \right)M_{m,x}(x,y)g_{xm} + \sum_{m=1}^{N_S}\left(y - \sum_{n=1}^{N_T}(N_n(x,y)y_n) \right)M_{m,x}(x,y)g_{ym}
$$

(3.24)

$$\tilde{f}_{,y}(x,y) = \sum_{n=1}^{N_T} N_{n,y}(x,y) f_n$$

$$+\sum_{m=1}^{N_S} \left(1 - \sum_{n=1}^{N_T}(N_{n,y}(x,y)y_n)\right) M_m(x,y)g_{ym} - \sum_{m=1}^{N_S}\left(\sum_{n=1}^{N_T}(N_{n,y}(x,y)x_n)\right) M_m(x,y)g_{ym}$$

$$+\sum_{m=1}^{N_S}\left(x - \sum_{n=1}^{N_T}(N_n(x,y)x_n)\right) M_{m,y}(x,y)g_{xm} + \sum_{m=1}^{N_S}\left(y - \sum_{n=1}^{N_T}(N_n(x,y)y_n)\right) M_{m,y}(x,y)g_{ym}$$

$$(3.25)$$

where the variable in the subscript after a comma indicates the partial differentiation with respect to that variable. After considering Equations (3.17) to (3.22), both the equations above can be simplified to the auxiliary conditions as follows,

$$\sum_{n=1}^{N_T} N_{n,x}(x,y)f_n - \sum_{m=1}^{N_S}\left(\sum_{n=1}^{N_T}(N_{n,x}(x,y)x_n)\right) M_m(x,y)g_{xm}$$

$$-\sum_{m=1}^{N_S}\left(\sum_{n=1}^{N_T}(N_{n,x}(x,y)y_n)\right) M_m(x,y)g_{ym} = 0$$

$$(3.26)$$

$$\sum_{n=1}^{N_T} N_{n,y}(x,y)f_n - \sum_{m=1}^{N_S}\left(\sum_{n=1}^{N_T}(N_{n,y}(x,y)y_n)\right) M_m(x,y)g_{ym}$$

$$-\sum_{m=1}^{N_S}\left(\sum_{n=1}^{N_T}(N_{n,y}(x,y)x_n)\right) M_m(x,y)g_{ym} = 0$$

$$(3.27)$$

The above auxiliary conditions (3.28) and (3.29) are required in the implementation of the presently developed Hermite-cloud method.

So far the Hermite-cloud method has been developed by the above formulation. In brief, it is based on the Hermite interpolation theorem, composed of the approximation $\tilde{f}(x,y)$ of unknown function $f(x,y)$ by Equation (3.23), and the approximations $\tilde{g}_x(x,y)$ and $\tilde{g}_y(x,y)$ of the first-order derivatives $g_x(x,y)$ and $g_y(x,y)$ by Equations (3.21) and (3.22), and further associated with the auxiliary conditions by Equations (3.26) and (3.27).

3.2.2 Numerical implementation

As mentioned above, the presently developed true meshless Hermite-cloud method employs the point collocation technique for the discretisation of generic engineering partial differential boundary-value (PDBV) problems. For example,

$$Lf(x,y) = P(x,y) \quad \text{PDEs in computational domain } \Omega \tag{3.28}$$

$$f(x,y) = Q(x,y) \quad \text{Dirichlet boundary condition on } \Gamma_D \tag{3.29}$$

$$\partial f(x,y)/\partial n = R(x, y) \quad \text{Neumann boundary condition on } \Gamma_N \tag{3.30}$$

where L is the differential operator, and $f(x,y)$ is an unknown real function. By using the point collocation technique and taking $\tilde{f}(x,y)$ as the approximation of $f(x,y)$, this PDBV problem can be discretised approximately and written as

$$L\tilde{f}(x_i,y_i) = P(x_i,y_i) \qquad i = 1,2,\ldots, N_\Omega \tag{3.31}$$

$$\tilde{f}(x_i,y_i) = Q(x_i,y_i) \qquad i = 1,2,\ldots, N_D \tag{3.32}$$

$$\partial \tilde{f}(x_i,y_i)/\partial n = R(x_i,y_i) \qquad i = 1,2,\ldots, N_N \tag{3.33}$$

where N_Ω, N_D and N_N are the numbers of randomly scattered points in the interior computational domain and along the Dirichlet and Neumann edges, respectively, in which the total number of scattered points is thus $N_T = (N_\Omega + N_D + N_N)$.

By substituting the approximation solutions by Equations (3.21) to (3.23) into Equations (3.33) to (3.35), by constructing the present Hermite-cloud method in the discretised PDBV problem, imposing the auxiliary conditions of Equations (3.26) to (3.27) and then rearranging the resulting equations, a set of discrete algebraic equations with respect to the unknown point values f_i, g_{xi} and g_{yi} is thus obtained and written in the matrix form as

$$[H_{ij}]_{(N_T+2N_s)\times(N_T+2N_s)}\{F_i\}_{(N_T+2N_s)\times1} = \{d_i\}_{(N_T+2N_s)\times1} \tag{3.34}$$

where $\{d_i\}$ and $\{F_i\}$ are $(N_T + 2N_s)$ order column vectors,

$$\{F_i\}_{(N_T+2N_s)\times1} = \{\{f_i\}_{1\times N_T}, \{g_{xi}\}_{1\times N_S}, \{g_{yi}\}_{1\times N_S}\}^T \tag{3.35}$$

$$\{d_i\}_{(N_T+2N_S)\times1} = \{\{P(x_i,y_i)\}_{1\times N_\Omega}, \{Q(x_i,y_i)\}_{1\times N_D}, \{R(x_i,y_i)\}_{1\times N_N}, \{0\}_{1\times 2N_S}\}^T$$

$$\tag{3.36}$$

$[H_{ij}]$ is a $(N_T + 2N_S) \times (N_T + 2N_S)$ coefficient square matrix

$[H_{ij}] =$

$$
\begin{bmatrix}
[LN_j(x_i,y_i)]_{N_Q \times N_T} & \left[L((x_i - \sum\limits_{n=1}^{N_T} N_n(x_i,y_i)x_n)M_j(x_i,y_i)) \right]_{N_Q \times N_S} & \left[L((y_i - \sum\limits_{n=1}^{N_T} N_n(x_i,y_i)y_n)M_j(x_i,y_i)) \right]_{N_Q \times N_S} \\
[N_j(x_i,y_i)]_{N_D \times N_T} & [0]_{N_D \times N_S} & [0]_{N_D \times N_S} \\
[0]_{N_N \times N_T} & [M_j(x_i,y_i)]_{N_N \times N_S} & [M_j(x_i,y_i)]_{N_N \times N_S} \\
[N_{j,x}(x_i,y_i)]_{N_S \times N_T} & \left[-\sum\limits_{n=1}^{N_T} N_{n,x}(x_i,y_i)x_n)M_j(x_j,y_i) \right]_{N_S \times N_S} & \left[-\sum\limits_{n=1}^{N_T} N_{n,x}(x_i,y_i)y_n)M_j(x_i,y_i) \right]_{N_S \times N_S} \\
[N_{j,y}(x_i,y_i)]_{N_S \times N_T} & \left[-\sum\limits_{n=1}^{N_T} N_{n,y}(x_i,y_i)x_n)M_j(x_j,y_i) \right]_{N_S \times N_S} & \left[-\sum\limits_{n=1}^{N_T} N_{n,y}(x_i,y_i)y_n)M_j(x_i,y_i) \right]_{N_S \times N_S}
\end{bmatrix}
$$

(3.37)

After the complete set of linear algebraic equation given in Equation (3.36) is solved numerically, $(N_T + 2N_S)$ point values $\{F_i\}$ are obtained; they consist of the N_T point values $\{F_i\}$ and $2N_S$ point values $\{g_{xi}\}$ and $\{g_{yi}\}$. Finally the approximate solution $\tilde{f}(x,y)$ and the corresponding first-order derivatives $\tilde{g}_x(x,y)$ and $\tilde{g}_y(x,y)$ of the PDBV problem can be computed through Equations (3.23), (3.21), and (3.22), respectively.

3.2.3 Examples for validation

In order to validate the computational accuracy and examine the numerical convergence of the present true meshless Hermite-cloud method, numerical comparisons are conducted for several classical 2-D partial differential boundary-value (PDBV) problems including include the 2-D Laplace equation with various mixed boundary conditions, the steady state heat conduction with a high gradient, and the 2-D Poisson equation with a local high gradient (Figure 3.1 through Figure 3.5). Using a refined version of the definition of the standard error, a measure of the global error ξ is defined below for the present numerical comparisons (Mukherjee and Mukherjee, 1997):

$$
\xi = \frac{1}{|f_{max}|} \sqrt{\frac{1}{N_T} \sum_{i=1}^{N_T} (\tilde{f}_i - f_i)^2}
$$

(3.38)

Example 3.1: Two-dimensional Laplace equation with various mixed boundary conditions

Let us consider a classical 2-D Laplace equation in a unit-square computational domain as follows

$$
\frac{\partial^2 f(x,y)}{\partial x^2} + \frac{\partial^2 f(x,y)}{\partial y^2} = 0 \qquad 0 < x < 1 \quad \text{and} \quad 0 < y < 1
$$

(3.39)

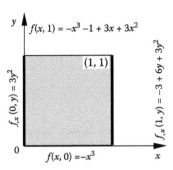

(a) Computational domain and boundary conditions.

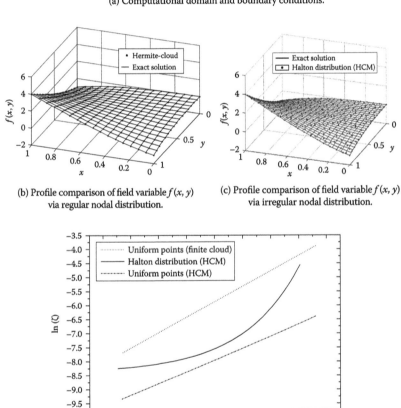

(b) Profile comparison of field variable $f(x, y)$ via regular nodal distribution.

(c) Profile comparison of field variable $f(x, y)$ via irregular nodal distribution.

(d) Convergence comparison (ξ-Global error, Δ Point distance).

Figure 3.1 Two-dimensional Laplace equation with mixed Dirichlet and Neumann boundary conditions. (From H. Li, T.Y. Ng, J.Q. Cheng. et al. (2003). *Computational Mechanics*, 33, 30–41. With permission.)

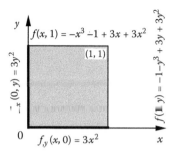

(a) Computational domain and boundary conditions.

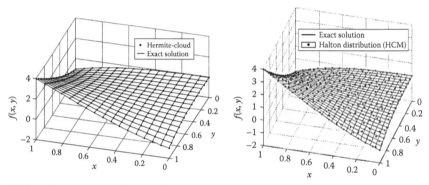

(b) Profile comparison of field variable $f(x, y)$ via regular nodal distribution.

(c) Profile comparison of field variable $f(x, y)$ via irregular nodal distribution.

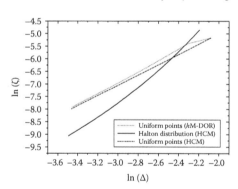

(d) Convergence comparison (Global error, point distance).

Figure 3.2 Two-dimensional Laplace equation with mixed Dirichlet and Neumann boundary conditions. (From H Li, T.Y. Ng, J.Q. Cheng et al. (2003). *Computational Mechanics*, 33, 30–41. With permission.)

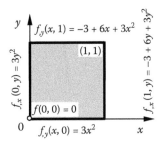

(a) Computational domain and boundary conditions.

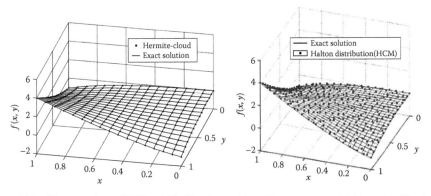

(b) Profile comparison of field variable $f(x, y)$ via regular nodal distribution.

(c) Profile comparison of field variable $f(x, y)$ via irregular nodal distribution.

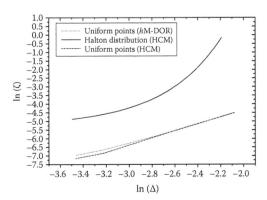

(d) Convergence comparison (Global error, point distance).

Figure 3.3 Two-dimensional Laplace equation with modified Neumann boundary conditions. (From H Li, T.Y. Ng, J.Q. Cheng et al. (2003). *Computational Mechanics*, 33, 30–41. With permission.)

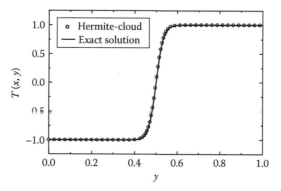

(a) Profile comparison of field variable $T(x, y)$.

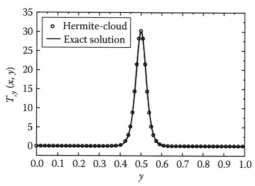

(b) Profile comparison of field variable $T_y(x, y)$.

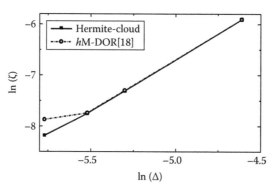

(c) Convergence comparison (Global error, point distance).

Figure 3.4 Steady-state heat conduction with high gradient. (From H Li, T.Y. Ng, J.Q. Cheng et al. (2003). *Computational Mechanics*, 33, 30–41. With permission.)

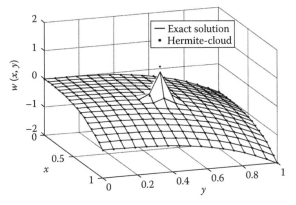

(a) Profile comparison of field variable $w(x, y)$.

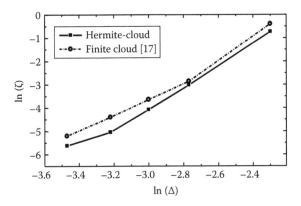

(b) Convergence comparison (Global error, point distance).

Figure 3.5 Two-dimensional Poisson equation with local high gradient. (From H Li, T.Y. Ng, J.Q. Cheng et al. (2003). *Computational Mechanics*, 33, 30–41. With permission.)

with the three mixed Dirichlet and Neumann boundary conditions along the four edges, as shown. The first two boundary cases, as shown in Figure 3.1 and Figure 3.2 are for mixed Dirichlet and Neumann boundary conditions. Figure 3.3 is for modified Neumann boundary conditions. The exact solution in the cubic form for the given 2-D Laplace problem with all the three boundary condition cases is given as

$$f(x, y) = -x^3 - y^3 + 3xy^2 + 3x^2y \tag{3.40}$$

The corresponding first-order derivatives with respect to x and y, respectively, are given as

$$f_{,x}(x, y) = -3x^2 + 3y^2 + 6xy, \quad f_{,y}(x, y) = -3y^2 + 3x^2 + 6xy \tag{3.41}$$

Table 3.1 Comparisons of Convergence with Exact Solution for
2-D Laplace Equation Mixed Boundary Conditions

Boundary Condition	Methods	Point/Cell Distributions	Global Error (ξ) (%)
Case 1	Present Hermite-cloud	9 × 9 Points	0.166
		11 × 11 Points	0.101
		17 × 17 Points	0.037
		26 × 26 Points	0.015
		33 × 33 Points	0.009
	MLS*	4 × 4 Cells, 9 × 9 Points	11.81
	EFG*	4 × 4 Cells, 9 × 9 Points	3.05
Case 2	Present Hermite-cloud	9 × 9 Points	0.5725
		11 × 11 Points	0.3636
		17 × 17 Points	0.1378
		26 × 26 Points	0.05575
		33 × 33 Points	0.03375
	MLS*	4 × 4 Cells, 9 × 9 Points	3.06
	EFG*	4 × 4 Cells, 9 × 9 Points	0.5
Case 3	Present Hermite-cloud	9 × 9 Points	1.105
		11 × 11 Points	0.7075
		17 × 17 Points	0.2775
		26 × 26 Points	0.1063
		31 × 31 Points	0.07775
	MLS*	4 × 4 Cells, 9 × 9 Points	3.01
	EFG*	4 × 4 Cells, 9 × 9 Points	0.9

Source: H Li, T.Y. Ng, J.Q. Cheng et al. (2003). *Computational Mechanics*, 33,
30–41. With permission.

*   Y.X. Mukherjee and S. Mukherjee. (1997). *Computational Mechanics* 19,
264–270.

The numerical analysis and computational accuracy comparison of the
different meshless techniques are shown in Table 3.1 for solving the 2-D
Laplace equation with the three cases of the different boundary-conditions.
The methods compared include the present Hermite-cloud method, the
classical moving least squares (MLS) method, and *h*-refinement of the element-free Galerkin (EFG) method (Mukherjee and Mukherjee, 1997). The
global errors decrease rapidly in a monotonic manner, with the increase
of regularly scattered point. They are generally less than 0.07775% for the
convergence solutions, for example, 0.00883% for case 1, 0.03375% for
case 2 and 0.07775% for case 3.

By the comparison of the present Hermite-cloud method with the existing MLS method and the *h*-refinement of the EFG method, it is observed that the computational accuracy of the Hermite-cloud method is much better than those of the MLS and EFG methods (Mukherjee and Mukherjee, 1997) when the distributive density of the regularly scattered points becomes larger than an 11 × 11 distribution. For example, the computational accuracy of the Hermite-cloud method for the case one is refined about 15 times that of the MLS and EFG methods with the same distribution density of regularly scattered points.

It is also noted that the present Hermite-cloud method as a true meshless technique is still able to achieve much better computational accuracy than the existing meshless MLS and EFG methods that are not true meshless techniques and thus the corresponding computations are more expensive due to the requirement of background meshing. Therefore, higher efficiency and elegance are achieved by the present Hermite-cloud method. For the three cases of different boundary conditions, the profiles of the field variable solution f(x, y) of the 2-D Laplace equation using the uniform nodal distribution are plotted in the (b) sections of Figures 3.1 through 3.3 while those using the Halton type of non-uniform irregular point distribution are plotted in the (c) sections of the figures. All the results mentioned are obtained through the present Hermite-cloud method, and the plotted results in the figures demonstrate very good agreement with the published solutions (Mukherjee an d Mukherjee, 1997). The comparison of the convergence properties for the three cases with the different boundary conditions are shown in the (d) sections of Figures 3.1 through 3.3 by the present Hermite-cloud method with regular and irregular nodal distributions, the finite cloud method (Aluru and Li, 2001), and the *h*M-DOR method developed by Ng et al. (2003). For the uniform nodal distribution, the convergence rates of the present Hermite-cloud approximate solutions are 2.17 for the case one, 2.04 for the case two, and 2.33 for the case three. It is concluded from all the three figures that the convergence properties of the present Hermite-cloud method are generally better than those of the finite cloud and *h*M-DOR methods, if the nodes are distributed uniformly.

Example 3.2: Steady-state heat conduction with high gradient

Let us consider a steady-state heat conduction problem defined in a two-dimensional rectangular plate domain with a heat source, which is governed via the partial differential equation of the temperature field as follows,

$$\frac{\partial^2 T(x,y)}{\partial x^2} + \frac{\partial^2 T(x,y)}{\partial y^2} = -2s^2 \sec h^2[s(y-0.5)] \tanh[s(y-0.5)]$$

$$(3.42)$$

$$0 < x < 0.5 \quad and \quad 0 < y < 1$$

The boundary conditions are given as

$$T(x,y)|_{y=0} = -\tanh(s/2), \quad T(x,y)|_{y=1} = \tanh(s/2) \tag{3.43}$$

$$\left.\frac{\partial T(x,y)}{\partial x}\right|_{x=0} = 0, \qquad \left.\frac{\partial T(x,y)}{\partial x}\right|_{x=0.5} = 0 \tag{3.44}$$

where s is a free parameter. The field variable $T_y(x,y)$ will have an increasing gradient, as s increases. For this steady-state heat conduction problem with a high gradient occurring near $y = 0.5$, the exact solution of temperature field $T(x,y)$ is given as

$$T(x,y) = \tanh[s(y-0.5)] \tag{3.45}$$

and

$$T_y(x,y) = s \cdot \sec h^2[s(y-0.5)] \tag{3.46}$$

By the presently developed Hermite-cloud method, the meshless numerical computation is conducted for the above 2-D steady-state heat conduction problem with a local gradient, and the corresponding computational results with comparisons are illustrated in both Table 3.2 and Figure 3.4. The temperature field $T(x,y)$ and its corresponding first-order derivative $T_y(x,y)$ are plotted in (a) and (b), respectively, and then compared with the exact solutions. It can be seen that the meshless Hermite-cloud method captures the characteristic of the high gradient near $y = 0.5$ very well.

In (c), the convergence of the present method is compared with that of the hM-DOR method (Ng et al., 2003), where it is noted that the convergence rate of the Hermite-cloud solution is 1.99 while that of the hM-DOR method is 1.82. It is also known that the global error of the Hermite-cloud method decreases linearly with an increase in the number of scattered points, and is generally better than that of the hM-DOR

Table 3.2 Comparisons of convergence with exact solution for steady-state heat conduction with high gradient

Point distribution	Global Error (ξ) for T(x,y) (%)		Global Error (ξ) for T$_y$(x,y) (%)	
	Present Hermite-cloud	hM-DOR (Ng et al.,2003)	Present Hermite-cloud	hM-DOR (Ng et al.,2003)
3 × 51	1.12	1.12	0.467	0.627
3 × 101	0.271	0.271	0.104	0.121
3 × 201	0.0671	0.0673	0.0276	0.0287
3 × 251	0.0429	0.0432	0.0178	0.0185
3 × 321	0.0279	0.0381	0.0109	0.0112

Source: H Li, T.Y. Ng, J.Q. Cheng et al. (2003). *Computational Mechanics*, 33, 30–41. With permission.

method, which experiences an adverse change in gradient when the relative distance between the points becomes smaller than a certain value. As a result, it is concluded that probably the present Hermite-cloud method possesses better computational stability even for the partial different equation with a local high gradient.

Table 3.2 provides detailed numerical information for convergence comparisons of the present Hermite-cloud method and the hM-DOR method. It is found that the difference between the global error of the two methods increases with the increase of the scattered points. This is in concordance with the results in (c). It is further observed that the Hermite-cloud method can achieve the very good characteristics of the convergence of the first-order derivatives. Probably this explains why the present Hermite-cloud method can achieve good computational accuracy not only for the approximate field-variable solutions but also for their first-order derivatives. It is also known that the global error incurred by the Hermite-cloud method decreases rapidly, as the number of the scattered points increases, for both the temperature field $T(x, y)$ and its corresponding first-order derivative $T_{,y}(x, y)$, with the minor errors of 0.0279% for $T(x, y)$, and 0.0109% for $T_{,y}(x, y)$, respectively.

Example 3.3: Two-dimensional Poisson equation with local high gradient

As the third example for the numerical case study, a 2-D Poisson equation with a local high gradient is considered in a unit-square computational domain, given as follows

$$\frac{\partial^2 w(x, y)}{\partial x^2} + \frac{\partial^2 w(x, y)}{\partial y^2} = -6(x + y) - \frac{4}{a^4}(a^2 - (x - b)^2 - (y - b)^2)$$

(3.47)

$$\exp\left[-\left(\frac{x - b}{a}\right)^2 - 4\left(\frac{y - b}{a}\right)^2\right] \quad 0 < x < 1 \quad and \quad 0 < y < 1$$

with the boundary conditions below

$$w(x, y)|_{x=0} = \exp[-(b^2 + (y - b)^2/a^2] - y^3 \tag{3.48}$$

$$w(x, y)|_{x=1} = \exp[-((1 - b)^2 + (y - b)^2)/a^2] - (1 + y^3) \tag{3.49}$$

$$w_{,y}(x, y)|_{y=0} = 2b \exp[-(b^2 + (x - b)^2)/a^2]/a^2 \tag{3.50}$$

$$w_{,y}(x, y)|_{y=1} = -3 - 2(1 - b)\exp[-((x - b)^2 + (1 - b)^2/a^2]/a^2 \tag{3.51}$$

The exact solution of the given 2-D Poisson equation with a local high gradient is written as

$$w(x, y) = -(x^3 + y^3) + \exp\left[-\left(\frac{x - b}{a}\right)^2 - \left(\frac{y - b}{a}\right)^2\right] \tag{3.52}$$

and

$$w_{,x}(x,y) = -3x^2 - 2(x-b)\exp[-((x-b)^2 + (y-b)^2/a^2]/a^2 \qquad (3.53)$$

$$w_{,y}(x,y) = -3y^2 - 2(y-b)\exp[-((x-b)^2 + (y-b)^2)/a^2]/a^2 \qquad (3.54)$$

It is noted that the local high gradient occurs near the centre point (0.5, 0.5) of the unit-square computational domain, if $a = 0.05$ and $b = 0.5$ are taken respectively. By the Hermite-cloud method, the present problem is solved numerically and the computational results are shown in Figure 3.5. Figure 3.5(a) illustrates the profiles of the field variable $w(x,y)$ by both the present Hermite-cloud method and the exact solutions. The numerical results by Hermite-cloud method very well match the exact solution, especially the local high gradient near the centre point is captured satisfactorily.

The global error of the field variable $w(x,y)$ is 0.724% and the errors of the first-order derivatives $w_{,x}(x,y)$ and $w_{,y}(x,y)$ are both 0.326%. This example again validates the numerical computational accuracy of the developed Hermite-cloud method for both the field-variable solutions and the corresponding first-order derivatives. Figure 3.5(b) demonstrates the comparison of the convergence characteristics for this problem between the present Hermite-cloud method and the finite cloud method (Aluru and Li, 2001), in which the convergence characteristics of the Hermite-cloud method are distinctly better.

3.2.4 Remarks

The presently developed Hermite-cloud method combines the point collocation technique for the discretisation of partial differential governing equations and employs the Hermite theorem for the construction of interpolation functions. The method is based on the classical RKPM, except that a fixed kernel is employed as the kernel function. The Hermite types of the interpolation functions developed require certain auxiliary conditions for the complete set of the discretised partial differential governing equations with the mixed Dirichlet and Neumann boundary conditions, as derived above. It is validated via the numerical case studies that the Hermite-cloud method is numerically stable and efficient. It is also shown that the computational accuracy of the present Hermite-cloud method is significantly refined at the scattered discrete points in the domain, not only for approximate field variable solutions, but also for the corresponding first-order derivatives of approximate solutions. Therefore, it is concluded that the presently developed Hermite-cloud method is a very efficient numerical technique for solving the partial differential boundary value problems, especially with high gradient fields.

3.3 POINT WEIGHTED LEAST-SQUARES METHOD

In this section, as a truly meshless technique, the point weighted least-squares (PWLS) method is developed. Two sets of distributed points are adopted and they are called the field nodes and the collocation points. The field nodes are used for the construction of trial functions, in which the radial point interpolation technique, based on the locally supported radial base function, is employed. The collocation points are independent of the field nodes, and are adopted to form the total residuals of the problem. The weighted least-squares technique is then used to obtain the solution of the problem by minimizing the functional of the summation of residuals. The present PWLS method possesses advantages over the conventional collocation methods. For example, it is very stable, the boundary conditions can be easily enforced, and the final coefficient matrix is symmetric. Several numerical examples are presented to validate the computational accuracy and demonstrate the performance of the method through 1-D and 2-D ordinary and partial differential equations. It is finally concluded after the implementation that the presently developed PWLS method can achieve good computational accuracy and works efficiently.

3.3.1 Formulation of PWLS method

In this subsection, several mathematical theories and techniques are introduced first as preliminary background information. They include the construction of the meshless shape function, the weighted least-squares technique, and the direct collocation (DC) method. After that, the point weighted least-squares (PWLS) method is formulated.

3.3.1.1 Construction of meshless shape function

Several techniques have been developed to construct the shape functions in the development of the meshless methods, for example, the moving least-squares (MLS) approximation (Lancaster and Salkauskas, 1986), the radial point interpolation method (RPIM; Liu et al., 2006) and the Kriging interpolation (Krige, 1975; Li et al., 2004c). In these approximation techniques, the Kriging interpolation is a form of the generalized linear regression for the formulation of an optimal estimator in the minimum mean square error sense. The RPIM employs the radial basis function and passing nodal value interpolation for construction of the shape functions. Dai et al. (2003) have shown that the Kriging interpolation and the RPIM can produce the same shape function, although they are derived via different mathematical approaches. However, the algorithm of the RPIM is simpler than that of the Kriging interpolation. Therefore, the RPIM is employed in this work.

Let us consider a random function $u(\mathbf{x})$ defined in the domain Ω discretised by a set of scattered nodes $x_i (1 \leq i \leq N)$, where N is the number of field nodes in the domain Ω. It is assumed that only the n nodes surrounding the point $\mathbf{x}$ have the effect on $u(\mathbf{x})$. The domain containing these surrounding nodes is called the influence or interpolation domain. The estimated value of function $u(\mathbf{x})$ at point $\mathbf{x}$ can be formulated by

$$u(\mathbf{x}) = \sum_{i=1}^{n} R_i(\mathbf{x})a_i + \sum_{j=1}^{m} p_j(\mathbf{x})b_j = \{R^T(\mathbf{x}), p^T(\mathbf{x})\} \left\{ \begin{array}{c} a \\ b \end{array} \right\} \tag{3.55}$$

where $R_i(\mathbf{x})$ is the radial basis function, n is the number of the nodes in the interpolation of $\mathbf{x}$, and $p_j(\mathbf{x})$ is the polynomial basis function that has the monomial terms. For example, in 2-D domain $\mathbf{x}^T = \{x, y\}, \mathbf{p}^T(\mathbf{x}) = \{1, x, y, x^2, xy, y^2, ...\}$, m is the number of $p_j(\mathbf{x})$ and usually $m < n$. a_i and b_j are the coefficients for $R_i(\mathbf{x})$ and $p_j(\mathbf{x})$, respectively. The vectors in Equation (3.55) are defined as

$$\mathbf{R}^T(\mathbf{x}) = \{R_1(\mathbf{x}), R_2(\mathbf{x}), ..., R_n(\mathbf{x})\}, \qquad \mathbf{P}^T(\mathbf{x}) = \{p_1(\mathbf{x}), p_2(\mathbf{x}), ..., p_m(\mathbf{x})\},$$

$$\mathbf{a}^T(\mathbf{x}) = \{a_1, a_2, ..., a_n\}, \quad \mathbf{b}^T(\mathbf{x}) = \{b_1, b_2, ..., b_m\}$$

$$\tag{3.56}$$

In the radial basis function (RBF) expressed by $R_i(\mathbf{x})$, the variable is the distance r_i only between the interpolation point $\mathbf{x}$ and a node $\mathbf{x}_i$, where $r_i = \sqrt{(x - x_i)^2 + (y - y_i)^2}$ for the 2-D space. Different types of radial basis functions are developed. The characteristics of the radial basis functions were widely investigated (Schaback and Wendland, 2000; Liu, 2003). The multi-quadrics (MQ) function is one of the most widely used radial basis functions and it is used in this work, although the other radial basis functions can also be used similarly. The form of the MQ-RBF is written as

$$R_i(r) = \left(r_i^2 + C^2\right)^p \tag{3.57}$$

where p and C are two unknown parameters that need to be determined. Usually they satisfy

$$p = I/2, \quad C > 0 \tag{3.58}$$

where I is an odd integer. The parameters p and C influence the performance of the MQ-RBF. However, so far no successful rigorous methods have been developed to achieve the theoretical best values for these parameters. In general, the parameters p and C can be determined by numerical examination. Detailed investigation of these parameters were done by Liu (2003) who considered p an opening parameter. They reported that $p = 1.03$

and $C = \alpha_c d_c (\alpha_c = 1.0 \sim 6.0)$, where d_c is the nodal spacing. These values can lead to good results for many engineering problems considered. As a result, $p = 1.03$ and $C = d_c (\alpha_c = 1.0)$ are used in the present work.

Polynomial basis functions are used in the function approximation expressed by Equation (3.55). In order to ensure that the interpolation matrix of the radial basis function is invertible, the polynomials added into the radial basis function cannot be arbitrary (Schaback and Wendland, 2000). Usually the polynomial with low degree is adopted to augment the radial basis function to guarantee the non-singularity of the matrix. For simplicity, the linear polynomial is always added into the radial basis function so that it can ensure the linear consistence (Liu, 2003). Therefore, the linear polynomials are used in the present numerical examinations, namely $m = 2$ for the 1-D cases and $m = 3$ for the 2-D problems. The coefficients a_i and b_j in Equation (3.55) are determined through enforcing the interpolation passing the n nodes surrounding the point $\mathbf{x}$. In addition, the following constraint condition is required to be satisfied in order to obtain unique solution (Golberg et al., 1999).

$$\sum_{i=1}^{n} p_j(\mathbf{x}_i) a_i = 0 \qquad j = 1, 2, \ldots, m \tag{3.59}$$

Equations (3.55) and (3.59) are rewritten here in the matrix form as follows

$$\mathbf{u}_s = \left\{ \begin{array}{c} \mathbf{u}_e \\ 0 \end{array} \right\} = \left[\begin{array}{cc} \mathbf{R}_0 & \mathbf{P}_m \\ \mathbf{P}_m^T & 0 \end{array} \right] \left\{ \begin{array}{c} \mathbf{a} \\ \mathbf{b} \end{array} \right\} = \mathbf{G} \left\{ \begin{array}{c} \mathbf{a} \\ \mathbf{b} \end{array} \right\} \tag{3.60}$$

where

$$\mathbf{u}_e = \left\{ u_1, u_2, \ldots, u_n \right\}^T \tag{3.61}$$

$$\mathbf{R}_0 = \left[\begin{array}{cccc} R_1(r_1) & R_2(r_1) & \cdots & R_n(r_1) \\ R_1(r_2) & R_2(r_2) & \cdots & R_n(r_2) \\ R_1(r_n) & R_2(r_n) & \cdots & R_n(r_n) \end{array} \right] \tag{3.62}$$

$$\mathbf{P}_m = \left[\begin{array}{cccc} 1 & 1 & \cdots & 1 \\ x_1 & x_2 & \cdots & x_n \\ y_1 & y_2 & \cdots & y_n \end{array} \right]^T \tag{3.63}$$

Since the distance is directionless, namely $R_j(r_i) = R_i(r_j)$, the matrix $\mathbf{R}_0$ is symmetric, and thus the matrix $\mathbf{G}$ is also symmetric. By solving Equation (3.60), the solution is obtained as

$$\left\{ \begin{array}{c} \mathbf{a} \\ \mathbf{b} \end{array} \right\} = \mathbf{G}^{-1}\mathbf{u}_s \tag{3.64}$$

By substituting the above solution expression into Equation (3.55), we have

$$u(\mathbf{x}) = \{\mathbf{R}^T(\mathbf{x}), \mathbf{p}^T(\mathbf{x})\}\mathbf{G}^{-1}\mathbf{u}_s = \mathbf{\Phi}^T(\mathbf{x})\mathbf{u}_s \tag{3.65}$$

where the shape function $\mathbf{\Phi}(\mathbf{x})$ is defined as

$$\mathbf{\Phi}^T(\mathbf{x}) = \{\phi_1(\mathbf{x}), \phi_2(\mathbf{x}), \ldots, \phi_{n+m}(\mathbf{x})\} = \{\mathbf{R}^T(\mathbf{x}), \mathbf{p}^T(\mathbf{x})\}\mathbf{G}^{-1} \tag{3.66}$$

The derivatives of $u(\mathbf{x})$ can be easily obtained by

$$u_{,l}(\mathbf{x}) = \mathbf{\Phi}_{,l}^T(\mathbf{x})\mathbf{u}_s = \left\{\mathbf{R}_{,l}^T(\mathbf{x}), \mathbf{p}_{,l}^T(\mathbf{x})\right\}\mathbf{G}^{-1}\mathbf{u}_s \tag{3.67}$$

where l denote the coordinates x and y. A comma designates a partial derivative with respect to the indicated spatial variable. It is known from the above formulation that the shape functions possess the delta function property (Gu and Liu, 2003), namely

$$\phi_i(x_j) = \delta_{ij} = \left\{ \begin{array}{ll} 1 & (i = j, \quad i = 1 \sim n) \\ 0 & (i \neq j, \quad i, j = 1 \sim n \end{array} \right. \tag{3.68}$$

and they satisfy the partition of unity, namely

$$\sum_{i=1}^{n} \phi_i(x_i) = 1 \tag{3.69}$$

3.3.1.2 Weighted least-squares technique

Let us consider the following (partial) differential equation

$$A(u) + q = 0 \text{ in problem domain } \Omega, \tag{3.70}$$

with the Neumann boundary condition given by

$$B(u) = \bar{t} \text{ on the boundary } \Gamma_t \tag{3.71}$$

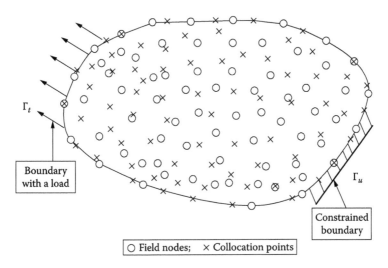

Boundary with a load

Γ_t

Γ_u

Constrained boundary

○ Field nodes;   × Collocation points

Figure 3.6 Problem domain discretised by field nodes and collocation points. (Modified from Q.X. Wang, H. Li, and K.Y. Lam. (2005). *Computational Mechanics,* 35, 170–181.)

and the Dirichlet (essential) boundary condition given by

$$u = \bar{u} \text{ on the boundary } \Gamma_u \tag{3.72}$$

where A, B are (partial) differential operators, and q represents an external source or force imposed over the problem domain.

As shown in Figure 3.6, the problem domain and boundaries are discretised by the N field nodes. Apart from the field nodes, the collocation points are also employed in the problem domain and along the boundaries for the construction of system equations. The total number of collocation points is M, and it includes the M_d internal collocation points, the M_u collocation points on the Dirichlet boundary, and the M_t collocation points on the Neumann boundary, namely

$$M = M_d + M_u + M_t \tag{3.73}$$

The approximate function u at a collocation point $\mathbf{x}_k$ is computed through the interpolation Equation (3.65) as

$$u(\mathbf{x}_k) = \sum_{j=1}^{n} \phi(\mathbf{x}_k) u_j \qquad (k = 1, 2, \ldots, M) \tag{3.74}$$

where n is the number of the filed nodes used in the local interpolation domain.

3.3.1.3 Direct collocation (DC) method

By substituting Equation (3.74) into Equations (3.70) to (3.72), we can have a system of equations for the direct collocation method as follows. For the M_d internal collocation points,

$$A(u_k) + q = \sum_{j=1}^{n} A(\phi_j)\, u_j + q(\mathbf{x}_k) = 0 \qquad (k = 1, \dots, M_d) \tag{3.75}$$

For the M_t collocation points on the Neumann boundary,

$$B(u_k) = \sum_{j=1}^{n} B(\phi_j)\, u_j = \overline{t}(\mathbf{x}_k) \quad (k = 1, 2, \dots, M_t) \tag{3.76}$$

For the M_u collocation points on the Dirichlet boundary,

$$u(\mathbf{x}_k) = \sum_{j=1}^{n} \phi_j u_j = \overline{u}(\mathbf{x}_k) \qquad (k = 1, 2, \dots, M_u) \tag{3.77}$$

By the assembly of Equations (3.75) to (3.77), we can obtain the following system of equations

$$\mathbf{K}\mathbf{U} = \mathbf{F} \tag{3.78}$$

The dimension of the matrix $\mathbf{K}$ is $(M \times N)$. In order to solve Equation (3.78), the following two techniques can be employed.

1. When $M = N$, namely the number of the collocation points is same as the number of the field nodes, the implementation becomes easy by simply considering the field nodes as the collocation points.
2. When $M > N$, the least square technique is employed to solve the system of equations (Zhang et al. 2001).

The direct collocation method is easy to implement. However, it is often numerically unstable and less accurate. Probably one of the reasons is that in the method the governing equations and boundary conditions are considered only on the collocation points by a point-by-point manner, while they are neglected on the other points. In addition, it is difficult for the direct collocation method to accurately enforce boundary conditions, especially the derivative (Neumann) boundary conditions. In the direct collocation method, the Neumann boundary conditions are enforced through a series of the separate Equations (3.76) at the M_t boundary points. Furthermore, it can be easily found that usually the matrix $\mathbf{K}$ in Equation (3.78) is asymmetric.

3.3.1.4 Formulation of point weighted least-squares (PWLS) method

When the approximation by Equation (3.74) is used, usually the governing Equation (3.70) and the derivative boundary conditions (3.71) cannot be satisfied exactly. In addition, if the RPIM shape functions with the delta function property are considered, the collocation points can be arranged to coincide with the field nodes on the Dirichlet boundary. This will always make the Dirichlet boundary condition (3.72) exactly satisfied. By substituting Equation (3.74) into Equations (3.70) and (3.71), the following residual functions $R^{(d)}$ and $R^{(t)}$ are obtained for the system of equations defined in the problem domain and the Neumann boundary, respectively.

$$R_k^{(d)} = A(u_k) + q(\mathbf{x}_k) = \sum_{j=1}^{n} A(\phi_j) u_j + q(\mathbf{x}_k) \qquad (k = 1 \sim M) \tag{3.79}$$

$$R_k^{(t)} = B(u_k) - \bar{t}(\mathbf{x}_k) = \sum_{j=1}^{n} B(\phi_j) u_j - \bar{t}(\mathbf{x}_k) \qquad (k = 1 \sim M_t) \tag{3.80}$$

The following weighted functional of all the residuals is formulated at all the collocation points,

$$
\begin{aligned}
J &= \sum_{k=1}^{M} \left[W_k^{(d)} \, R_k^{(d)} \right]^2 + \sum_{k=1}^{M_t} \left[W_k^{(t)} \, R_k^{(t)} \right]^2 \\
&= \sum_{k=1}^{M} \left[W_k^{(d)} \left(\sum_{j=1}^{n} A(\phi_j) u_j + q(\mathbf{x}_k) \right) \right]^2 + \sum_{k=1}^{M_t} \left[W_k^{(t)} \left(\sum_{j=1}^{n} B(\phi_j) u_j - \bar{t}(\mathbf{x}_k) \right) \right]^2
\end{aligned}
\tag{3.81}
$$

where $W_k^{(d)}$ and $W_k^{(t)}$ are the weight coefficients that will be discussed in detail in the subsequent subsection. By minimizing the functional J, one can have

$$\frac{\partial J}{\partial u_1} = \frac{\partial}{\partial u_1} \left\{ \sum_{k=1}^{M} \left[W_k^{(d)} \left(\sum_{j=1}^{n} A(\phi_j) u_j + q(\mathbf{x}_k) \right) \right]^2 + \sum_{k=1}^{M_t} \left[W_k^{(t)} \left(\sum_{j=1}^{n} B(\phi_j) u_j - \bar{t}(\mathbf{x}_k) \right) \right]^2 \right\} = 0$$

$$\frac{\partial J}{\partial u_2} = \frac{\partial}{\partial u_2} \left\{ \sum_{k=1}^{M} \left[W_k^{(d)} \left(\sum_{j} A(\phi_j) u_j + q(\mathbf{x}_k) \right) \right]^2 + \sum_{k=1}^{M_t} \left[W_k^{(t)} \left(\sum_{j=1}^{n} B(\phi_j) u_j - \bar{t}(\mathbf{x}_k) \right) \right]^2 \right\} = 0$$

$$\vdots$$

$$\frac{\partial J}{\partial u_N} = \frac{\partial}{\partial u_N} \left\{ \sum_{k=1}^{M} \left[W_k^{(d)} \left(\sum_{j} A(\phi_j) u_j + q(\mathbf{x}_k) \right) \right]^2 + \sum_{k=1}^{M_t} \left[W_k^{(t)} \left(\sum_{j=1}^{n} B(\phi_j) u_j - \bar{t}(\mathbf{x}_k) \right) \right]^2 \right\} = 0$$

$$\tag{3.82}$$

Equation (3.82) can also be rewritten in the following matrix form as

$$\mathbf{K}\mathbf{U} = \mathbf{F}$$

(3.83)

where

$$
\begin{aligned}
\mathbf{K}_{ij} &= \mathbf{K}_{ij}^{(d)} + \mathbf{K}_{ij}^{(t)} \\
&= \sum_{k=1}^{M} W_k^{(d)} A(\phi_i) A(\phi_j) + \sum_{k=1}^{M_t} W_k^{(t)} B(\phi_i) B(\phi_j)
\end{aligned}
$$

(3.84)

$$
\mathbf{F}_i = \sum_{k=1}^{M_t} B(\phi_i) W_k^{(t)} \, \overline{t}(\mathbf{x}_k) - \sum_{k=1}^{M} A(\phi_i) W_k^{(d)} q(\mathbf{x}_k)
$$

(3.85)

The dimension of the matrix $\mathbf{K}$ is $(N \times N)$ in Equation (3.83). After enforcing the Dirichlet boundary condition in the system of equations, the nodal values of the N field nodes are obtained. It is also observed from Equation (3.84) that the final coefficient matrix $\mathbf{K}$ in the PWLS method is symmetric, namely

$$
\begin{aligned}
\mathbf{K}_{ij} &= \sum_{k=1}^{M} W_k^{(d)} \, A(\phi_i) \, A(\phi_j) + \sum_{k=1}^{M_t} W_k^{(t)} B(\phi_i) \, B(\phi_j) \\
&= \sum_{k=1}^{M} W_k^{(d)} \, A(\phi_j) \, A(\phi_i) + \sum_{k=1}^{M_t} W_k^{(t)} B(\phi_j) \, B(\phi_i) \\
&= \mathbf{K}_{ji}
\end{aligned}
$$

(3.86)

3.3.2 Numerical implementation of PWLS method

In this subsection, several features for the numerical implementation of the point weighted least-squares (PWLS) method are discussed, for example, how to determine the number of collocation points, and how to compute the weight coefficients $W_k^{(d)}$ and $W_k^{(t)}$. The advantages or properties of the PWLS method are also discussed.

3.3.2.1 Number of collocation points

As mentioned earlier, the number of collocation points M may be different from the number of field nodes N. Here M can be selected arbitrarily as long as it is large enough to make Equation (3.82) non-singular. Theoretically, the more collocation points used, the more accurate is the solution. If fewer collocation points are scattered, the computational error of the solution will increase. However, as the collocation points M increase, the computational cost will increase so a reasonable M should be selected. Unfortunately, to date no theoretical best value is available for the number of collocation points M. The optimized value of M may be achieved through numerical examinations.

3.3.2.2 Weight function

The weight functions $W_k^{(d)}$ and $W_k^{(t)}$ employed in the functional Equation (3.81) become the weight coefficients associated with the residual values at the kth collocation point in the problem domain and the Neumann boundary, respectively. The weight coefficients characterize the different influence on the residuals by different collocation points. In fact, the weight functions can be selected by considering the properties of the problem. For example, if a sub-domain or sub-boundary is more important in the problem when compared with other sub-domains or sub-boundaries, one can employ the larger weight coefficients for the collocation points in this sub-domain or on this sub-boundary. The weight function employed is very useful while solving practical engineering applications. For example, in order to solve a crack problem by the present PWLS method, the larger weight coefficients may be imposed in the sub-domain near the crack tip, because the tip region is the most important area site for this kind of crack problem.

   In the present work, the conventional partial differential equations are solved as examples and the weight coefficients are computed by simple weight functions. Several 1-D weight functions used in the present work are listed as follows.

Weight function I (WI): constant weight function:

$$W(x_k) = 1.0, \qquad (x_k = 0 \sim L) \tag{3.87}$$

Weight function II (WII): quadric weight function 1:

$$W(x_k) = 4x_k^2 - 4x_k + 2, \qquad (x_k = 0 \sim L) \tag{3.88}$$

Weight function III (WIII): quadric weight function 2:

$$W(x_k) = -4x_k^2 - 4x_k + 1, \qquad (x_k = 0 \sim L) \tag{3.89}$$

The three weight functions are plotted in Figure 3.7. WI represents that the residuals of all the collocation points in the problem domain have the same influences on the functional by Equation (3.81). WII means that the boundary regions are more important than the internal region, and WIII means that the internal region has more important influence than the boundary

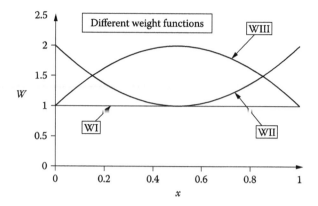

Figure 3.7 Weight functions used in PWLS method. (Modified from Q.X. Wang, H. Li, and K.Y. Lam. (2005). *Computational Mechanics, 35*, 170–181.)

regions. These three weight functions will be investigated in the following numerical examples.

3.3.2.3 Properties of PWLS method

The present PWLS method possesses several advantages.

1. The functional is formed by the weighted residuals of the collocation points and the least square technique is adopted to obtain the
   solution. These make the PWLS method more stable and numerically
   accurate.
2. The derivative boundary conditions can be easily imposed by adding the weighted boundary residuals in the functional. The residuals
   of the governing equations for boundary points are also considered
   simultaneously.
3. The weight function can flexibly adjust the influences of the residuals
   for different sub-domains or sub-boundaries in the problem. This is
   very useful to obtain more accurate numerical solution for practical
   engineering applications.
4. The final coefficient matrix is symmetric, and this makes the PWLS
   method computationally efficient.

In addition, the collocation points in the PWLS method can be independent of the field nodes. This makes an easy and efficient implementation
of PWLS to obtain more accurate solutions at reasonable computational
cost.

3.3.3 Examples for validation

In this subsection, several 1-D and 2-D examples of ordinary and partial differential equations that are always associated with mechanics problems are solved to demonstrate the performance of the PWLS method. The following norms are defined as the error indicators,

$$e_0 = \frac{1}{N} \frac{\sum_{i=1}^{N} |u_i^{\text{exact}} - u_i^{\text{num}}|}{\sum_{i=1}^{N} |u_i^{\text{exact}}|} \tag{3.90}$$

where u_i^{exact} and u_i^{num} are the exact and numerical approximate solutions of the function, respectively. The errors for the first-order derivatives of the function are defined as e_{1x} and e_{1y}, respectively

$$e_{1x} = \frac{1}{N} \frac{\sum_{i=1}^{N} |u_{i,x}^{\text{exact}} - u_{i,x}^{\text{num}}|}{\sum_{i=1}^{N} |u_{i,x}^{\text{exact}}|}, \quad e_{1y} = \frac{1}{N} \frac{\sum_{i=1}^{N} |u_{i,y}^{\text{exact}} - u_{i,y}^{\text{num}}|}{\sum_{i=1}^{N} |u_{i,y}^{\text{exact}}|} \tag{3.91}$$

where $u_{i,x}^{\text{exact}}$ and $u_{i,y}^{\text{exact}}$ are the first-order derivatives of the exact solution in x and y directions, respectively. $u_{i,x}^{\text{num}}$ and $u_{i,y}^{\text{num}}$ are the first-order derivatives obtained by the numerical approximate methods. The convergence rate of the relative error norm is defined as

$$\tilde{R}(e) = \frac{\text{Log}_{10}(e_{i+1}/e_i)}{\text{Log}_{10}(h_{i+1}/h_i)} \tag{3.92}$$

where e should be e_0, e_{1x} or e_{1y}. e_i and e_{i+1} are the relative errors, and h_i and h_{i+1} are the uniform nodal spacings for two different node distributions, respectively.

Example 3.4: 1-D Poisson equation

Let us consider the 1-D mechanics problem of a bar described by the following Poisson equation,

$$E\tilde{A}\frac{d^2u}{dx^2} + q(x) = 0, \quad (0 < x < l) \tag{3.93}$$

where E is Young's modulus, $\tilde{A}$ is the cross-section area, u and $q(x)$ are the axial displacement and the body force in the x direction, respectively. For simplicity, let us take $E = 1.0$, $\tilde{A} = 1.0$. The source term $q(x) = -(3.4\pi)^2 \sin(3.4\pi x)$ is given in this example. The following two types of boundary conditions are considered.

Boundary condition I (BC-I): Dirichlet boundary conditions at both ends of the problem domain:

$$u|_{x=0}= 0, \quad u|_{x=1}=-\sin(3.4\pi) \tag{3.94}$$

Boundary condition II (BC-II): Dirichlet boundary condition at left end $(x = 0)$ and Neumann boundary condition at right end $(x = 1)$:

$$u|_{x=0} = 0 \tag{3.95}$$

$$\frac{du}{dx}\bigg|_{x=1} = u_{,x}|_{x=1}= -3.4\pi\cos(3.4\pi) \tag{3.96}$$

The exact solution of the problem is easily obtained by integrating the differential governing Equation (3.93) with the boundary conditions (3.94) to (3.96), namely

$$u^{exact}(x) = -\sin(3.4\pi x) \tag{3.97}$$

Figure 3.8 demonstrates the numerical results computed by the PWLS method, in which the problem domain is discretised by 21 uniformly distributed nodes and 3 nodes are chosen in the local interpolation domain.

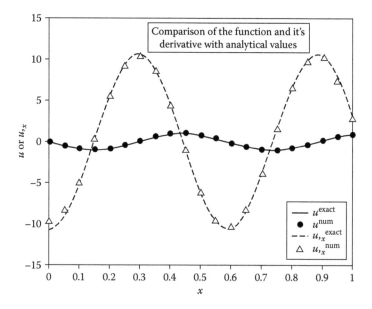

Figure 3.8 Results of u and corresponding derivatives by PWLS method. (Modified from Q.X. Wang, H. Li, and K.Y. Lam. (2005). Computational Mechanics, 35, 170–181.)

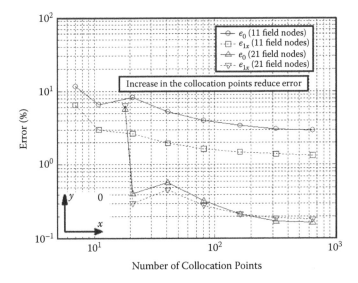

Figure 3.9 Influence of number of collocation points. (Modified from Q.X. Wang, H. Li, and K.Y. Lam. (2005). *Computational Mechanics*, 35, 170–181.)

The number of the collocation points is the same as that of the field nodes. The weight function WI given by Equation (3.87) is employed. The PWLS method produces very accurate results for both the function and the corresponding derivative, with the errors $e_0 = 0.31\%$ and $e_{1x} = 0.2\%$, respectively.

Influence of number of collocation points — The variation of the error computed with the number of collocation points M is plotted in Figure 3.9. Sufficient collocation points are necessary to obtain a non-singular solution. The accuracy is generally improved with the increase of the number of collocation points. However, more computational cost is required when a large value of M is used. A proper M should be chosen for practical engineering applications. It is also observed from Figure 3.9 that the computational result is usually acceptable when M is the same as the number of the field nodes N. Therefore, $M = N$ is used in the following case studies for simplicity.

Influence of weight functions — To examine the influences of different weight functions given by Equations (3.87) to (3.89), the computed errors of Equation (3.93) with BC-I given by Equation (3.94) are obtained as plotted in Figure 3.10. The computational result by WIII is the most accurate among the three cases. From Figure 3.7, it is found that for WIII a larger weight coefficient is imposed to the residuals in the middle of the domain compared with the boundaries. Since the solution has three inflection points in the middle of the problem domain, it leads to larger residuals, and thus WIII is used in this example.

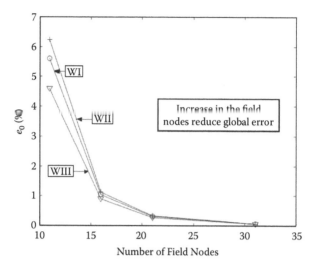

Figure 3.10 Influence of weight functions used in PWLS method. (Modified from Q.X. Wang, H. Li, and K.Y. Lam. (2005). *Computational Mechanics, 35,* 170–181.)

Convergence — To examine the convergence of the present PWLS method for this 1-D problem, the 11, 21, 41, and 81 nodes are uniformly distributed. In the interpolation, the 3-node, 4-node and 5-node are employed respectively in the local interpolation domains for the colloca-tion points. The computational errors e_0 and e_{1x} are listed in Table 3.3.

Table 3.3 Results of *u* computed by PWLS

Number of interpolation nodes	Number of field nodes	Boundary condition I		Boundary condition II	
		e_0 (%)	e_{1x} (%)	e_0 (%)	e_{1x} (%)
3	11	4.6	1.65	18.67	2.23
	21	0.27	0.18	2.03	0.28
	41	1.72e-2	1.93e-2	0.21	0.032
	81	1.77e-3	2.16e-3	2.32e-2	3.69e-3
4	11	1.18	1.09	11.86	1.84
	21	0.14	0.13	1.54	0..23
	41	1.78e-2	1.53e-2	0.13	0.02
	81	2.22e-3	1.81e-3	1.10e-2	2.05e-3
5	11	0.21	0.43	10.2	1.49
	21	5.62e-3	0.15	0.31	0.05
	41	1.56e-4	3.70e-4	7.01e-3	1.30e-3
	81	4.95e-6	1.01e-5	1.56e-4	3.01e-5

Source: Modified from Q.X. Wang, H. Li, and K.Y. Lam. (2005). *Computational Mechanics, 35,* 170–181.

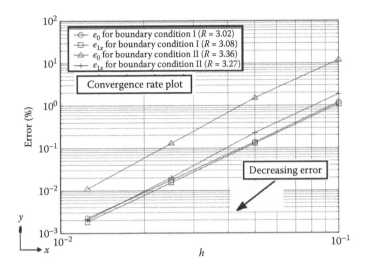

Figure 3.11 Convergence study of four-node interpolation of PWLS method. (Modified from Q.X. Wang, H. Li, and K.Y. Lam. (2005). *Computational Mechanics*, 35, 170–181.)

The convergence rates for the 4-node interpolation scheme are plotted in Figure 3.11, where the convergence rates for both the function value and their derivatives are about 3.1. It is thus concluded that the computational accuracy and numerical convergence of the PWLS method are very good for this example.

Distribution of irregular nodes — To examine the stability of the PWLS method subject to the irregular nodal distribution, 21 nodes are randomly scattered. The computational results with the 4-node interpolation scheme are shown in Figure 3.12. It shows that very good computational results are obtained by the PWLS method, where e_0 is equal to 0.21% for the BC-I case and e_0 is 1.36% for the BC-II case.

Example 3.5: 2-D Poisson equation with mixed boundary conditions

Let us consider a two-dimensional Poisson equation defined in a square domain,

$$\frac{\partial^2 u}{\partial x^2} + \frac{\partial^2 u}{\partial x^2} = \sin(\pi x)\sin(\pi y) \quad (0 < x < l, \ 0 < y < l) \tag{3.98}$$

Dirichlet boundary conditions:

$$u(x,y)|_{x=0} = 0,$$
$$u(x,y)|_{y=0} = 0 \tag{3.99}$$

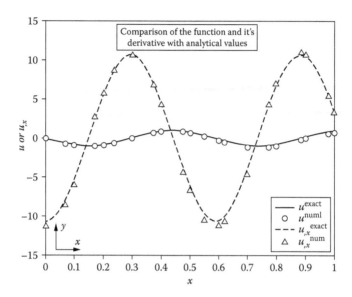

Figure 3.12 Results of PWLS method by irregular nodes for boundary condition II. (Modified from Q.X. Wang, H. Li, and K.Y. Lam. (2005). *Computational Mechanics*, 35, 170–181.)

Neumann (derivative) boundary conditions:

$$\left.\frac{\partial u}{\partial x}\right|_{x=1} = \frac{-1}{2\pi}\cos(\pi x)\sin(\pi y),$$

$$\left.\frac{\partial u}{\partial x}\right|_{y=1} = \frac{-1}{2\pi}\sin(\pi x)\cos(\pi y),$$

(3.100)

The exact solution of this problem is given as

$$u^{exact}(x, y) = \frac{-1}{2\pi^2}\sin(\pi x)\sin(\pi y)$$

(3.101)

The four arrangements of the regularly distributed field nodes are employed to discretize the square problem domain and boundaries. They are 36 nodes (6 × 6), 121 nodes (11 × 11), 441 nodes (21 × 21), and 1681 nodes (41 × 41). Figure 3.13(a) illustrates the arrangement of the 121 nodes, where WI is employed first and $M = N$ is used. In the radial basis interpolation, $p = 1.03$ and $C = d_c$ are adopted. In addition, the size of local interpolation is defined as $r_i = 3.0d_c$. The convergences of the result computed by the PWLS method are plotted in Figure 3.14, where h is the nodal spacing. The PWLS method has achieved very good convergence for both the

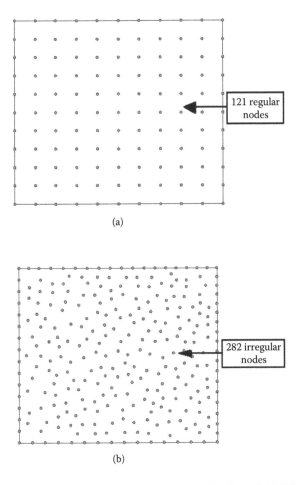

(a)

121 regular nodes

(b)

282 irregular nodes

Figure 3.13 Nodal distributions in square domain. (Modified from Q.X. Wang, H. Li, and K.Y. Lam. (2005). *Computational Mechanics*, 35, 170–181.)

function values and their derivatives for this 2-D Poisson equation, where both the convergence rates are about 3.0.

To demonstrate a comparison, this problem is also solved by the direct collocation method presented in the previous section. The results computed by the direct collocation method are also presented in Figure 3.14. It is clearly shown that the computational accuracy of the PWLS method is higher than that of the direct collocation method. However, the discrepancy of the convergence rates is small between the direct collocation method and the PWLS method. The stability of the PWLS method is examined by the randomly distributed nodes. Table 3.4 illustrates the arrangement of 282 nodes distributed irregularly. The computational results are listed in Table 3.4. The PWLS method produces very accurate results

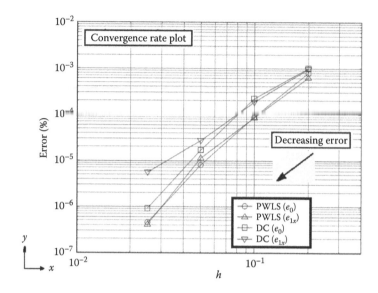

Figure 3.14 Convergence study for two-dimensional Poisson equation problem. (Modified from Q.X. Wang, H. Li, and K.Y. Lam. (2005). *Computational Mechanics*, 35, 170–181.)

even by randomly distributed nodes. The results of the direct collocation method are also listed in the same table for comparison. The errors of the direct collocation method are larger than those of the PWLS method. Influences of the weight functions are also examined. For this 2-D problem with a square domain, the weight functions are constructed simply by

$$W(x,y) = W_{(x)} \cdot W_{(y)} \tag{3.102}$$

where $W_{(x)}$ and $W_{(y)}$ are the 1-D weight functions given by Equations (3.87) to (3.89) in x and y directions, respectively. The results via the different weight functions with the regular distribution of 441 nodes are listed in Table 3.5. The results through the three weight functions are very close; WII gives slightly better results. This is because the solution has the inflection points along boundaries, while WII gives the larger weight coefficients

Table 3.4 Results of u computed by PWLS and DC methods with 282 irregular nodes

	e_0 (%)	e_{1x} (%)	e_{1y} (%)
PWLS	1.25e-3	1.09e-3	1.02e-3
DC	2.77e-3	3.16e-3	3.59e-3

Source: Modified from Q.X. Wang, H. Li, and K.Y. Lam. (2005). *Computational Mechanics*, 35, 170–181.

Table 3.5 Results of u computed by PWLS method with 441
irregular nodes under different weight functions

	e_0 (%)	e_{Ix} (%)
$WI_{(x)} \cdot WI_{(y)}$	8.17e-04	1.12e-03
$WII_{(x)} \cdot WII_{(y)}$	7.53e-04	1.06e-03
$WIII_{(x)} \cdot WIII_{(y)}$	8.56e-04	1.61e-03

Source: Modified from Q.X. Wang, H. Li, and K.Y. Lam. (2005).
Computational Mechanics, 35, 170–181.

at boundaries. Therefore, WII is more reasonable than WI and WIII for
this example.

Example 3.6: 2-D Partial differential equation in circle domain with Dirichlet boundary condition

Let us consider the partial differential equation below

$$\Delta^2 u = ku^l \tag{3.103}$$

where k and l are two parameters. In the area of chemical engineering,
the square root of k is called the Thiele modulus, representing the ratio of
the kinetics to transport resistance in the domain considered, and l is the
order of reaction. The computational domain of this problem is defined as
a cylindrical domain idealized as a unit circle, and the following Dirichlet
condition is considered along the entire circumferential boundary,

$$u|_{r=1} = 1.0 \tag{3.104}$$

This example was also attempted by Balakrishnan and Ramachandran
(2001). Here $l = 1$ and $k = 9$ are taken in order to compare the computational results with the analytical solution by Balakrishnan and Ramachandran
(2001). Two types of nodal distributions with 84 nodes and 287 nodes, as
shown in Figure 3.15 are considered. In the numerical computation, the
sizes of the support domain are adjusted so that 20 nearest neighboring nodes are selected. The computational results are listed in Table 3.6.
Comparing with the results provided by Balakrishnan and Ramachandran
(2001), it is concluded that the PWLS method can perform excellently. The
results computed by the direct collocation method are also obtained and
listed in Table 3.6. The PWLS method can achieve higher accuracy than the
direct collocation method.

In brief, it is demonstrated via this example that the PWLS method using
the radial point interpolation method can work well even for a problem
with a circular domain. The solution of the problem is computationally
stable and numerically convergent.

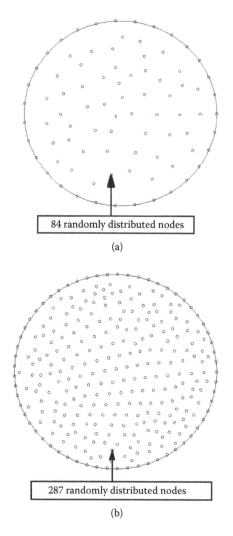

Figure 3.15 Nodal distributions in circle domain. (Modified from Q.X. Wang, H. Li, and K.Y. Lam. (2005). *Computational Mechanics*, 35, 170–181.)

3.3.4 Remarks

In our development of the PWLS method, the problem domain is discretised by the distributed field nodes that are used for the construction of trial function based on the local radial point interpolation. The collocation points that are independent of the field nodes are adopted to form the total residuals of the problem. The weighted least-squares technique is employed to obtain the solution of the problem by minimizing the summation of

Table 3.6 Results of u computed by PWLS and DC methods for $l = 1$ and $k = 9$

Location r	Reference solution	84 nodes		287 nodes	
		PWLS (Error %)	DC (Error %)	PWLS (Error %)	DC (Error %)
0	0.2048	0.2036 (−0.59)	0.2068 (0.98)	0.2048 (0.00)	0.2049 (0.05)
0.25	0.2347	0.2339 (−0.34)	0.2362 (0.64)	0.2347 (0.00)	0.2347 (0.00)
0.50	0.3373	0.3362 (−0.33)	0.3404 (0.92)	0.3373 (0.00)	0.3374 (0.03)
0.75	0.5581	0.5572 (−0.16)	0.5619 (0.68)	0.5587 (0.11)	0.5588 (0.13)

Source: Modified from Q.X. Wang, H. Li, and K.Y. Lam. (2005). *Computational Mechanics*, 35, 170–181.

residuals at all the collocation points. Both 1-D and 2-D examples are presented for examination of the convergence and performance of the present PWLS method. Through the implementations of these numerical examples, it is demonstrated that the PWLS method possesses the several advantages listed below.

1. It is a truly meshless method since no mesh or integration is required.
2. Both the Dirichlet and Neumann boundary conditions can be easily enforced.
3. The final coefficient matrix is symmetric.

All advantages of the PWLS method ensure that it is a very good potential meshless technique for a wide range of engineering computational applications.

3.4 LOCAL KRIGING (LOKRIGING) METHOD

The present local Kriging (LoKriging) method is based on the local weak forms of the partial differential governing equations and by the Kriging interpolation (Krige, 1975) for construction of meshless shape functions. The shape functions constructed by this interpolation possess the delta function property through the randomly distributed points such that the essential boundary conditions can be easily implemented. The local weak forms of the partial differential governing equations are formulated by the weighted residual method within the simple local quadrature domain. The spline function with high continuity is employed as the weight function. Therefore, the presently developed LoKriging method is truly meshless because it does not require mesh either for construction of the shape functions or for integration of the local weak form. Several examples of 2-D static structural problems are examined to illustrate the computational accuracy and performance of the method. It is concluded through the examples that the LoKriging method works efficiently and accurately during computation.

3.4.1 Formulation of Kriging Interpolation

Let us consider a random function $u(\mathbf{x})$ defined in the domain Ω discretised by a set of scattered nodes $\mathbf{x}_i$ ($1 \leq i \leq N$) where N is the number of the nodes distributed in the domain Ω. It is assumed that only the nodes surrounding the point $\mathbf{x}_0$ have effect on $u(\mathbf{x}_0)$. The domain covering these surrounding nodes is called the influence or interpolation domain. The estimated value of the function $u(\mathbf{x})$ at the point $\mathbf{x}_0$, u^b, can be obtained (Trochu,1993; Olea,1999)

$$u^b(\mathbf{x},\mathbf{x}_0) = \sum_{i=1}^{n} \lambda_i u(\mathbf{x}_i) \tag{3.105}$$

where $u(\mathbf{x}_i)$ is the value at $\mathbf{x}_i$ ($i = 1,2,\ldots,n$), and n is the number of the nodes in the influence domain of $\mathbf{x}_0$. λ_i is the weight assigned to the neighborhood nodes, and it is determined by minimizing the squared variance of the estimation error $E\{[u(\mathbf{x}_0) - u^b(\mathbf{x},\mathbf{x}_0)]^2\}$. Provided that the estimation Equation (3.105) is no-bias, namely the expected values of $u(\mathbf{x}_0)$ and $u^b(\mathbf{x},\mathbf{x}_0)$ must be identical, one can have

$$E[u(\mathbf{x}_0)] = E[u^b(\mathbf{x}, \mathbf{x}_0)] = \sum_{i=1}^{n} \lambda_i E[u(\mathbf{x}_i)] \tag{3.106}$$

In the Kriging system, the random function $u(\mathbf{x}_0)$ is decomposed into the sum of two portions,

$$u(\mathbf{x}_0) = Z_a(\mathbf{x}_0) + Z_b(\mathbf{x}_0) \tag{3.107}$$

where $Z_a(\mathbf{x}_0) = E[u(\mathbf{x}_0)]$ is called the drift. $Z_b(\mathbf{x}_0)$ represents a stationary fluctuation and $E[Z_b(\mathbf{x}_0)] = 0$. Since the drift represents the expected value of $u(\mathbf{x}_0)$, it may be written as

$$Z_a(\mathbf{x}_0) = \sum_{i=1}^{n} \lambda_i Z_a(\mathbf{x}_i) \tag{3.108}$$

Usually the drift can be chosen arbitrarily. If it is assumed that the drift belongs to a finite linear subspace S and it is taken as a linear polynomial, Equation (3.108) is rewritten as

$$\sum_{i=1}^{n} \lambda_i p_l(\mathbf{x}_i) = p_l(\mathbf{x}_0), \quad 1 \le l \le m \tag{3.109}$$

where the basis function $p_l(\mathbf{x})$ is the monomial in S. For example, $\mathbf{x}^T = [x \; y]$, $\mathbf{p}^T = [1 \; x \; y]$ for the two-dimensional domain. The squared variance of the estimation error is given by

$$E\left[u(\mathbf{x}_0) - \sum_{i=1}^{n} \lambda_i u(\mathbf{x}_i) \right]^2 = E[u(\mathbf{x}_0)]^2 - \sum_{i=1}^{n} 2\lambda_i E[u(\mathbf{x}_0)u(\mathbf{x}_i)]$$

$$+ \sum_{i=1}^{n} \sum_{j=1}^{n} \lambda_i \lambda_j E[u(\mathbf{x}_i)u(\mathbf{x}_j)] \tag{3.110}$$

By minimizing Equation (3.110) with respect to the coefficients λ_i subjected to m linear constraints that satisfy the no-bias conditions, the solution is characterized by a linear system of $(n+m)$ equations with respect to $(n + m)$ unknowns $\lambda_1, \lambda_2, ..., \lambda_n$ and $\mu_1, \mu_2, ..., \mu_m$,

$$\sum_{j=1}^{n} E[u(\mathbf{x}_i)u(\mathbf{x}_j)]\lambda_j + \sum_{i=1}^{m} \mu_l p_l(\mathbf{x}_i) = E[u(\mathbf{x}_0)u(\mathbf{x}_i)], \quad 1 \le i \le n \tag{3.111}$$

$$\sum_{j=1}^{n} \lambda_j p_l(\mathbf{x}_j) = p_l(\mathbf{x}_0), \quad 1 \le l \le m \tag{3.112}$$

The coefficients μ_l $(1 \le l \le m)$ are the Lagrange multipliers associated with the constraints. Since the squared variance of the estimation error is always positive, the quadratic form Equation (3.110) must be positive. Furthermore, if it is positive definite, the solution of the Kriging system given by Equations (3.111)) and (3.112) exists and it is unique (Phan and Trochu, 1998). Based on the intrinsic hypothesis (Trochu, 1993), the covariance $E[u(\mathbf{x}_0)u(\mathbf{x}_i)]$ is replaced by the semivariogram $\gamma(\mathbf{h})$ in the following form

$$\gamma(\mathbf{x}_0, \mathbf{x}_i) = \gamma(\mathbf{h}) = \frac{1}{2} E\{[u(\mathbf{x}_i) - u(\mathbf{x}_0)]^2\} \tag{3.113}$$

where $\mathbf{h}$ is the Euclidean distance between $\mathbf{x}_0$ and $\mathbf{x}_i$. Similarly, the covariance $E[u(\mathbf{x}_i)u(\mathbf{x}_j)]$ is replaced by $\gamma(\mathbf{x}_i,\mathbf{x}_j)$. As a result, the Kriging system given by Equations (3.111) and (3.112) is rewritten in the matrix form of

$$\mathbf{G}\,\mathbf{c} = \mathbf{g} \tag{3.114}$$

where

$$\mathbf{G} = \begin{bmatrix} \mathbf{R} & \mathbf{P} \\ \mathbf{P}^T & 0 \end{bmatrix} = \begin{bmatrix} \gamma(\mathbf{x}_1,\mathbf{x}_1) & \cdots & \gamma(\mathbf{x}_1,\mathbf{x}_n) & p_1(\mathbf{x}_1) & \cdots & p_m(\mathbf{x}_1) \\ \cdots & \cdots & \cdots & \cdots & \cdots & \cdots \\ \gamma(\mathbf{x}_n,\mathbf{x}_1) & \cdots & \gamma(\mathbf{x}_n,\mathbf{x}_n) & p_1(\mathbf{x}_n) & \cdots & p_m(\mathbf{x}_n) \\ p_1(\mathbf{x}_1) & \cdots & p_1(\mathbf{x}_n) & 0 & \cdots & 0 \\ \cdots & \cdots & \cdots & \cdots & \cdots & \cdots \\ p_m(\mathbf{x}_1) & \cdots & p_m(\mathbf{x}_n) & 0 & \cdots & 0 \end{bmatrix} \tag{3.115}$$

$$\mathbf{c} = [\lambda_1 \quad \lambda_2 \quad \cdots \quad \lambda_n \quad \mu_1 \quad \mu_2 \quad \cdots \quad \lambda_m]^T \tag{3.116}$$

$$\mathbf{g} = [\boldsymbol{\gamma}(\mathbf{x}_0) \quad \mathbf{p}(\mathbf{x}_0)]^T = [\gamma(\mathbf{x}_0,\mathbf{x}_1) \quad \cdots \quad \gamma(\mathbf{x}_0,\mathbf{x}_n) \quad p_1(\mathbf{x}_0) \quad \cdots \quad p_m(\mathbf{x}_0)^T] \tag{3.117}$$

In fact, there are several semivariogram models available for specific computational applications. For example, Olea (1999) listed about seven semivariogram models. Due to the simplicity and availability in geostatistical software packages, the first four models, namely the spherical, exponential, Gaussian and power models, are more widely used. However, each model has advantages as well as disadvantages. For solving different problems, an appropriate semivariogram model should be chosen. The Gaussian semivariogram is one of the most widely used models and produces good numerical results for problems in engineering computations (Dai et al., 2003; Gu, 2003). In this work therefore, the Gaussian model is employed and given by

$$\gamma(h) = c_0\left(1 - e^{-3\left(\frac{h}{a_0}\right)^2}\right) \tag{3.118}$$

where h is the lag, and c_0 and a_0 are the sill and range, respectively. The sill c_0 represents the average variance of the points at such a distance away from the considered point that there is no correlation between the points, and the range a_0 represents the distance at which there is no longer a correlation between the points. The range a_0 is taken as

$$a_0 = a \cdot r_i \tag{3.119}$$

where α is a coefficient, and

$$r_i = \beta \cdot d_{\min} \tag{3.120}$$

where β is a scaling parameter, and $d_{\min}$ is the shortest distance between the interpolation point and its neighboring points. By substituting the weights λ_i solved by Equation (3.114) into Equation (3.105), one can have the estimated value in the simple form (Stein, 1999):

$$u^b(\mathbf{x}, \mathbf{x}_0) = \Phi(\mathbf{x})\mathbf{u} \tag{3.121}$$

where $\mathbf{u} = [\mathbf{u}(\mathbf{x}_1)\ \mathbf{u}(\mathbf{x}_2)\ \dots\ \mathbf{u}(\mathbf{x}_n)]^T$. $\Phi(\mathbf{x})$ is defined as the shape function matrix, and is written as

$$\Phi(\mathbf{x}) = \gamma(\mathbf{x}_0)^T \mathbf{S} + \mathbf{p}(\mathbf{x}_0)^T \mathbf{Y} \tag{3.122}$$

$$\mathbf{S} = \mathbf{R}^{-1}(\mathbf{I} - \mathbf{PY}) \tag{3.123}$$

$$\mathbf{Y} = (\mathbf{P}^T \mathbf{R}^{-1} \mathbf{P})^{-1} \mathbf{P}^T \mathbf{R}^{-1} \tag{3.124}$$

Figure 3.16 plots the shape function and corresponding first-order derivatives in the x and y directions. It is known from the figure that the shape

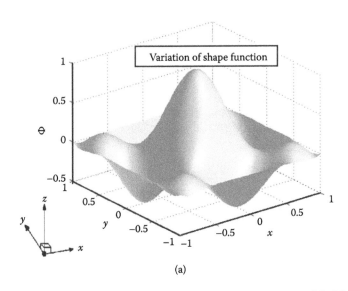

(a)

Figure 3.16 Shape function and corresponding first-order derivatives. (Modified from K.Y. Lam, Q.X. Wang, and H. Li. (2004). *Computational Mechanics*, 33, 235–244.)

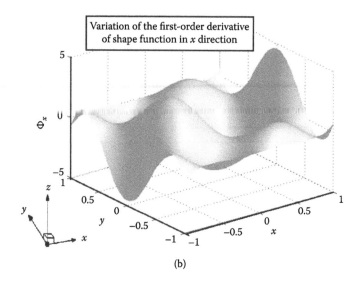

(b)

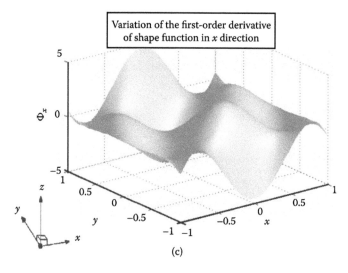

(c)

Figure 3.16 (continued) Shape function and corresponding first-order derivatives. (Modified from K.Y. Lam, Q.X. Wang, and H. Li. (2004). *Computational Mechanics,* 33, 235–244.)

functions derived from the Kriging interpolation possess the delta function property, namely

$$\phi_i(\mathbf{x}_j) = \delta_{ij} = \begin{cases} 1 & (i = j, i = 1 \sim n) \\ 0 & (i \neq j, i, j = 1 \sim n) \end{cases} \tag{3.125}$$

and satisfy the partition of unity,

$$\sum_{i=1}^{n} \phi(\mathbf{x}_i) = 1 \tag{3.126}$$

Kriging shape functions are able to reproduce any function exactly in the basis. In particular, if all the constants and linear terms are included, it will reproduce a general linear polynomial exactly, namely

$$\sum_{i=1}^{n} \phi_i \, \mathbf{x}_i = \mathbf{x}_j \quad (j = 1 \sim n) \tag{3.127}$$

Similar to the other interpolation methods, the above Kriging interpolation formalism is sensitive to the topological structure of a solid body, and is suitable for convex simply connected bodies with continuous boundaries. If the problem domain is non-convex or multi-connected, the traditional interpolation formulation will be unsuitable. The usual method for the numerical analysis of these non-convex or multi-connected domain problem is to transform to several sub-domains that are simply connected and convex. Furthermore, special numerical techniques are required to handle the continuity along the interfaces between the sub-domains, such that several strategies are developed in the meshless functional approximation, for example, adding the special basis function using specially shaped domains. However, these techniques cannot solve all the problems with non-convex boundaries. Further research for more efficient interpolation approximation is required if a body is non-convex.

3.4.2 Numerical implementation of LoKriging method

In this subsection, we use the classical governing equations of 2-D solid mechanics problems with linear elasticity theory as examples to describe how to implement the LoKriging method.

3.4.2.1 Local weak forms of elasto-static problems

Let us consider the following 2-D problem of linear elasticity in the domain Ω bounded by Γ,

$$\sigma_{ij,j} + b_i = 0 \quad \text{in} \quad \Omega \tag{3.128}$$

where σ_{ij} is the stress tensor corresponding to the displacement field u_i, b_i is the body force, and $(),_j$ denotes $\partial()/\partial x_j$. The corresponding boundary conditions are given as

$$u_i = \bar{u}_i \quad \text{on the essential boundary } \Gamma_u \tag{3.129}$$

$$t_i = \sigma_{ij}n_j = \bar{t}_i \quad \text{on the natural boundary } \Gamma_t \tag{3.130}$$

where $\bar{u}_i$ and $\bar{t}_i$ are the prescribed displacements and tractions, respectively, and n_j is the unit outward normal to the boundary $\Gamma(\Gamma = \Gamma_u \cup \Gamma_t)$.

In the present LoKriging method, a weak form is constructed first over a sub-domain Ω_s bounded by Γ_s, where Ω_s is located entirely inside the global domain Ω. Since the Kriging shape functions satisfy the delta function property, the essential boundary condition given by Equation (3.129) is imposed directly. By the local weighted residual method, the generalized local weak form of Equation (3.128) is written as

$$\int_{\Omega_s} w_i (\sigma_{ij,j} + b_i) d\Omega = 0 \tag{3.131}$$

where w_i is the weight function.

By integrating the first term of the left side of Equation (3.131) by parts, one can have

$$\int_{\Gamma_s} w_i \sigma_{ij} n_j \, d\Gamma - \int_{\Omega_s} (w_{i,j} \, \sigma_{ij} - w_i b_i) d\Omega = 0 \tag{3.132}$$

In the sub-domain Ω_s of the node $\mathbf{x}_i$, $w_i(\mathbf{x}) \neq 0$. Usually the shape of the sub-domain Ω_s can be arbitrary. However, it is taken conveniently to be a circle or rectangle in 2-D problems. If there is an intersection between the local boundary Γ_s and the global boundary Γ, usually the boundary Γ_s is composed of the three parts as shown in Figure 3.17. The internal boundary Γ_{si} and the boundaries Γ_{su} and Γ_{st} over which the essential and natural boundary conditions are specified, respectively. By imposing the natural boundary condition given by Equation (3.130) in Equation (3.132), one can have

$$\int_{\Gamma_{si}} w_i \, t_i \, d\Gamma + \int_{\Gamma_{su}} w_i \, t_i \, d\Gamma + \int_{\Gamma_{st}} w_i \bar{t}_i \, d\Gamma - \int_{\Omega_s} (w_{i,j} \, \sigma_{ij} - w_i b_i) d\Omega = 0 \tag{3.133}$$

If there is no intersection between Γ_s and Γ, $\Gamma_{si} = \Gamma_s$, the integrals along Γ_{su} and Γ_{st} don't exist, and Equation (3.133) becomes

$$\int_{\Gamma_{si}} w_i t_i \, d\Gamma - \int_{\Omega_s} (w_{i,j}\sigma_{ij} - w_i b_i) d\Omega = 0 \tag{3.134}$$

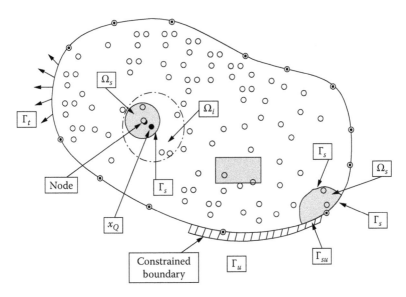

Figure 3.17 Local sub-domain Ωs and interpolation domain Ωi. (Modified from K.Y. Lam, Q.X. Wang, and H. Li. (2004). *Computational Mechanics, 33,* 235–244.)

Through the above deduction, Equation (3.133) or Equation (3.134) becomes a localized boundary value problem. As long as the union of all the local domains covers the global domain, satisfactory numerical results can be obtained (Atluri and Zhu, 1998).

It is known that the local weak form given by Equation (3.133) is constructed by the weighted residual method. As a result, the choice of weight function becomes important for the performance of the method. Several special weight functions may be employed to satisfy the conditions we require. These special weight functions always have the following advantages for the meshless methods.

1. The continuity of the weight function must satisfy the requirement in the local weak form. For example, if the highest order of the derivatives of the weight function is the first-order, the weight function must have at least first-order continuity.
2. The weight function is equal to zero along the boundary of quadrature domain. With this property, the local weak form can be simplified since the integration along the internal boundary vanishes.

The spline weight functions have the above advantages and are easily constructed (Atluri et al., 1999b). Furthermore, compared with the quadratic and cubic spline functions, the fourth order spline function has higher continuity, and its form is simpler than that of the cubic spline function. As a result, the fourth order spline function is employed here and given by

$$w_i(\mathbf{x}) = \begin{cases} 1 - 6\left(\dfrac{d_i}{r_s}\right)^2 + 8\left(\dfrac{d_i}{r_s}\right)^3 - 3\left(\dfrac{d_i}{r_s}\right)^4 & 0 \le d_i \le r_s \\ 0 & d_i > r_s \end{cases} \tag{3.135}$$

where $d_i = |\mathbf{x}-\mathbf{x}_i|$ is the distance from the node $\mathbf{x}_i$ to the point $\mathbf{x}$, r_s is the size of the local sub-domain, and r_s is computed by

$$r_s = c \cdot d_{min} \tag{3.136}$$

where d_{min} is the shortest distance between the node i and its neighboring nodes and c is a scaling parameter. Figure 3.18 illustrates the fourth order spline weight function and its corresponding derivative.

3.4.2.2 Discretisation of local weak form

By substituting the displacement expressed by Equation (3.121) into the local weak form given by Equation (3.133), one can have

$$\mathbf{Ku} = \mathbf{f} \tag{3.137}$$

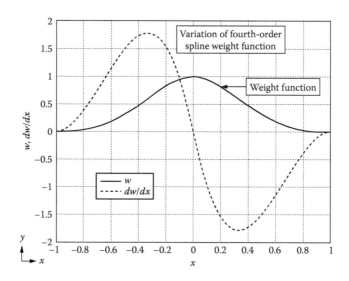

Figure 3.18 Fourth order spline weight function and corresponding first-order derivative. (Modified from K.Y. Lam, Q.X. Wang, and H. Li. (2004). *Computational Mechanics*, 33, 235–244.)

where $\mathbf{K}$ is the stiffness matrix, and $\mathbf{f}$ is the force vector, namely

$$K_{ij} = \int_{\Omega_s} \mathbf{v}_i^T \, \mathbf{DB}_j \, d\Omega - \int_{\Gamma_{si}} \mathbf{w}_i \, \mathbf{NDB}_j \, d\Gamma - \int_{\Gamma_{su}} \mathbf{w}_i \, \mathbf{NDB}_j \, d\Gamma \qquad (3.138)$$

$$\mathbf{f}_i = \int_{\Omega_s} \mathbf{w}_i \, \mathbf{b}_i \, d\Omega + \int_{\Gamma_{st}} \mathbf{w}_i \, \bar{\mathbf{t}}_i \, d\Gamma \qquad (3.139)$$

$$\mathbf{D} = \frac{E}{1-v^2} \begin{bmatrix} 1 & v & 0 \\ v & 1 & 0 \\ 0 & 0 & \dfrac{1-v}{2} \end{bmatrix} \quad \text{for plane stress} \qquad (3.140)$$

$$\mathbf{B}_j = \begin{bmatrix} \phi_{j,x} & 0 \\ 0 & \phi_{j,y} \\ \phi_{j,y} & \phi_{j,x} \end{bmatrix} \qquad (3.141)$$

$$\mathbf{v}_i = \begin{bmatrix} w_{i,x} & 0 \\ 0 & w_{i,y} \\ w_{i,y} & w_{i,x} \end{bmatrix} \qquad (3.142)$$

$$\mathbf{N} = \begin{bmatrix} n_x & 0 & n_y \\ 0 & n_y & n_x \end{bmatrix} \qquad (3.143)$$

where $\mathbf{w}_i$, $\mathbf{b}_i$, and $\bar{\mathbf{t}}_i$ are the weight function, body force, and traction matrixes of the corresponding node i, respectively and E is the Young's modulus and v the Poisson ratio.

The Gauss quadrature is employed here to perform the numerical integrations in Equations (3.138) and (3.139). For each Gauss quadrature point $\mathbf{x}_Q$, the Kriging interpolation is employed to compute the estimated value on it. Therefore, there exist the two local domains as shown in Figure 3.17, the sub-domain Ω_s (size r_s) and the interpolation domain Ω_i (size r_i). Usually these two domains are independent of each other. The sizes of the local sub-domain for the local integration and the interpolation domain have been defined by Equations (3.136) and (3.120), respectively.

3.4.3 Examples for validation

In this subsection, several case studies of the patch tests are examined for the validation of the local Kriging (LoKriging) method. The case studies also include the 1-D beam problem and the 2-D plate problem to examine the numerical convergence and computational performance of the LoKriging method.

Example 3.7: Standard patch test

Figure 3.19 is plotted for the standard patch test. The displacements along all the boundaries are characterized by a linear function of x and y on the patch of dimension $L_x = 2$ and $L_y = 2$. This linear function is given as

$$\begin{cases} u_x = x + y \\ u_y = x - y \end{cases} \tag{3.144}$$

The satisfaction of the patch test requires that the displacements of any interior nodes be given by the same linear functions and the computed strains and stresses are constants in the patch.

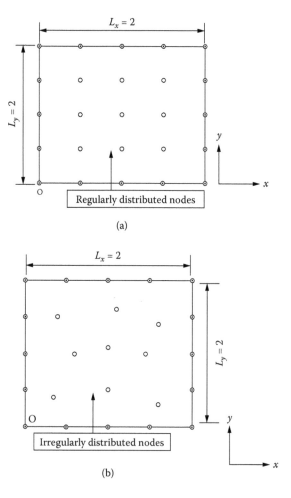

(a)

(b)

Figure 3.19 Standard patch test. (Modified from K.Y. Lam, Q.X. Wang, and H. Li. (2004). *Computational Mechanics*, 33, 235–244.)

Table 3.7 Results of interior nodes located irregularly in patch shown in figure 3.19(b)

Coordinate (x, y) for Interior Node	u_x	u_y	σ_{xx}	σ_{yy}	σ_{xy}
(0.35 ,0.4)	0.75000	−0.50000	0.76923077	-0.76923077	0.76923077
(0.6, 1.0)	1.60000	−0.40000	0.76923077	-0.76923077	0.76923077
(0.4, 1.5)	1.90000	−0.11000	0.76923077	-0.76923077	0.76923077
(1.0 ,0.5)	1.50000	0.50000	0.76923077	-0.76923077	0.76923077
(1.0, 1.08)	2.08000	−0.08000	0.76923077	-0.76923077	0.76923077
(1.1, 1.6)	2.70000	−0.50000	0.76923077	-0.76923077	0.76923077
(1.6, 0.3)	1.90000	1.30000	0.76923077	-0.76923077	0.76923077
(1.45, 1.0)	2.45000	0.45000	0.76923077	-0.76923077	0.76923077
(1.6, 1.4)	3.00000	0.20000	0.76923077	-0.76923077	0.76923077

Source: Modified from K.Y. Lam, Q.X. Wang, and H. Li. (2004). *Computational Mechanics, 33,* 235–244.

In this example, two kinds of nodal distributions are examined. The first is distribution of all 25 nodes regularly as shown in Figure 3.19(a). The second has 9 irregularly distributed interior nodes among the 25 nodes as shown in Figure 3.19(b). The essential boundary conditions are easily imposed since the Kriging shape function possesses delta function properties. In both the nodal distributions, $E = 1$ and $v = 0.3$ are taken. It was found that the present LoKriging method can exactly pass the patch test with regularly distributed nodes. Table 3.7 presents the coordinates and the numerically computed result of the displacements and stresses for the second nodal distribution including the 9 irregularly distributed nodes. The computational results demonstrate that the present meshless LoKriging method can pass the patch test even with irregularly distributed nodes.

Example 3.8: Higher order patch test

Two case studies of the higher order patches are presented, as shown in Figure 3.20. In the first case study, a uniform axial stress with unit intensity is

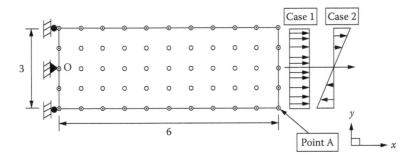

Figure 3.20 High order patch test. (Modified from K.Y. Lam, Q.X. Wang, and H. Li. (2004). *Computational Mechanics, 33,* 235–244.)

Table 3.8 Comparison of computed solution with exact solution at point A as shown in Figure 3.20

	u_x			u_y		
	Exact result	Computed result	Relative error (%)	Exact result	Computed result	Relative error (%)
Case I	6.000000	6.000859	0.0143	0.375000	0.374975	0.0066
Case 2	−6.000000	−6.018960	0.3160	−12.187500	−12.290670	0.8465

Source: Modified from K.Y. Lam, Q.X. Wang, and H. Li. (2004). *Computational Mechanics*, 33, 235–244.

imposed at the right end. The exact solution of this case with $E = 1$ and $v = 0.25$ is $u_x = x$ and $u_y = -y/4$. In the second case study, a linearly varying normal stress is imposed at the right end. The exact solution of this case with $E = 1$ and $v = 0.25$ is $u_x = 2xy/3$ and $u_y = -(x^2+y^2/4)/3$. The 55 nodes are scattered in this patch. The relative error of the displacements u_x and u_y at point A are examined for both the first and second cases, as shown in Table 3.8. The present meshless LoKriging method can pass the first case exactly where the analytical solution is a linear function. However, a computational error exists for the second case since the Kriging interpolation used here can reproduce a linear polynomial only; the exact solution of the second case is quadratic.

Example 3.9: Cantilever beam

The present LoKriging method is also employed for the cantilever beam problem, as shown in Figure 3.21(a), for which the exact solution is given by Timoshenko and Goodier (1970) as

$$u_x = -\frac{Py}{6EI}\left[(6L - 3x)x + (2 + v)\left(y^2 - \frac{D^2}{4}\right)\right] \tag{3.145}$$

$$u_y = -\frac{P}{6EI}\left[3vy^2(L - x) + (4 + 5v)\frac{D^2x}{4} + (3L - x)x^2\right] \tag{3.146}$$

The stresses corresponding to Equation (3.145) and (3.146) are given as

$$\sigma_{xx}(x,y) = -\frac{P(L - x)}{I} \tag{3.147}$$

$$\sigma_{yy}(x,y) = 0 \tag{3.148}$$

$$\sigma_{xy}(x,y) = -\frac{P}{2I}\left(y^2 - \frac{D^2}{4}\right) \tag{3.149}$$

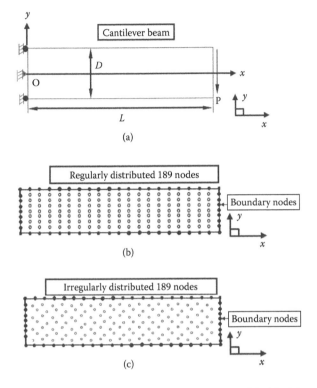

Figure 3.21 Cantilever beam with data nodes and end load. (Modified from K.Y. Lam, Q.X. Wang, and H. Li. (2004). *Computational Mechanics*, 33, 235–244.)

where the parameters $E = 3.0 \times 10^7$, $v = 0.3$, $L = 48$, $D = 12$, and $P = 1000$ are taken. Before the numerical results of this example are computed, several parameters, such as α in Equation (3.119) and β in Equation (3.120), are studied first by the energy norm because the energy norm can reflect comprehensively the computational accuracy of strain and stress. The energy norm is defined as

$$\text{Energy Norm: } e_e = \left\{ \frac{1}{2} \int_\Omega (\varepsilon^{Num} - \varepsilon^{Exact})^T D(\varepsilon^{Num} - \varepsilon^{Exact}) d\Omega \right\}^{\frac{1}{2}} \quad (3.150)$$

Figure 3.22 illustrates the energy errors for different α and β values. This figure shows that with a larger β, the energy errors decrease with increasing the interpolation domain. However, it is observed from Equation (3.120) that the interpolation domain is determined by the parameter β and the shortest

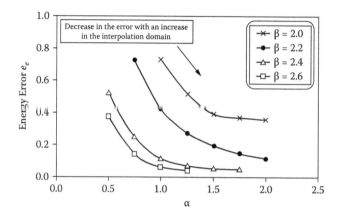

Figure 3.22 Parameter study for cantilever beam. (Modified from K.Y. Lam, Q.X. Wang, and H. Li. (2004). *Computational Mechanics*, 33, 235–244.)

distance d_{min} between the interpolation point and its neighboring points, where d_{min} is determined by the node distribution. Therefore, when the node distribution remains unchanged, the interpolation domain is directly related to β. Theoretically, the larger β is good for the computed results. However, if the interpolation domain becomes larger, more field nodes are required for the interpolation, and the dimension of the interpolation matrix will be enlarged. In addition, increasing the interpolation domain also enlarges the band width of the final stiffness matrix. As a result, a larger interpolation domain or a larger β will increase the computational cost. However, based on the numerical experiment results shown in Figure 3.22, β = 2.4 ~ 2.6 is a good choice. Similarly, the influence of the parameter α that controls the range of the semivariogram model, can also be obtained from Figure 3.22. If α is taken as 1.0 ~ 1.5, the energy errors become smaller.

The case studies of the convergence rates at β = 2.4 are demonstrated, as shown in Figure 3.23(a), where the nodes are distributed regularly. The mesh size *d* in the figure is defined as the distance in *x*-direction between the two neighboring nodes. It is seen from the figure that the convergence rates are different for different values of α. The appropriate choice of α and β greatly influences the convergence rate. The comparison of the convergence rate with other meshless methods, such as LRPIM and MLPG, is presented in Figure 3.23(b) at β = 2.4 and α = 1.25. It is known that the convergence rate of the LoKriging method is very good, such that β = 2.4 and α = 1.25 are taken in this problem.

The above numerical experiment technique is quite common in the development of meshless methods to determine the corresponding parameters, e.g., α and β, associated with the size of the interpolation domain or scale change of the semivariogram model. To date, no successfully rigorous approaches have achieved theoretical values for these parameters. The present work aims to develop s meshless LoKriging method by which a

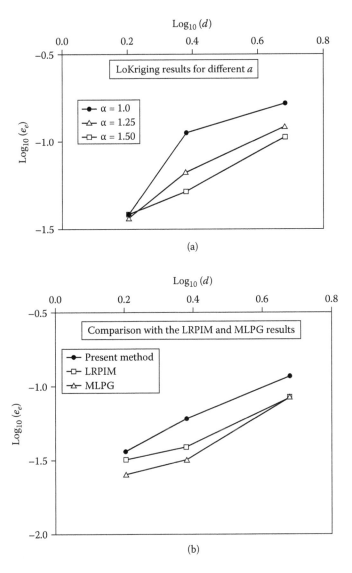

Figure 3.23 Convergence study for cantilever beam ($\beta = 2.4$). (Modified from K.Y. Lam, Q.X. Wang, and H. Li. (2004). *Computational Mechanics*, 33, 235–244.)

simple numerical experiment technique is employed to study these parameters. Obviously future efforts will be required for quantitative studies of the parameters and convergence.

As shown in Figure 3.21(b) and Figure 3.21(c), both the regularly distributed nodes and the irregularly distributed nodes are investigated for this

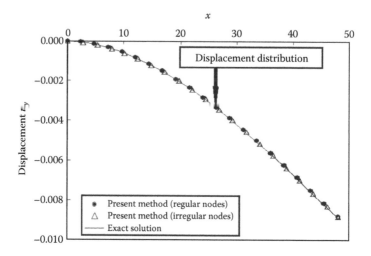

Figure 3.24 Displacement distribution of cantilever beam along x-axis. (Modified from K.Y. Lam, Q.X. Wang, and H. Li. (2004). *Computational Mechanics*, 33, 235–244.)

cantilever beam. The comparison of the numerical results computed by the present LoKriging method with the exact solutions of the displacement u_y along x-axis is shown in Figure 3.24. The numerical results agree well with the exact solution. Figure 3.25 and Figure 3.26 illustrate the normal stress σ_{xx} and the shear stress σ_{xy} at the section $x = L/2$ of the beam,

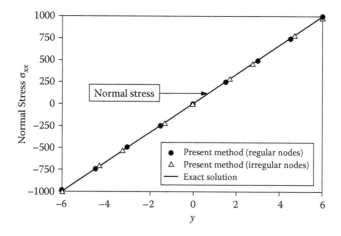

Figure 3.25 Normal stress at section $x = L/2$ of cantilever beam. (Modified from K.Y. Lam, Q.X. Wang, and H. Li. (2004). *Computational Mechanics*, 33, 235–244.)

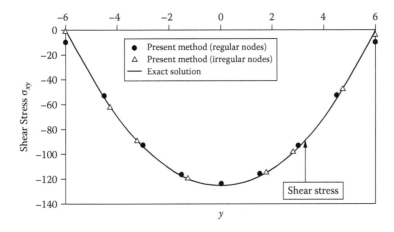

Figure 3.26 Shear stress at section $x = L/2$ of cantilever beam. (Modified from K.Y. Lam, Q.X. Wang, and H. Li. (2004). *Computational Mechanics*, 33, 235–244.)

respectively. The computed results of the normal and shear stresses are almost equal to the exact solutions in these two case studies with different nodal distributions. Therefore, it is can be said that very good numerical results can be obtained by the present LoKriging method for irregular nodal distribution.

Example 3.10: Infinite plate with circular hole

Let us consider an infinite plate with a central hole as $x^2 + y^2 \le a^2$, where a is the radius of the hole. The plate is subjected to a unit uniform tension of $\sigma = 1.0$ in x direction at infinity. Due to the symmetry, only a quarter of the plate is modeled as shown in Figure 3.27. For this plane–strain problem, the material properties $E = 1000$ MPa and $v = 0.3$ are taken. The symmetry conditions are imposed along the left and bottom edges, and the inner boundary at $x^2 + y^2 = a^2$ is traction free. The exact stress solution of this problem is given as

$$\sigma_{xx} = \sigma\left[1 - \frac{a^2}{r^2}\left(\frac{3}{2}\cos 2\theta + \cos 4\theta\right) + \frac{3a^4}{2r^4}\cos 4\theta\right] \qquad (3.151)$$

$$\sigma_{yy} = \sigma\left[-\frac{a^2}{r^2}\left(\frac{1}{2}\cos 2\theta - \cos 4\theta\right) - \frac{3a^4}{2r^4}\cos 4\theta\right] \qquad (3.152)$$

$$\sigma_{xy} = \sigma\left[-\frac{a^2}{r^2}\left(\frac{1}{2}\sin 2\theta + \sin 4\theta\right) + \frac{3a^4}{2r^4}\sin 4\theta\right] \qquad (3.153)$$

where r and θ are the polar coordinates, and θ is measured in a counter-clockwise direction from the positive x-axis. The traction boundary

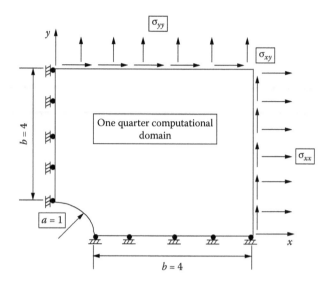

Figure 3.27 Infinite plate with central hole on fourfold symmetry. (Modified from K.Y. Lam, Q.X. Wang, and H. Li. (2004). *Computational Mechanics, 33, 235–244.*)

conditions given by the exact solution in Equations (3.151) to (3.153) are imposed on the right ($x = a + b$) and top ($y = a + b$) edges. Three kinds of nodal distributions are examined here, as shown in Figure 3.28 in which (a) and (b) correspond to cases with 54 and 165 nodes, respectively. The nodes are scattered regularly in θ direction and irregularly in r direction. In (c), the 165 nodes are scattered irregularly in both the θ and r directions.

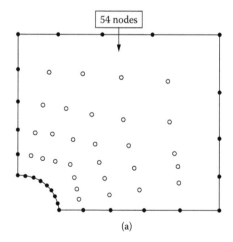

Figure 3.28 Nodal distribution in plate with central hole. (Modified from K.Y. Lam, Q.X. Wang, and H. Li. (2004). *Computational Mechanics, 33, 235–244.*)

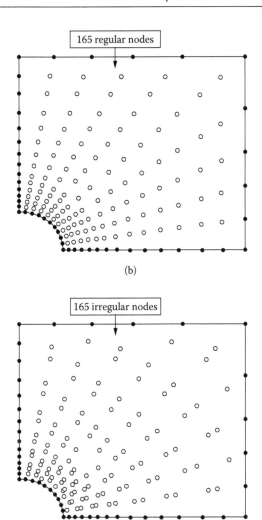

(b)

(c)

Figure 3.28 (continued) Nodal distribution in plate with central hole. (Modified from K.Y. Lam, Q.X. Wang, and H. Li. (2004). *Computational Mechanics*, 33, 235–244.)

Figure 3.29 presents a comparison of numerically computed results with the exact solution for the stress σ_{xx} along $x = 0$. It is observed from (a) that the numerical solutions by the 54 and 165 nodes are satisfactory. Furthermore, the results by the 165 nodes are more accurate than that by the 54 nodes, as shown. In Figure 3.29(b), the numerical results by both the regular and irregular distributions of the 165 nodes agree well with the exact solution.

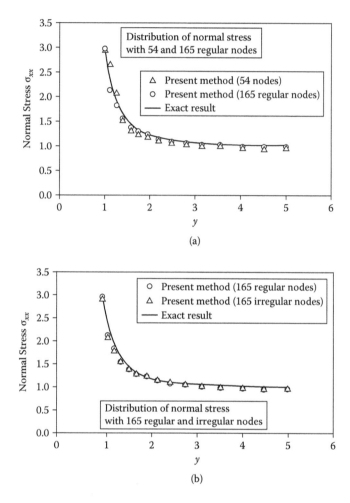

Figure 3.29 Normal stress σ_{xx} along x = 0 for plate with central hole. (Modified from K.Y. Lam, Q.X. Wang, and H. Li. (2004). *Computational Mechanics*, **33**, 235–244.)

3.4.4 Remarks

The local Kriging method has been developed in this section and termed the LoKriging method in which Kriging interpolation is employed for the construction of shape functions with the delta function property using scattered points. This property of the shape function makes it easy to impose the essential boundary conditions. The local weak form is achieved by the local weighted residual method in which the spline function is employed as the weight function since spline weight functions are accurate and convenient.

The choice of the parameters for the interpolation domain and semivariogram model is examined by numerical cases studies. The convergence studies demonstrate that the presently developed LoKriging method possesses good convergence for the problem considered. The numerical case studies show that the LoKriging method presented is easy to implement and accurate for computing approximate displacements and stress fields.

3.5 VARIATION OF LOCAL POINT INTERPOLATION METHOD (VLPIM)

In the variation of local point interpolation method (vLPIM), the meshless polynomial interpolation technique is used for the construction of trial functions with the delta function property. The spline function is employed as the weight function. Using the spline function can easily construct a weight function with desired order and shape and ensure numerical accuracy and computational efficiency, especially for high order differential equations. In order to validate the meshless vLPIM, static problems are investigated first to demonstrate the efficiency of the method. In the dynamic analysis, the Newmark technique is adopted for the integration of time. At each time step, an iteration technique is required to achieve the equilibrium position. It is demonstrated via the numerical examples that the present vLPIM is able to provide very good computational results for the approximate solutions of the problems.

3.5.1 Meshless point interpolation

For the development of meshless vLPIM, the trial function is constructed by interpolation or approximation. The polynomial point interpolation technique developed by Liu and Gu (2001) is employed. One of the advantages of this technique is that the constructed shape functions possess the delta function property, thus it becomes easy to impose the essential boundary conditions.

Let us consider a function $w(\mathbf{x})$ defined in the domain Ω discretised by a set of field nodes. The interpolation function $w(\mathbf{x})$ based on the nodes surrounding a point $\mathbf{x}$ using the polynomial basis can be written as

$$w(\mathbf{x}) = \sum_{i=1}^{n} p_i(\mathbf{x})\, a_i = \mathbf{p}^T(\mathbf{x})\mathbf{a} \tag{3.154}$$

where $p_i(\mathbf{x})$ is a monomial in the space coordinates $\mathbf{x}^T = \{x \quad y\}$, n is the number of nodes in the local interpolation domain for a point $\mathbf{x}$, and a_i is the coefficient for $p_i(\mathbf{x})$ corresponding to the interpolation point $\mathbf{x}$. Therefore, one can write in the matrix form as

$$\mathbf{a} = \left\{ a_1 \quad a_2 \quad \cdots \quad a_n \right\}^T \tag{3.155}$$

$$\mathbf{p}^T(\mathbf{x}) = \left\{ 1 \quad \mathbf{x} \quad \mathbf{x}^2 \quad \cdots \quad \mathbf{x}^{n-1} \right\} \tag{3.156}$$

The coefficients a_i in Equation (3.154) are determined by enforcing Equation (3.154) at n nodes surrounding the point $\mathbf{x}$, and written in the following matrix form

$$\mathbf{w}_e = \mathbf{P}_0 \mathbf{a} \qquad (3.157)$$

where

$$\mathbf{w}_e = \{w_1 \quad w_2 \quad w_3 \quad \cdots w_n\}^T \qquad (3.158)$$

$$\mathbf{p}_0 = \{\mathbf{p}(\mathbf{x}_1) \quad \mathbf{p}(\mathbf{x}_2) \quad \mathbf{p}(\mathbf{x}_3) \quad \cdots \quad \mathbf{p}(\mathbf{x}_n)\} = \begin{Bmatrix} 1 & \mathbf{x}_1 & \mathbf{x}_1^2 & \cdots & \mathbf{x}_1^{n-1} \\ 1 & \mathbf{x}_2 & \mathbf{x}_2^2 & \cdots & \mathbf{x}_2^{n-1} \\ \vdots & \vdots & \vdots & \vdots & \vdots \\ 1 & \mathbf{x}_n & \mathbf{x}_n^2 & \cdots & \mathbf{x}_n^{n-1} \end{Bmatrix}$$

$$(3.159)$$

By Equation (3.157), the coefficients are obtained as

$$\mathbf{a} = \mathbf{P}_0^{-1} \mathbf{w}_e \qquad (3.160)$$

After that, we have

$$w(\mathbf{x}) = \mathbf{\Phi}(\mathbf{x}) \mathbf{w}_e \qquad (3.161)$$

where $\mathbf{\Phi}(\mathbf{x})$ is the shape function defined as

$$\mathbf{\Phi}(\mathbf{x}) = \mathbf{p}^T(\mathbf{x}) \mathbf{P}_0^{-1} = \{\phi_1(\mathbf{x}) \quad \phi_2(\mathbf{x}) \quad \phi_3(\mathbf{x}) \quad \cdots \quad \phi_n(\mathbf{x})\} \qquad (3.162)$$

It is noted that the shape functions possess the delta function property, namely

$$\phi_i(\mathbf{x} = \mathbf{x}_i) = 1, \quad i = 1, 2, \ldots, n \qquad (3.163)$$

$$\phi_j(\mathbf{x} = \mathbf{x}_i) = 0, \quad j \neq i \qquad (3.164)$$

$$\sum_{i=1}^{n} \phi_i(\mathbf{x}) = 1 \qquad (3.165)$$

In the polynomial-based PIM, one of the major challenges is the possible singularity of the $\mathbf{P}_0$ matrix. Several techniques have been proposed to overcome this problem (Liu, 2003; Liu and Gu, 2001). For 1-D problems however, the $\mathbf{P}_0$ matrix is usually not singular. In the present work, the numerical examples of MEMS devices are simplified to 1-D beam structures and then solved.

3.5.2 Numerical implementation of vLPIM

In this subsection, let us take the classical governing equation of the Timoshenko beam problem with the linear theory of elasticity as an example to describe how to implement the vLPIM.

3.5.2.1 Weak form of governing equation

For analysis of the Timoshenko beam, the normality assumption is not made. In other words, the plane sections remain plane, but not necessarily normal to the longitudinal axis after deformation, and the transverse shear strain is no longer zero. The beam theory based on these relaxed assumptions is called the shear deformation beam theory (Reddy, 1993). In general, the non-damping motion equations of a Timoshenko beam are written as

$$
\begin{cases}
\rho A \dfrac{\partial^2 w}{\partial t^2} - \dfrac{\partial}{\partial x}\left[GAk_s \left(\dfrac{\partial w}{\partial x} + \theta \right) \right] - f = 0 \\[4mm]
\rho I \dfrac{\partial^2 \theta}{\partial t^2} - \dfrac{\partial}{\partial x}\left(EI \dfrac{\partial \theta}{\partial x} \right) + GAk_s \left(\dfrac{\partial w}{\partial x} + \theta \right) = 0
\end{cases}
\quad \text{in domain } \Omega \qquad (3.166)
$$

where w is the deflection of the beam, θ the rotation, ρ the mass density, E the modulus of elasticity, I the moment of inertia, A the cross-section area, G the shear modulus, k_s the shear correction coefficient, and f the external force. The auxiliary conditions are

$$
\begin{cases}
w(x_0) = \bar{w}, & \text{on } \Gamma_w \\[2mm]
\theta(x_0) = \bar{\theta}, & \text{on } \Gamma_\theta
\end{cases}
\qquad (3.167)
$$

$$
\begin{cases}
M(x_0) = EI \left. \dfrac{\partial \theta}{\partial x} \right|_{x=x_0} = \bar{M}, & \text{on } \Gamma_M \\[4mm]
V(x_0) = GAk_s \left(\theta + \dfrac{\partial w}{\partial x} \right) \Big|_{x=x_0} = \bar{V}, & \text{on } \Gamma_V
\end{cases}
\qquad (3.168)
$$

$$
\begin{cases}
w(x,t_0) = w_0(x) \\[2mm]
\theta(x,t_0) = \theta_0(x)
\end{cases}
\quad \text{in } \Omega
\qquad (3.169)
$$

$$
\begin{cases}
\dfrac{\partial w(x,t_0)}{\partial t} = v_0(x) \\[4mm]
\dfrac{\partial \theta(x,t_0)}{\partial t} = \gamma_x(x)
\end{cases}
\quad \text{in } \Omega
\qquad (3.170)
$$

where Γ_w, Γ_θ, Γ_M and Γ_v are the boundaries of w, θ, M, and V imposed, respectively, and t is the time and t_0 is the initial time. The local weak form of the differential Equation (3.166) over a local support domain Ω_s bounded by Γ_s can be constructed by the weighted residual method as

$$
\begin{cases}
\displaystyle \int_{\Omega_s} v\left\{\rho A \frac{\partial^2 w}{\partial t^2} - \frac{\partial}{\partial x}\left[GAk_s\left(\frac{\partial w}{\partial x}+\theta\right)\right]-f\right\}d\Omega =0 \\[3mm]
\displaystyle \int_{\Omega_s} v\left\{\rho I \frac{\partial^2 \theta}{\partial t^2} - \frac{\partial}{\partial x}\left(EI\frac{\partial\theta}{\partial x}\right)+GAk_s\left(\frac{\partial w}{\partial x}+\theta\right)\right\}d\Omega =0
\end{cases}
\tag{3.171}
$$

where v is the weight function. Integrating Equation (3.171) by parts, one can have

$$
\begin{cases}
\displaystyle \int_{\Omega_s} v\rho A \frac{\partial^2 w}{\partial t^2}d\Omega + \int_{\Omega_s}\left[\frac{dv}{dx}GAk_s\left(\frac{\partial w}{\partial x}+\theta\right)-vf\right]d\Omega -\left[\bar{n}vGAK_s\left(\frac{\partial w}{\partial x}+\theta\right)\right]\Big|_{\Gamma_s}=0 \\[3mm]
\displaystyle \int_{\Omega_s} v\rho I \frac{\partial^2 \theta}{\partial t^2}d\Omega + \int_{\Omega_s}\left[\frac{dv}{dx}\left(EI\frac{\partial\theta}{\partial x}\right)+GAK_s\left(\frac{\partial w}{\partial x}+\theta\right)\right]d\Omega -\left[\bar{n}v\left(EI\frac{\partial\theta}{\partial x}\right)\right]\Big|_{\Gamma_s}=0
\end{cases}
\tag{3.172}
$$

where $\bar{n}$ is the unit outward normal to domain Ω_s. Note that the boundary Γ_s for the support domain is usually composed of the five portions, namely the internal boundary Γ_{si}, the boundaries Γ_{sw}, $\Gamma_{s\theta}$, Γ_{sM}, and Γ_{sV}, over which the essential boundary conditions w, θ and the natural boundary conditions M, V are specified, respectively. The boundaries Γ_{sw} with Γ_{sV} and $\Gamma_{s\theta}$ with Γ_{sM} are mutually disjointed. By imposing the natural boundary condition given by Equation (3.168), one can have

$$
\begin{cases}
\displaystyle \int_{\Omega_s} v\rho A \frac{\partial^2 w}{\partial t^2}d\Omega + \int_{\Omega_s}\left[\frac{dv}{dx}GAk_s\left(\frac{\partial w}{\partial x}+\theta\right)-vf\right]d\Omega -[\bar{n}v\bar{V}]_{\Gamma_{sV}} -\left[\bar{n}vGAK_s\left(\frac{\partial w}{\partial x}+\theta\right)\right]\Big|_{\Gamma_s-\Gamma_{sV}}=0 \\[3mm]
\displaystyle \int_{\Omega_s} v\rho I \frac{\partial^2 \theta}{\partial t^2}d\Omega + \int_{\Omega_s}\left[\frac{dv}{dx}\left(EI\frac{\partial\theta}{\partial x}\right)+vGAK_s\left(\frac{\partial w}{\partial x}+\theta\right)\right]d\Omega -[\bar{n}v\bar{M}]_{\Gamma_{sM}} -\left[\bar{n}v\left(EI\frac{\partial\theta}{\partial x}\right)\right]\Big|_{\Gamma_s-\Gamma_{sM}}=0
\end{cases}
\tag{3.173}
$$

It is observed from Equation (3.173) that the two types of functions, the weight and trial functions, are required in this method. Since the method is regarded as a weighted residual technique, the weight function plays an important role in its performance. There is no special requirement for a weight function as long as the condition of continuity is satisfied. In Equation (3.173), however, the terms along the boundary Γ_{si} of the local support domain will vanish if the weight function is equal to zero. In the local meshless methods, most boundary integrations come from Γ_{si}. Therefore, if the integrations along the boundary Γ_{si} are removed, the computational time can be reduced significantly.

In the vLPIM, the Petrov–Galerkin approach is employed. In other words, the weight function is constructed by the same interpolation form as the trial function. Figure 3.30(a) and (b) illustrate the weight functions of the

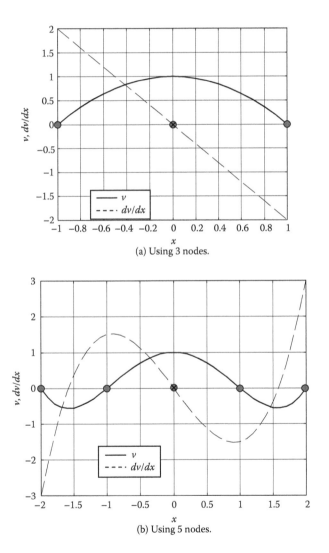

(a) Using 3 nodes.

(b) Using 5 nodes.

Figure 3.30 Weight function and corresponding first-order derivative constructed by point interpolation.

point interpolation form, and the corresponding first-order derivatives by the polynomial point interpolations of the three and five nodes, respectively. The weight function by point interpolation possess the following disadvantages.

1. When fewer nodes are scattered, the order of the weight function will be lower. For example, if three nodes are scattered, the weight function only has second order continuity.
2. When more nodes (e.g., five) are scattered, the weight function has the negative values in some domains.
3. To ensure the weight function is equal to zero along Γ_s, there must be nodes scattered on Γ_s. However, this condition cannot always be satisfied.

In order to avoid the shortcomings of the weight function mentioned above, several special weight functions may be adopted. In the present work however, the fourth order spline function is adopted. It is widely used in MLSA and possess a simpler form than the cubic spline function (Atluri et al., 1999b),

$$
v_i(x) = \begin{cases} 1 - 6\left(\dfrac{d_i}{r_v}\right)^2 + 8\left(\dfrac{d_i}{r_v}\right)^3 - 3\left(\dfrac{d_i}{r_v}\right)^4 & 0 \le d_i \le r_v \\ 0 & d_i \ge r_v \end{cases} \tag{3.174}
$$

where $d_i = |x - x_i|$ is the distance from the node x_i to the point x, and r_v is the size of the local support domain. Plots of the fourth order spline weight function and its derivative are given in Figure 3.31. It is observed that this weight function has the higher order continuity up to the fourth order. The most important advantage of this weight function is that it is easy to construct with the zero value on the boundary of the local support domain. When the weight function becomes zero on the local support boundary Γ_s or Γ_s that does not intersect with the global boundaries (Γ_w, Γ_θ, Γ_M and Γ_V), the local weak form given by Equation (3.172) can be simplified to

$$
\begin{cases} \displaystyle\int_{\Omega_s} v\rho A \frac{\partial^2 w}{\partial t^2}\, d\Omega + \int_{\Omega_s}\left[\frac{dv}{dx}GAk_s\left(\frac{\partial w}{\partial x}+\theta\right)-vf\right]d\Omega = 0 \\[4mm] \displaystyle\int_{\Omega_s} v\rho I \frac{\partial^2 \theta}{\partial t^2}\, d\Omega + \int_{\Omega_s}\left[\frac{dv}{dx}\left(EI\frac{\partial \theta}{\partial x}\right)+vGAk_s\left(\frac{\partial w}{\partial x}+\theta\right)\right]d\Omega = 0 \end{cases} \tag{3.175}
$$

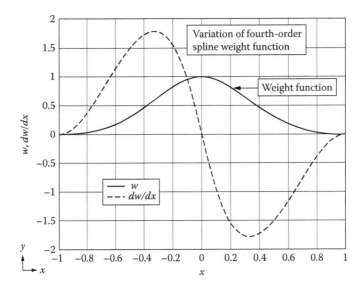

Figure 3.31 Fourth order spline weight function and first-order derivative. (Modified from K.Y. Lam, Q.X. Wang, and H. Li. (2004). *Computational Mechanics,* 33, 235–244.)

3.5.2.2 Discretised equations of dynamic system

In the Timoshenko beam theory, the deflection w and the rotation θ are independent variables. Only the space domain is discretised. Utilizing the local point interpolations, one can have

$$
\begin{cases}
w(x,t) = \Phi^w(x)\mathbf{w}_e(t) \\
\theta(x,t) = \Phi^\theta(x)\theta_e(t)
\end{cases}
\tag{3.176}
$$

where $\Phi^w(x)$ and $\Phi^\theta(x)$ are the shape functions of the deflection and rotation, respectively. They are constructed by the point interpolation given by Equation (3.162), and $\mathbf{w}_e(t)$ and $\theta_e(t)$ are the deflection and rotation of the node at time t. The discretised system of governing equations is written as

$$
\mathbf{M}\ddot{\mathbf{u}}(t) + \mathbf{K}\mathbf{u}(t) = \mathbf{f}(t)
\tag{3.177}
$$

where $\mathbf{M}$ and $\mathbf{K}$ are the mass matrix and the stiffness matrix, respectively, $\mathbf{u}(t)$ is the vector of the nodal deflection and rotation, $\ddot{\mathbf{u}}(t)$ is the second order derivative of $\mathbf{u}(t)$, and $\mathbf{f}(t)$ is the vector of the external force. As a result,

$$
\mathbf{u}(t) = \{w_1, \theta_1, \dots, w_n, \theta_n\}^T
\tag{3.178}
$$

The elements of $\mathbf{M}$, $\mathbf{K}$ and $\mathbf{f}$ are written as

$$m_{ij}^{11} = \int_{\Omega_s} \rho A v_i \Phi_j^w \, d\Omega \tag{3.179}$$

$$m_{ij}^{12} = m_{ij}^{21} = 0 \tag{3.180}$$

$$m_{ij}^{22} = \int_{\Omega_s} \rho I v_i \Phi_j^\theta \, d\Omega \tag{3.181}$$

$$k_{ij}^{11} = \int_{\Omega_s} GAk_s \frac{dv_i}{dx} \frac{d\Phi_j^w}{dx} d\Omega + \left[\bar{n} GAk_s v_i \frac{d\Phi_j^w}{dx} \right]_{\Gamma_{si}+\Gamma_{sw}+\Gamma_{s\theta}+\Gamma_{sM}} \tag{3.182}$$

$$k_{ij}^{12} = \int_{\Omega_s} GAk_s \frac{dv_i}{dx} \phi_j^\theta \, d\Omega + \left[\bar{n} GAk_s v_i \Phi_j^\theta \right]_{\Gamma_{si}+\Gamma_{sw}+\Gamma_{s\theta}+\Gamma_{sM}} \tag{3.183}$$

$$k_{ij}^{21} = \int_{\Omega_s} GAk_s v_i \frac{d\Phi_j^w}{dx} d\Omega \tag{3.184}$$

$$k_{ij}^{22} = \int_{\Omega_s} \left(EI \frac{dv_i}{dx} \frac{d\Phi_j^\theta}{dx} + GAk_s v_i \Phi_j^\theta \right) d\Omega + \left[\bar{n} EI v_i \frac{d\Phi_j^\theta}{dx} \right]_{\Gamma_{si}+\Gamma_{sw}+\Gamma_{s\theta}+\Gamma_{sV}} \tag{3.185}$$

$$f_i^w = \int_{\Omega_s} v_i \, f(t) \, d\Omega + \left[\bar{n} v_i \, \bar{V} \right]_{\Gamma_{sV}} \tag{3.186}$$

$$f_i^\theta = [\bar{n} v_i \, \bar{M}]_{\Gamma_{sM}} \tag{3.187}$$

The dynamics of Equation (3.177) is solved by the Newmark method, which is an unconditionally stable implicit technique of temporal discretisation (Reddy, 1993).

3.5.3 Examples for validation

In this subsection, two case studies of the beam problems are used for validation of the vLPIM. The case studies include a fixed–fixed thin beam and a cantilever thick beam for the examination of numerical convergence rate and computational performance of the vLPIM.

Example 3.11: Static analysis of fixed–fixed thin beam

In order to examine the computational efficiency of the present method, a fixed–fixed thin beam subject to uniformly distributed load is considered, as shown in Figure 3.32(a). The parameters of this beam are given as $E = 1.2 \times 10^7$, Poisson ratio $\bar{v} = 0.3$, length $L = 10$, depth $D = 0.1$, width $\bar{t} = 1.0$, and uniform force $f(x) = 1.0$. As well known, there is a shear locking phenomenon in the analysis of thin beam using the thick beam theory.

Several studies have been conducted to get rid of shear locking phenomena such as adding transverse shear strain as an additional variable in MLPG (Cho et al., 2000). Liu (2003) concluded that the shear locking phenomena could be avoided easily by simply using high order polynomial point interpolation such as the four or more interpolation nodes in local meshless methods. There is no problem to meet this requirement in 1-D polynomial point interpolation. In the present example, the four nodes are employed in each of the interpolation domains to get rid of the shear locking phenomena.

A total of 21 nodes are scattered in the domain for the discretisation of the beam. The deflection and rotation are computed and plotted in

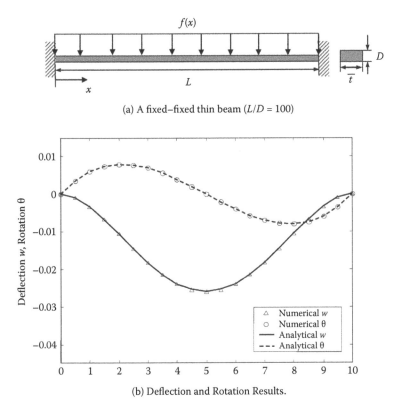

(a) A fixed–fixed thin beam ($L/D = 100$)

(b) Deflection and Rotation Results.

Figure 3.32 Static analysis of fixed–fixed thin beam.

Figure 3.32(b). The negative value of the deflection means a downward deformation. The analytical solution of the Bernoulli-Euler beam theory is also plotted in the same figure for comparison. The figure indicates that the present method is able to produce satisfactory computational accuracy for the problem. In the meantime, the shear locking is also overcome although the ratio L/D is very large.

Example 3.12: Static analysis of cantilever thick beam

In this example, a cantilever thick beam is analyzed, as shown in Figure 3.33(a). The parameters of the thick beam are given as $E = 3.0 \times 10^7$, $\bar{v} = 0.3$ $L = 48$, $D = 12$, $\bar{t} = 1.0$, and the concentrated load $f(L) = 1000$, especially $L/D = 4$ in this example for which the thick beam theory must be considered. The analytical solution of the problem is available (Timoshenko and Goodier, 1970).

$$w(x) = -\frac{f(L)}{6EI}\left[(4+5\bar{v})\frac{D^2x}{4} + (3L-x)x^2\right] \tag{3.188}$$

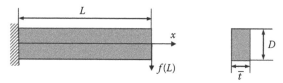

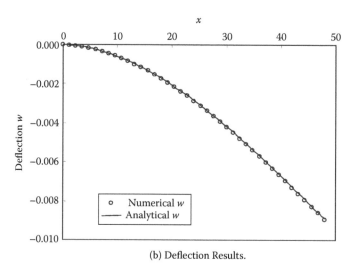

(a) A cantilever thick beam ($L/D = 4$)

(b) Deflection Results.

Figure 3.33 Static analysis of cantilever thick beam.

The deflection computed by the vLPIM is plotted in Figure 3.33(b) in which the negative value means a downward deformation. The numerical results obtained by the present meshless method agree very well with the analytical solution.

3.5.4 Remarks

In the presently developed meshless technique termed the variation of local point interpolation method, the concepts of other local meshless methods are followed similarly and the meshless polynomial interpolation technique is employed. The local weak form of the governing equations of a thick beam is constructed by the partial differential equation developed by Timoshenko beam theory. In the local weak form, the spline function is employed as the weight or test function since spline weight functions are more accurate and convenient to use in the Galerkin formulation. The discretised system of dynamic equations of Timoshenko beam theory is based on the local weighed residual method and the meshless polynomial interpolation technique. Static problems are examined first to demonstrate the computational efficiency of the method. The Newmark scheme is implemented to solve the dynamic equation directly, and an iteration technique is used to solve the nonlinear equation at each time step. In the subsequent chapter, the vLPIM will be used to simulate the fundamental characteristics of beam-based microelectromechanical systems (MEMS).

3.6 RANDOM DIFFERENTIAL QUADRATURE (RDQ) METHOD

A novel strong-form meshless RDQ method is developed in this section. First, several references are studied for the background of fixed RKPM interpolation function and DQ method. Then the motivation for the development of RDQ method is elaborated. The formulations of the fixed RKPM and DQ method are discussed in detail followed by the development of the RDQ method.

The originally proposed SPH method (Lucy, 1977) is unable to reproduce the higher order terms well or satisfy the consistency condition in solving the problems with finite boundaries. Hence, it was enhanced over the last decade to improve its consistency and stability (Liu et al., 1995, 1996; Liu and Jun, 1998; Bonet and Lok, 1999). Liu et al. (1995, 1996) modified the SPH window function by introducing a correction function term called the RKPM by which the SPH window function is called the modified window function. The RKPM is more widely used now than the original SPH method because of higher order reproducibility. Wong and Shie (2008) performed large deformation analysis by the SPH-based Galerkin method with the moving least-squares (MLS) approximation.

Wu et al. (2008) proposed a differential RKPM in which the separate sets of the differential reproducing condition were developed to compute the derivative shape functions instead of conventionally computing them by directly taking the derivative of RKPM approximation. Li et al. (2008) used the RKPM method to perform ductile fracture simulations and achieved very good agreement with the FEM and existing experimental data.

Following the idea of an integral quadrature, Bellman et al. (1972) proposed the DQ method in which the derivative at any grid point is approximated by the weighted sum of function values in whole domain provided that all the grid points are always collinear. The important task in the DQ method is to determine the weighting coefficients. Bellman et al. (1972) suggested two approaches to compute the weighting coefficients. The first is the use of polynomial function as a test function. The second uses the test function in which the coordinates of the grid points are chosen as the roots of shifted Legendre polynomials.

However, these approaches have a problem when the order of the system of algebraic equations is very high, making the resulting matrix highly ill conditioned. To overcome this problem, Quan and Chang (1989a and b) proposed another approach in which the weighting coefficients are computed by Lagrange interpolation polynomials. Shu et al. (1994) proposed a general approach combining both Bellman (1972) and Quan and Chang (1989a and b) approaches.

Shu (2000) proved by the analysis of linear vector space that the polynomials used in Bellman et al. (1972) and Quan and Chang (1989a and b) approaches are nothing but different sets of the base polynomial vectors of the function approximation. If one of the base vectors satisfies the approximate function equation, the other base vectors do so. This means that all the approaches will lead to the same values of the weighting coefficients, and in turn similar function approximations.

Shan et al. (2008) coupled the local multiquadric-based radial basis functions (RBFs) with the DQ method to solve fluid flow problems with 3-D curved boundaries. Liew et al. (2003b) applied the DQ method to model elastic bonding in 3-D composite laminates. Liew et al. (2003a) also applied the MLS based DQ (MLSDQ) to solve the moderately thick plate for shear deformation, and to solve the fourth order bending differential equation of thin plate over irregular boundaries (Liew et al., 2004). More work about the DQ method can be found in Naadimuthu et al. (1984) and Ding et al. (2006).

The main objective of the development of RDQ is to extend the applicability of the DQ method over the irregular and non-uniform domains discretised by either uniform or random field nodes. Two types of nodal distributions are used in the RDQ method. First, the background (virtual) nodes are created in a fixed pattern (uniform or Chebyshev-Gauss-Lobatto nodes). Second, the field nodes are created either in the uniform or random manner,

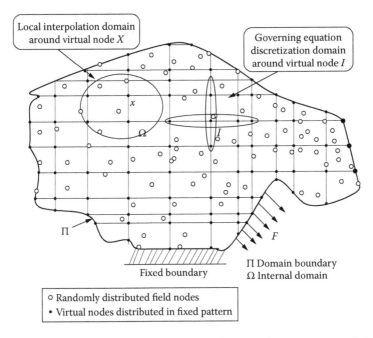

Figure 3.34 Working principle of RDQ method for irregular computational domain discretised by random field nodes coupled with virtual nodes distributed in fixed pattern. (Modified from H. Li and S. Mulay. (2011). *Computational Mechanics*, DOI: 10.1007/s004466-011-0622-5.)

as shown in Figure 3.34. The virtual nodes are interpolated by surrounding field nodes with the fixed RKPM interpolation function, and a linear transformation matrix is developed to relate the values of the field variable at the virtual nodes with the values of the nodal parameter at the field nodes.

The governing differential equations and boundary conditions are discretised at the virtual nodes located within the domain and along the boundaries, respectively, by the locally applied DQ method. The local domains are created around each virtual node in the x and y directions in which the virtual nodes falling are considered for the discretisation of derivative terms from the governing PDEs at the concerned virtual node in the respective directions as shown in Figure 3.34.

In the discretised equations, the terms related to the values of the function at the virtual nodes are replaced by the corresponding interpolation equations using the linear transformation matrix developed earlier. A system of algebraic equations is obtained after simplifying it, which is solved for the unknown values of the nodal parameter u_i at the field nodes. Since the fixed RKPM function does not possess the Dirac delta function property, the values of nodal parameter u_i at the field nodes are not equal to the values of function approximation f_i^b at the field nodes, namely $u_i \neq f_i^b$. As a result, the values of approximate function f_i^b at the field nodes are computed by interpolating each field node by the nearby field nodes with the fixed RKPM interpolation function as $f_i^b = \sum_{j=1}^{NP} N_j \, u_j$, where N_j and u_j are the values of shape function and nodal parameter at the jth field node, respectively.

Total NP field nodes surrounding the ith field node are included in the interpolation domain of the ith field node, such that $NP \subseteq N$, where N represents the total field nodes in the domain. This is broadly the working principle of the RDQ method.

One of the important requirements of the DQ method is that all the discretised field nodes should be always arranged in a collinear manner and the domain should be a regular one. Suppose, while solving a dynamic problem by the DQ method, all the field nodes are scattered in a specific pattern at the beginning of computation. After the first time increment, the field nodes are randomly distributed with a modified boundary. This situation restricts the application of the DQ technique for the further time increments. As a result, the DQ method is restricted to problems with regular domains discretised by field nodes distributed in a fixed pattern. This limitation is overcome by the present RDQ method via the fixed RKPM interpolation function to construct a novel strong form meshless method.

Compared with other existing methods, the merit of the RDQ method is the combination of both the Eulerian and Lagrangian grids. The virtual nodes represent a Eulerian grid fixed in a computational domain. The field nodes represent a Lagrangian grid that moves freely in a domain as per the computation. No interconnectivity information among the field nodes is required while computing the values of approximate functions and thus they are free to move anywhere in the domain. The governing equation is easily discretised by the virtual nodes at every time increment, as they are fixed in a space. As a result, the proposed RDQ method is applicable to solve large deformation and moving boundary problems, namely crack propagation and hydrogel simulation. The classical FEM faces many challenging issues due to the distortion of mesh and singularity in elements. As compared with the other meshless methods, especially the weak form methods, the RDQ method is also capable of capturing the local high gradients.

The fixed RKPM interpolation function and the formulations of DQ method are presented in the following subsections, and consequently the working principle of the RDQ method is elaborated.

3.6.1 Formulation of fixed reproducing kernel particle method

The approximate equation by the fixed RKPM interpolation function is derived in this section. The fixed RKPM interpolation function is a special case of RKPM function in which the kernel is fixed at the central node as explained by Aluru and Li (2001). The fixed RKPM interpolation function has several advantages over the other forms, such as the classical, moving, and multiple fixed functions. For example, it has the constant moment matrix due to the fixed kernel and it follows the partition of unity principle. The approximate value of function as per the fixed RKPM interpolation is given as

$$f^h(x,y) = \int_\Omega C(x,y,u,v) \, K(x_k - u, y_k - v) \, f(u,v) \, du \, dv \qquad (3.189)$$

where $C(x,y,u,v)$ is the unknown correction function, $f^h(x,y)$ is an approximation of the function $f(x,y)$ at a node (x, y), and $K(x_k - u, y_k - v)$ is the kernel function fixed at the node (x_k, y_k) . Order of the approximate function $f^h(x,y)$ is determined by the order of the monomial basis functions used in the correction function $C(x,y,u,v) = P^T(u,v) \, c(x,y)$, where $P^T(u,v) = \{b_1(u,v), b_2(u,v),..., b_m(u,v)\}$ is the mth order monomial column vector and $c(x,y)$ is the mth order unknown row vector. The unknown correction functions are computed by the consistency condition, such that the approximate function $f^h(x,y)$ passes through all the nodes located in the domain Ω as

$$C(x,y,u,v) = P^T(u,v) \, c(x,y) \qquad (3.190)$$

where $c(x,y)_{(1\times m)} = \{c_1(x,y), c_2(x,y),..., c_m(x,y)\}$ is the mth order row vector of the unknown correction function coefficients. The 1-D bases of the linear monomials in the polynomial basis function are given as

$$P^T(u,v) = \{1, \; u, \; v\}, \quad (m = 3) \qquad (3.191)$$

The 2-D bases of the quadratic monomials are given as

$$P^T(u,v) = \{1, \; u, \; v, \; u^2, \; uv, \; v^2\}, \quad (m = 6) \qquad (3.192)$$

The coefficients of unknown correction function $c(x,y)$ are determined by the consistency or reproducing condition, as the approximate function is required to satisfy every monomial of the polynomial basis function as

$$b_i(x,y) = \int_\Omega P^T(u,v) \, c(x,y) \, K(x_k - u, y_k - v) \, b_i(u,v) \, du \, dv, \quad i = 1,2,...,m$$

$$(3.193)$$

This equation is rewritten in a discretised or particle form as

$$\{b_i(x,y)\}_{(m\times 1)} = \sum_{I=1}^{NP} P^T(x_I,y_I) \, c(x,y) \, K(x_k - x_I, y_k - y_I) \, b_i(x_I,y_I) \, dV_I$$

$$(3.194)$$

where NP represents the total field nodes in a local interpolation domain of the virtual node (x_k, y_k) and dV_I is a nodal volume at the Ith field node. Equation (3.194) is written in a matrix form as

$$\{b_i(x,y)\} = M \, c(x,y) \tag{3.195}$$

where M is an $m \times m$ moment matrix given in the discretised form as

$$M_{ij} = \sum_{I=1}^{NP} \{b_i(x_I,y_I)\} \, K(x_k - x_I,\, y_k - y_I) \, b_j(x_I,y_I) dV_I, \quad i \text{ and } j = 1, 2,..., m \tag{3.196}$$

It is seen from Equation (3.196) that the moment matrix is constant. This is an advantage of the fixed RKPM interpolation function over its other variants such as the moving or multiple fixed RKPM functions because their moment matrices are not constant and must be computed for each local domain. Equation (3.196) is written in generalized form as

$$M = F W F^T \tag{3.197}$$

where F is am $m \times NP$ matrix given as

$$F = \begin{bmatrix} b_1(x_1,y_1) & b_1(x_2,y_2) & \cdots & b_1(x_{NP},y_{NP}) \\ b_2(x_1,y_1) & b_2(x_2,y_2) & \cdots & b_2(x_{NP},y_{NP}) \\ \vdots & \vdots & \ddots & \vdots \\ b_m(x_1,y_1) & b_m(x_2,y_2) & \cdots & b_m(x_{NP},y_{NP}) \end{bmatrix} \tag{3.198}$$

W is an $NP \times NP$ diagonal matrix given as

$$W = \begin{bmatrix} K(x_k - x_1, y_k - y_1)dv_1 & 0 & \cdots & 0 \\ 0 & K(x_k - x_2, y_k - y_2)dv_2 & \cdots & 0 \\ \vdots & \vdots & \ddots & \vdots \\ 0 & 0 & \cdots & K(x_k - x_{NP}, y_k - y_{NP})dv_{NP} \end{bmatrix} \tag{3.199}$$

The vector $c(x,y)$ is obtained from Equation (3.195) as

$$c(x,y)_{(m\times1)} = M^{-1}\{b_i(x,y)\}_{(m\times1)} \tag{3.200}$$

Substituting Equations (3.200) and (3.190) into Equation (3.189) results in

$$f^h(x,y) = \int_\Omega P^T(x,y) \, M^{-1}\{b_i(u,v)\} \, K(x_k - u, y_k - v) \, f(u,v) \; du \; dv \tag{3.201}$$

This equation is rewritten in the discretised form as

$$f^h(x,y) = \sum_{I=1}^{NP} P^T(x,y)\, M^{-1} B(x_I,y_I)\, K(x_k - x_I, y_k - y_I)\, f(u,v)\, dV_I \quad (3.202)$$

$$f^h(x,y) = \sum_{I=1}^{NP} N_I(x,y)\, u_I \quad\quad\quad (3.203)$$

where $N_I(x,y)$ and u_I are the values of the shape function and unknown nodal parameter at the Ith field node, respectively. The $N_I(x,y)$ is given as

$$N_I(x,y) = P(x,y)_{1\times m}\, M^{-1}_{m\times m}
\begin{bmatrix}
b_1(x_I,y_I)\, K(x_k - x_I,\ y_k - y_I) \\
b_2(x_I,y_I)\, K(x_k - x_I,\ y_k - y_I) \\
\vdots \\
b_m(x_I,y_I)\, K(x_k - x_I,\ y_k - y_I)
\end{bmatrix}_{m\times 1}
\Rightarrow P(x,y)C_{II}^{-1}$$

$$(3.204)$$

where C_{II}^{-1} denotes the Ith column of C^{-1} such that $I \subseteq NP$. Equation (3.203) is a final expression for the approximation of function by the fixed RKPM interpolation. The kernel function is a window function that approximates the distribution of field variables over a local domain and therefore has a non-zero value over a certain sub-domain and zero outside of it.

$$K(x_k - u, y_k - v) = \frac{1}{d_x} w\left(\frac{x_k - u}{d_x}\right) \frac{1}{d_y} w\left(\frac{y_k - v}{d_y}\right) \quad (3.205)$$

$$w(z_I) = \begin{cases}
0, & z_I < -2 \\[4pt]
\dfrac{1}{6}(z_I + 2)^3, & -2 \le z_I \le -1 \\[4pt]
\dfrac{2}{3} - z_I^2(1 + \dfrac{z_I}{2}), & -1 \le z_I \le 0 \\[4pt]
\dfrac{2}{3} - z_I^2(1 - \dfrac{z_I}{2}), & 0 \le z_I \le 1 \\[4pt]
\dfrac{-1}{6}(z_I - 2)^3, & 1 \le z_I \le 2 \\[4pt]
0, & z_I > 2
\end{cases} \quad (3.206)$$

where $z_I = [(x_k - u)/d_x]$, and d_x and d_y are the cloud sizes in the x and y directions, respectively. The cubic spline function $w(z_I)$ is used in the RDQ method, as given in Equation (3.206). It is observed from Equation (3.189) that the fixed RKPM interpolation function does not possess the delta function property, but with the cubic spline function $w(z_I)$ in Equation (3.206), it has the partition of unity property such that the summation of the shape function values at all the NP field nodes is equal to one as given by

$$\sum_{I=1}^{NP} N_I(x,y) = 1 \tag{3.207}$$

Jin et al. (2001) and Atluri and Shen (2002) discussed the equivalence of the RKPM and MLS shape functions. Atluri and Shen (2002) showed that the shape functions obtained by the RKPM and MLS are identical if the same kernel and window functions are chosen in both with the same consistency order k.

3.6.2 Formulation of differential quadrature method

The key task in the DQ method is to compute the DQ weighting coefficients. Several approaches have been proposed to compute the weighting coefficients (Bellman et al., 1972; Quan and Chang, 1989a-b; Shu et al., 1994). One is Shu's general approach (Shu et al., 1994) based on the Lagrange interpolation polynomials adopted in the RDQ method. A locally applied DQ method is used in the RDQ method, in which the derivative terms from the governing differential equation are discretised at the concerned virtual node by the nearby surrounding virtual nodes only, as shown in Figure 3.34. The first-order derivative in the x direction by locally applying the DQ method is given as

$$f_{,x}^{(1)}(x_i) = \sum_{j=1}^{N_x} a_{ij} \, f(x_j) \tag{3.208}$$

where $f_{,x}^{(1)}(x_i)$ is the approximate first-order derivative at the virtual node (x_i) in the x direction and $f(x_j)$ is the approximate function value at the jth virtual node where $j \in [1, \, N_x]$. a_{ij} are the DQ weighting coefficients of the first-order derivative, evaluated by Shu's general approach as

$$a_{ij}^x = \frac{1}{x_i - x_j} \prod_{m=1, m \neq i, j}^{N_x} \left(\frac{x_i - x_m}{x_j - x_m} \right), \quad \text{and} \quad a_{ii}^x = -\sum_{k=1, k \neq i}^{N_x} a_{ik}^x \tag{3.209}$$

The approximate derivatives of the first-order for a 2-D problem are similarly given by

$$f_x^{(1)}(x_i, y_j) = \sum_{k=1}^{N_x} a_{ik}^x f(x_k, y_j) \tag{3.210}$$

$$f_y^{(1)}(x_i, y_j) = \sum_{k=1}^{N_y} a_{jk}^y f(x_i, y_k) \tag{3.211}$$

where a_{ik}^x and a_{jk}^y are the DQ weighting coefficients in the x and y direction, respectively. They are computed by Equation (3.209), and N_x and N_y are the total virtual nodes in the local DQ domain of the concerned virtual node (x_i, y_j) in the x and y directions, respectively. Similarly, the approximate derivatives of the second order are as

$$\frac{d^2 f(x_i, y_j)}{dx^2} = \sum_{k=1}^{N_x} b_{ik} f^b(x_k, y_j) \tag{3.212}$$

$$\frac{d^2 f(x_i, y_j)}{dy^2} = \sum_{k=1}^{N_y} b_{jk} f^b(x_i, y_k) \tag{3.213}$$

$$\frac{d^2 f(x_i, y_j)}{dx\,dy} = \sum_{l=1}^{N_x} a_{il} \sum_{k=1}^{N_y} a_{jk} f^b(x_l, y_k) \tag{3.214}$$

where b_{ik} and b_{jk} are the DQ weighting coefficients computed as

$$b_{ij} = 2\, a_{ij} \left[a_{ii} - \frac{1}{x_i - x_j} \right], \text{ for } j = 1, 2, \ldots, N_x, \; \forall \; i \neq j : b_{ij} \;\; b_{ii} = -\sum_{j=1_{j \neq i}}^{N_x} b_{ij}$$
$$\tag{3.215}$$

A recurring formula for the computation of higher order derivatives was developed (Shu, 2000).

It is difficult to ensure a good numerical solution by accurately computing the weighting coefficients. Quan and Chang observed that the accuracy of a numerical solution depends on the values of the weighting coefficient and the distribution of grid points in a domain (Quan and Chang, 1989a).

They showed that the general collocation method is actually same as the DQ method, but the grid points placed at the roots of the first kind of Chebyshev polynomial consistently gives better solution by the DQ method than that obtained by the orthogonal collocation method. They implemented the general procedure to solve a PDE (Quan and Chang, 1989a-b) and presented the numerical results with the DQ weighting coefficients computed by Equations (3.216) to (3.218) as

$$a_{ij} = \frac{1}{x_j - x_i} \prod_{\substack{m=1 \\ m \neq i,j}}^{n} \frac{x_i - x_m}{x_j - x_m}, \quad \text{for } i \neq j \quad \text{and} \quad a_{ii} = \sum_{\substack{k=1 \\ k \neq i}}^{n} \frac{1}{x_i - x_k}, \quad \text{for } i = j$$

(3.216)

$$b_{ij} = \frac{2}{x_j - x_i} \left(\prod_{\substack{m=1 \\ m \neq i,j}}^{n} \frac{x_i - x_m}{x_j - x_m} \right) \left(\sum_{\substack{k=1 \\ k \neq i,j}}^{n} \frac{1}{x_i - x_k} \right), \quad \text{for } i \neq j$$

(3.217)

$$b_{ii} = 2 \sum_{\substack{k=1 \\ k \neq i}}^{n-1} \left[\frac{1}{x_i - x_k} \left(\sum_{\substack{l=k+1 \\ l \neq i}}^{n} \frac{1}{x_i - x_l} \right) \right], \quad \text{for } i = j$$

(3.218)

where a_{ij} and a_{ii}, and b_{ij} and b_{ii} are the weighting coefficients of the first and second order derivatives, respectively, computed at the ith node using the total n domain nodes. Even though both, Shu (Shu, 2000) and Quan and Chang (Quan and Chang, 1989a-b), assumed Lagrange interpolation polynomial as a test function, the final weighting coefficient equations are little different. Using the linear vector space analysis, Shu et al. (1994) proved that different polynomials used in Bellman et al. (1972) and Quan and Chang (1989a and b) approaches are nothing but different sets of base polynomial vectors of a function approximation such that all the base vectors satisfy the equation of function approximation if one of them does so.

In summary, the fixed RKPM function and DQ method are discussed as preliminary works. Among several approaches to compute the DQ weighting coefficients, Shu's general approach (Shu et al., 1994; Shu, 2000) is adopted in the RDQ method. Development of the RDQ method is explained in the next subsection.

3.6.3 Development of RDQ method

As explained earlier, two types of distributive nodes are used in the RDQ method. The first group are the virtual nodes used for the discretisation of the governing equation, and the second group are called field nodes and used to determine the field variable distributions within the computational domain as shown in Figure 3.34. The detailed working principle of the RDQ method along with the implementation of boundary conditions is discussed in the subsequent subsections.

3.6.3.1 Bridge to link discretisation of governing differential equation with approximation of function value at field nodes

The objective of the development of RDQ method is to extend the use of DQ method for a regular computational domain discretised by the random field nodes and for an irregular computational domain discretised by uniform or random field nodes. This is achieved by creating the virtual and field nodes in a computational domain as shown in Figure 3.34. A local domain is created around every virtual node in the x and y directions, and the virtual nodes falling in it are considered for the discretisation of derivative term at the concerned virtual node; this is called the DQ local domain.

The derivative terms from the governing differential equation are approximated at the virtual nodes located in the internal computational domain, by applying the DQ method with the DQ local domain of the concerned virtual node. Another local domain is created around each virtual node, in which the field nodes (uniform or random) falling are considered for the approximation of field variable values at the concerned virtual node by the fixed RKPM interpolation function as shown in Figure 3.34.

When all the equations of the approximation of field variables at the virtual nodes are combined, a transformation matrix comprising the values of shape function is obtained that associates the values of field variables at the virtual nodes with the values of nodal parameters at all the field nodes. In the discretised governing equation, the terms of the function values at the virtual nodes are replaced by their corresponding transformation equations in the form of the shape functions developed earlier. The discretised governing equation in this way is represented in the form of the unknown values of nodal parameters at all the field nodes.

Any shape can be used to create the local interpolation domain. A circular shape is adopted in the RDQ method for convenience. Let us consider $d_x = \alpha_x \Delta_x$ as the domain size in the x direction (Aluru and Li, 2001) where α_x is the size of local domain and Δ_x is an average nodal spacing in the x direction. Two approaches are studied to define the value of α_x. The first is numerical analysis and the second is developing a positivity condition. The value of α_x defined by the positivity condition ensures that the kernel is non-zero over a local domain of interpolation and zero outside it.

The positivity condition is derived based on the fact that the fixed RKPM interpolation function is non-zero only over a certain local domain. As per Equation (3.206), the kernel function is zero if $z_I > 2$ or $z_I < -2$, which means that to have a non-zero value of a kernel function, the α_x and α_y values should satisfy $z_I < 2$ or $z_I > -2$ conditions, as given by

$$\frac{(x_k - x_i)}{(\alpha_x \, \Delta_x)} \geq -2, \quad \text{and} \quad \frac{(x_k - x_i)}{(\alpha_x \, \Delta_x)} \leq 2 \tag{3.219}$$

$$\therefore \alpha_x \in \left[\left(\frac{x_k - x_i}{2 \, \Delta_x} \right), \ -\left(\frac{x_k - x_i}{2 \, \Delta_x} \right) \right], \quad \text{for the domain size } [2, -2] \tag{3.220}$$

where the index k refers the fixed virtual node, and the index i refers the field node in the local interpolation domain of the kth virtual node. The highest absolute value of α_x is selected from all the values obtained from Equation (3.220). From the numerical analysis approach, α_x values equal to 1.17 and 2.23 are found to give good results. Comparing the results obtained from both the approaches, α_x of 1.17 (Aluru and Li, 2001) is used in the RDQ method.

Let Equation (3.212) be discretised at the virtual node (x_i, y_j) by the RDQ method as

$$\frac{d^2 f(x_i, y_j)}{dx^2} = \{ b_{i1} \quad b_{i2} \quad \cdots \quad b_{iN_x} \}_{1 \times N_x} \begin{bmatrix} f^b(x_1, y_j) \\ f^b(x_2, y_j) \\ \vdots \\ f^b(x_{N_x}, y_j) \end{bmatrix}_{N_x \times 1}$$

$$\tag{3.221}$$

The terms $f^b(x_{N_x}, y_j)$ in Equation (3.221) are replaced by the fixed RKPM interpolation functions from the linear transformation matrix as

$$\begin{bmatrix} f^b(x_1, y_j) \\ f^b(x_2, y_j) \\ \vdots \\ f^b(x_{N_x}, y_j) \end{bmatrix}_{(N_x \times 1)} = \begin{bmatrix} N_1(x_1, y_j) & N_2(x_1, y_j) & \cdots & N_{NP}(x_1, y_j) \\ N_1(x_2, y_j) & N_2(x_2, y_j) & \cdots & N_{NP}(x_2, y_j) \\ \vdots & \vdots & \ddots & \vdots \\ N_1(x_{N_x}, y_j) & N_2(x_{N_x}, y_j) & \cdots & N_{NP}(x_{N_x}, y_j) \end{bmatrix}_{(N_x \times NP)} \begin{bmatrix} u_1 \\ u_2 \\ \vdots \\ u_{NP} \end{bmatrix}_{(NP \times 1)}$$

$$\tag{3.222}$$

Substituting Equation (3.222) in Equation (3.221) results in

$$
\frac{d^2f(x_i,y_j)}{dx^2} = \{b_{i1} \quad \cdots \quad b_{iN_x}\}_{1\times N_x}
\begin{bmatrix}
N_1(x_1,y_j) & \cdots & N_{NP}(x_1,y_j) \\
\vdots & \ddots & \vdots \\
N_1(x_{N_x},y_j) & \cdots & N_{NP}(x_{N_x},y_j)
\end{bmatrix}_{(N_x\times NP)}
\begin{bmatrix}
u_1 \\
u_2 \\
\vdots \\
u_{NP}
\end{bmatrix}_{(NP\times1)}
$$

(3.223)

Equation (3.223) is further simplified to

$$
\frac{d^2f(x_i,y_j)}{dx^2} = \{\; k_1 \quad k_2 \quad \cdots \quad k_{NP} \;\}_{1\times NP}
\begin{bmatrix}
u_1 \\
u_2 \\
\vdots \\
u_{NP}
\end{bmatrix}_{NP\times1}
$$

(3.224)

where

$$
k_m = \{b_{i\,N_x}\}_{1\times N_x} \{N_m\}_{N_x\times m}, \text{ where } m \in [1,\ NP]
$$

(3.225)

As a result, Equation (3.224) is the final linear algebraic expression for Equation (3.212) in terms of the unknown values of the nodal parameters at the field nodes. Note from Equation (3.222) that some values of the shape function may be zero; therefore the final stiffness matrix is banded and sparse. Similarly, Equation (3.212) is discretised at all the virtual nodes located in the internal computational domain of the problem and corresponding Equation (3.224) is obtained at all the virtual nodes.

All the equations via the discretisation and boundary conditions are combined to form the final stiffness matrix. Therefore, the numbers of the total rows and columns of the final stiffness matrix are given by the total virtual and field nodes, respectively. The total virtual nodes should be at least equal to or greater than the total field nodes, resulting in a non-symmetric and rectangular stiffness matrix. The detailed procedure to solve the final stiffness matrix is elaborated next.

3.6.3.2 Imposing Dirichlet boundary conditions

The fixed RKPM interpolation function has no Kronecker delta function property; thus $u_I \neq f^h(x_I,y_I)$, namely the value of nodal parameter u_I at the field node, is not equal to the value of approximate function $f^h(x_I,y_I)$. But, it is still possible to exactly impose the Dirichlet boundary condition.

As the governing differential equations and boundary conditions are imposed at the virtual nodes, extra virtual nodes are created at the locations of field nodes scattered along the Dirichlet boundary to ensure that the boundary values at all the field nodes located on the Dirichlet boundary are imposed. It is assumed for all the virtual nodes (original and extra created) located along the Dirichlet boundary that $u_I = f^h(x_I, y_I)$. For example, in a given 1-D domain discretised by a total of four virtual nodes, the first and fourth virtual nodes are on the Dirichlet boundary, and the corresponding function values imposed are b_1 and b_4. The modified global matrix after imposing $u_I = f^h(x_I, y_I)$ for the first and fourth virtual nodes is given as

$$
\begin{bmatrix}
1 & 0 & 0 & 0 \\
k_{21} & k_{22} & k_{23} & k_{24} \\
k_{31} & k_{32} & k_{33} & k_{34} \\
0 & 0 & 0 & 1
\end{bmatrix}
\begin{Bmatrix}
u_1 \\
u_2 \\
u_3 \\
u_4
\end{Bmatrix}
=
\begin{Bmatrix}
b_1 \\
F_2 \\
F_3 \\
b_4
\end{Bmatrix}
\tag{3.226}
$$

The values of u_1 and u_4 are obtained as b_1 and b_4, respectively, by solving Equation (3.226).

3.6.3.3 Imposing Neumann boundary conditions

The Neumann boundary condition is imposed in two ways. By the first approach, all the equations of Neumann boundary condition are discretised by the RDQ method at the virtual nodes located on the Neumann boundary. By the second approach, the Neumann boundary condition is converted to the Dirichlet boundary condition, and then imposed as the Dirichlet boundary condition. Let us take Equation (3.210) as an example that needs to be satisfied at the virtual node (x_i, y_i) located on the Neumann boundary.

$$
\frac{df(x_i, y_i)}{dx} = 0
\tag{3.227}
$$

Equation (3.227) is modified as

$$
\left. \frac{df}{dx} \right|_i = \frac{f_{next} - f_i}{b} = 0 \quad \Rightarrow \quad f_{next} = f_i
\tag{3.228}
$$

where f_i and f_{next} are the values of function at the virtual node (x_i, y_i) and the immediately next virtual node in the x direction, respectively, and b

is the nodal spacing as $h = x_{next} - x_i$. As a result, the Neumann boundary condition in Equation (3.227) at the virtual node (x_i, y_i) is converted to the Dirichlet boundary condition. Finally, these boundary equations are assembled into the global stiffness matrix.

It is observed by solving several test problems via both the approaches that the first using the RDQ method gives good results for problems with complex equations of Neumann boundary conditions such as an elasticity problem that has different stresses as the Neumann boundary conditions. The second approach gives good results for problems with simple equations for the Neumann boundary condition.

3.6.3.4 Solving final system of equations and computing approximate function values

It is necessary to have the virtual nodes at least equal to the field nodes to achieve a unique solution, and the solution is improved as the virtual nodes are further increased. Suppose the final system of equations is $K_{mxn} U_{nx1} = F_{mx1}$, where m and n are the numbers of total virtual and field nodes, respectively, in the computational domain such that $m \geq n$. In order to get a square stiffness matrix K, if $m > n$, it is solved in a least-squares sense by multiplying K^T on both sides to get $K'_{nxn} U_{n \times 1} = F'_{nx1}$. As a result, the solution vector is computed by minimizing the residual error $(E = K'U - F')$. This approach is similar to the LS approximation.

The final system of equations is solved by the LS approach as discussed earlier, and the values of the nodal parameters at the field nodes are then computed. The fixed RKPM interpolation function does not possess the delta function property. Thus, $u_I \neq f^h(x_I, y_I)$, namely the value of the nodal parameter u_I at the Ith field node, is not equal to the value of the approximate function $f^h(x_I, y_I)$. Therefore, a procedure discussed earlier is used to correctly impose the Dirichlet boundary conditions in the RDQ method. However, for field nodes not located on the Dirichlet boundary, $u_I \neq f^h(x_I, y_I)$ is still true. Therefore, a second level of interpolation is performed, and the value of approximate function at each field node is computed by interpolating each field node by the surrounding field nodes with the fixed RKPM interpolation function as

$$f^h(x_i, y_j) = \sum_{I=1}^{NP} N_I(x_i, y_j)\, u_I \tag{3.229}$$

where $f^h(x_i, y_j)$ is an approximate value of the function at the field node (x_i, y_j), NP represents the total field nodes surrounding the field node (x_i, y_j), such that $NP \subseteq N$, where N denotes total field nodes in the domain and N_I and u_I are the values of shape function and nodal parameter at the Ith field node, respectively.

3.6.3.5 Computing approximate derivatives of field variables

Two approaches are employed in the RDQ method to compute the approximate derivatives of the field variables. The first one is the diffused derivative of a standard MLS approximation (Nayroles et al., 1992), and the second is the weighted derivatives approach (Mulay et al., 2009).

3.6.3.5.1 Diffused derivative approach

Approximate derivatives at the field nodes are computed after the evaluation of approximate function values by Equation (3.229). Two approaches are studied to compute the values of approximate derivates at the field nodes. The first approach is by taking the derivative of a shape function in Equation (3.229) given as

$$\frac{df^b(x_i,y_j)}{dx} = \sum_{I=1}^{NP} \frac{dN_I(x_i,y_j)}{dx} u_I \tag{3.230}$$

$$\frac{d^2f^b(x_i,y_j)}{dx^2} = \sum_{I=1}^{NP} \frac{d^2N_I(x_i,y_j)}{dx^2} u_I \tag{3.231}$$

This approach is similar to a diffused derivative of standard MLS approximation (Nayroles et al., 1992). It is seen from Equation (3.204) that the only variable in the equation of shape function is an array $b_I(x,y)$. As a result, an appropriate derivative of the array $b_I(x,y)$ is taken while computing the derivative of shape function. If the maximum complete second order of monomial is considered in the shape function, then

$$N_{I,x}(x,y) = \begin{bmatrix} 0 & 1 & 0 & 2x & y & 0 \end{bmatrix} C_{II}^{-1} \tag{3.232}$$

$$N_{I,y}(x,y) = \begin{bmatrix} 0 & 0 & 1 & 0 & x & 2y \end{bmatrix} C_{II}^{-1} \tag{3.233}$$

$$N_{I,xx}(x,y) = \begin{bmatrix} 0 & 0 & 0 & 2 & 0 & 0 \end{bmatrix} C_{II}^{-1} \tag{3.234}$$

$$N_{I,yy}(x,y) = \begin{bmatrix} 0 & 0 & 0 & 0 & 0 & 2 \end{bmatrix} C_{II}^{-1} \tag{3.235}$$

$$N_{I,xy}(x,y) = \begin{bmatrix} 0 & 0 & 0 & 0 & 1 & 0 \end{bmatrix} C_{II}^{-1} \tag{3.236}$$

As a result, the appropriate equations from (3.232) to (3.236) are substituted in Equations (3.230) and (3.231) to compute the values of approximate function derivatives at the field nodes. Based on convergence studies, this approach does not give correct results if the order of monomials included in the shape function basis is less than the actual order of the field variable distribution. As a result, the second approach called the weighted derivatives approach is developed.

3.6.3.5.2 Weighted derivative approach

This approach is formulated by computing the weighted derivatives at a concerned field node x_k by the weighted sum of derivative values with the field nodes located in the local domain of the concerned node x_k. For example, the first-order derivative at the node x_k in the x direction is given by $f(x_k)_{,x} = \sum_{i=1}^{NP} N_i(x_k) f(x_i)_{,x}$, where NP is the number of total field nodes in the local domain of node x_k, and $N_i(x_k)$ and $f(x_i)_{,x} = \{[f(x_i) - f(x_k)] / (x_i - x_k)\}$ are the values of shape function and approximate derivative at the ith field node, respectively. The values of shape function are computed by the fixed RKPM interpolation function.

Similarly, the second order derivative at the field node x_k in the x direction is computed by $f(x_k)_{,xx} = \sum_{i=1}^{NP} N_i(x_x) f(x_i)_{,xx}$ where $N_i(x_k)$ and $f(x_i)_{,xx} = \{[f(x_i)_{,x} - f(x_k)_{,x}] / (x_i - x_k)\}$ are the values of shape function and derivative at the ith node, respectively. The derivatives in the y direction are similarly computed.

It is observed after solving several test problems that the weighted derivatives approach gives good results when compared with the diffused derivative approach. As a result, it is adopted while computing the convergence rates of the RDQ method. Furthermore, the values of the approximate derivatives computed by the weighted derivative approach are more accurate for field nodes distributed uniformly as compared with field nodes distributed randomly. This is because the randomness of the field nodes is not handled in the weighted derivative approach. Therefore, the improved weighted derivatives approach is developed to increase the accuracy of the weighted derivative approach for the field nodes distributed randomly.

3.6.3.5.3 Improved weighted derivative approach

A new method called the improved weighted derivative approach is developed in this subsection to improve the accuracy of weighted derivative approach for the field nodes distributed randomly in the computational domain.

Let us consider the ith concerned field node with the NP field nodes distributed randomly in the interpolation domain and the field node j as one of the NP nodes such that $j \in [1, NP]$. The relative position of the i to j nodes is shown in Figure 3.35. The Euclidean length between them is computed

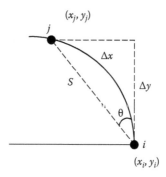

Figure 3.35 Relative positioning of field nodes *i* and *j*. (Modified from H. Li and S. Mulay. (2007). *Computational Mechanics*, DOI: 10.1007/s004466-011-0622-5.)

by $S = \sqrt{(\Delta x)^2 + (\Delta y)^2}$ and the gradient between them in the direction S is given as

$$\left(\frac{\partial f}{\partial S}\right)_i = \frac{f_j - f_i}{S} \qquad (3.237)$$

Therefore, the gradient in the y direction is given as

$$\left(\frac{\partial f}{\partial y}\right)_i = \left(\frac{\partial f}{\partial S}\right)_i \cos(\theta) \Rightarrow \left\{ -\frac{\cos(\theta)}{S} \quad \frac{\cos(\theta)}{S} \right\} \left\{ \begin{array}{c} f_i \\ f^j \end{array} \right\} \qquad (3.238)$$

All the individual values of $(\partial f/\partial y)_i$ are computed, depending on the relative locations of the field nodes i and j. The $(\partial f/\partial y)_i$ is thus given as

$$\left(\frac{\partial f}{\partial y}\right)\bigg|_i = \sum_{j=1}^{NP} N_j \left(\frac{\partial f}{\partial y}\right)_j \qquad (3.239)$$

Similarly, any equation of the approximate derivative can be developed in the x or y direction. It is observed from Equation (3.239) that the shape function N_j smooths the solution and the angle θ captures the variation of the gradient due to the random location of the field nodes. The accuracy of the improved weighted derivative approach is better than the weighted derivative approach.

3.6.4 Remarks

First, the formulations of the fixed RKPM and DQ methods are discussed and the principle of the RDQ method is explained. The stiffness matrix in the RDQ method is developed by discretizing the governing differential equation at the virtual nodes and imposing the appropriate boundary conditions. The final system of equations is computed by the LS approach, as the stiffness matrix may be rectangular.

Two approaches are discussed to evaluate the approximate derivatives of the function. The weighted derivatives approach is found to give more accurate solutions. As a result, it is adopted in the RDQ method. To correctly compute the approximate derivatives of the function for field nodes distributed randomly, a novel approach called the improved weighted derivative is proposed. It is observed from the solution of test problems that the improved weighted derivative approach correctly captures the gradients for field nodes distributed randomly.

Further, the final system of equations is solved by the direct or iterative methods. Therefore, during further numerical analyses, the Gauss elimination with a partial pivoting is implemented as a direct method, and portable extensible toolkit for the scientific computing (PETSC) is implemented as an iterative method. The PETSC is a free library with different iterative solvers; it has the major advantages noted below.

It works with only the nonzero elements of the stiffness matrix, which drastically reduces the time of computation.

It can be run in a parallel mode.

It uses the LU pre-conditioner by default and simplifies the sparse stiffness matrix.

It is good practice to get a converged solution even for big matrices with less computational time. In general, the convergence of iterative method degrades with an increase in the number of matrices. Therefore, the pre-conditioner reduces the condition number of the matrix, thereby increasing the accuracy of results. The computational time taken by PETSC is considerably lower than that for the Gauss elimination and the solution obtained by PETSC exactly matches that by the Gauss elimination for most problems.

3.7 SUMMARY

As well known, the current meshless methods continue to be developed. Many novel meshless methods have been proposed. In this chapter, the five novel meshless methods presented were all developed recently by the present authors. Three methods are in the strong form and called the

Hermite-cloud method, the RDQ method, and the point weighted least-squares (PWLS) method. The remaining two are in the weak form and designated the local Kriging (LoKriging) method and the variation of local point interpolation method (νLPIM).

In fact, a number of new meshless methods have been developed. More details can be found in relevant published books and journal papers. Meshless methods continue to serve as one of the hottest research areas of computational science and engineering because no mesh is required and these methods can overcome some shortcomings of traditional methods such as the FEM and FDM. The authors believe that the meshless methods will find more and more applications in simulation and analysis of engineering and science problems as time goes by.

Chapter 4

Convergence and consistency analyses

4.1 INTRODUCTION TO CONVERGENCE ANALYSIS

The meshless strong form RDQ method is introduced in the previous chapter. The primary focus of this chapter is to perform its convergence and consistency analyses. The DQ method is one of the efficient techniques for derivative approximation, but it always requires a regular domain with all the nodes distributed along the straight lines only. This severely restricts the DQ, while solving problems with irregular domains discretised by random field nodes. This limitation of the DQ method is overcome in the novel strong form RDQ method.

The RDQ method extends the applicability of the DQ technique over irregular/non-uniform and regular domains discretised by random or uniformly distributed field nodes by approximating a function value with the fixed RKPM method and discretizing a governing differential equation by the locally applied DQ method.

A superconvergence condition for the RDQ method is developed in this chapter. It gives the convergence rate of function more than $O(h^{p+1})$ for uniform and random field nodes scattered in the domain, where p is the highest order of the monomials used in the approximation of function. Finally, the convergence analysis of the RDQ method is performed and the superconvergence condition is verified by solving several 1-D, 2-D, and elasticity problems. The applicability of the RDQ method in solving the nonlinear governing differential equations is successfully demonstrated by the fixed–fixed and cantilever microswitches under the nonlinear electrostatic load. It is concluded that the RDQ method effectively handles the irregular and regular domains discretised by either uniform or random field nodes with good rates of convergence.

The main objective of this chapter is to perform convergence and consistency analyses of the RDQ method, with its application for the analysis of the MEMS switches. The superconvergence condition is developed while studying the convergence analysis, which always gives the convergence rate of function value greater than $O(h^{p+1})$, where p is the highest order of the monomials used in the approximation of function. For the test problems solved in this chapter by the superconvergence condition, the convergence rates of $O(h^{p+\alpha})$, where $\alpha \geq 1$ for the function value approximation and $\alpha \approx 0.7$ to 1 for the derivative approximation, are obtained for the uniform along with randomly distributed field nodes.

The advantage of the RDQ method as compared with the existing strong form meshless methods is that the fixed RKPM function is used only to approximate the values of field variable distribution at virtual nodes, and not to approximate the derivative terms from the governing PDE. The derivative terms from the governing PDE are approximated by the locally applied DQ method. As a result, the accuracy of the derivative approximation in the RDQ method is independent of the order of monomial used in the fixed RKPM interpolation. This differentiates the RDQ method from other collocation-based strong form methods in which the derivative terms from the governing PDE are approximated by the derivatives of shape function. In comparison with other weak form methods, the RDQ method is capable of well capturing the local high gradients as will be demonstrated by convergence analysis.

Compared with the classical FEM, the merit of the RDQ method is that it can effectively handle a problem with moving boundaries, which is still difficult for FEM due to the problems of mesh distortion and singularities in elements. The field nodes act as Lagrangian grid points and the virtual nodes act as Eulerian grid points in the RDQ method. As a result, the field nodes are free to move anywhere in the domain, and no interconnectivity information among them is required. The virtual nodes are fixed in a space so that it is always possible to discretize the governing PDE over the virtual nodes by locally applying the DQ method.

In the subsequent portions of this chapter, the superconvergence condition is derived and several test problems are solved by the RDQ method. The MEMS microswitches of the type fixed–fixed and cantilever are then solved by the RDQ method.

4.2 DEVELOPMENT OF SUPERCONVERGENCE CONDITION

Let us consider N_v and N_r as the numbers of total virtual and field nodes, respectively, distributed in a given computational domain. In order to identify the relationship between them for a given problem, suppose the two graphs are plotted: $\ln(E)$ versus $\ln(h_r)$ and $\ln(E)$ versus $\ln(h_r/h_v)$, where E is a global error and h_r and h_v are the inter-particle spacings corresponding to field and virtual nodes, respectively. Let m_1 and m_2 be the

Table 4.1 Possibilities of m_1, m_2, and λ signs

m_1	m_2	λ	Is Valid?
+ve	+ve	+ve	No
+ve	−ve	−ve	Yes
−ve	−ve	+ve	No
−ve	+ve	−ve	No

Source: H. Li, S. Mulay, and S. See. (2009a). *Computer Modeling in Engineering and Sciences*, 48, 43–82. With permission.

slopes of these graphs, respectively. The equations of the m_1 and m_2 are then written as

$$\ln(E) = m_1 \ln(h_r) \quad \text{and} \quad \ln(E) = m_2 \ln(h_r/h_v) \tag{4.1}$$

$$\therefore \lambda = \frac{m_1}{m_2} = \frac{\ln(h_r/h_v)}{\ln(h_r)} \tag{4.2}$$

where $\lambda = (m_1/m_2)$, $N_v \geq N_r$ for a unique solution, and $h_r = L/(N_r - 1)$ and $h_v = L/(N_v - 1)$, where L is a domain length. It is stated by observation that $\ln(h_r)$ has a zero or negative value as $h_r \leq 1$, $\ln(h_r/h_v)$ always has a zero or positive value as $h_r \geq h_v$, and $ln(E)$ has either a positive or negative value as $E > 0$ or $E < 0$. Based on this input, four conditions are possible for the signs of m_1, m_2, and λ. Therefore, a careful consideration of these four conditions results in only one valid possibility, as given in Table 4.1. As a result, m_1, m_2, and λ should have +ve, −ve, and −ve signs, respectively. For the different number of field and fixed number of virtual nodes during the convergence studies, m_2 will not be a constant but will have a different value at each (h_r/h_v) location on the graph. Therefore, the third graph (λ) versus (h_r) is plotted to maintain a constant value of λ with respect to the different values of N_r, where the origin of this graph is at $(h_v, 0)$, as the lowest values of $h_r = h_v$ and $\lambda = 0$. The slope of the third graph by the origin along with any two points, (λ_1, h_{r1}) and (λ_2, h_{r2}), on the graph is given as

$$\lambda_1 - 0 = h_{r1} - h_v \quad \text{and} \quad \lambda_2 - 0 = h_{r2} - h_v \tag{4.3}$$

$$\therefore \quad \frac{\lambda_1}{\lambda_2} = \frac{(h_{r1} - h_v)}{(h_{r2} - h_v)} \tag{4.4}$$

Equation (4.4) is simplified for a fixed value of m_1 as

$$\frac{(m_2)_2}{(m_2)_1} = \frac{(h_{r1} - h_v)}{(h_{r2} - h_v)} \tag{4.5}$$

If the values of m_1 and λ are fixed, the successive values of m_2 can be computed, and by further fixing the values of h_v and h_{r1}, the successive values of h_{r2} can be computed by Equation (4.5). If a test problem is solved by these successive values of h_r (corresponding to fixed value of m_1) and a fixed value of N_v, the m_1 convergence rate should be achieved as per Equation (4.5). It is concluded therefore that if a test problem is solved with the field nodes obtained by Equation (4.5) and a function value is approximated with the pth order monomials, a convergence rate higher than $O(h^{p+1})$ is possibly obtained. Therefore, Equation (4.5) is called the superconvergence condition. Its application is demonstrated in the next section by solving several 1-D and 2-D test problems.

4.3 CONVERGENCE ANALYSIS

The convergence analysis of the RDQ method is performed in this section by solving several test problems of 1-D, 2-D, and elasticity. The convergence rates for all the presented problems are evaluated by computing the global error as (Aluru and Li, 2001; Mukherjee and Mukherjee, 1997)

$$\varepsilon = \frac{1}{\left| f^e \right|_{max}} \sqrt{\frac{1}{NP} \sum_{I=1}^{NP} \left[f_I^{(e)} - f_I^{(n)} \right]^2} \tag{4.6}$$

where ε is a global error in the solution, $f_I^{(e)}$ and $f_I^{(n)}$ are the exact and numerical values of function at the Ith field node, respectively, and NP is the number of total field nodes in a computational domain. Equation (4.6) is same as the L^2 error norm but averaged over the total field nodes, and normalized by an absolute maximum value of exact solution.

The rate of the convergence is computed as described (Liu and Quek, 2003; Zienkiewicz et al., 2005). Taylor series is a mathematical representation of a continuous function as the sum of infinite terms computed by the values of derivatives at a single point as

$$f(x+h) = f(x) + h \frac{\partial f(x)}{\partial x} + \frac{h^2}{2} \frac{\partial^2 f(x)}{\partial x^2} + \cdots + \frac{h^n}{n!} \frac{\partial^n f(x)}{\partial x^n} + O(h^{n+1}) \tag{4.7}$$

While computing the shape functions, if the monomials up to the pth order from the Pascal's triangle are included, the first $(p + 1)$ terms from the Taylor series will be exactly reproduced. The difference between the exact f_e and numerical f_n values of function is an error of the order $O(h^{p+1})$; therefore the order of the rate of convergence is $O(h^{p+1})$ as

$$(f_e - f_n) = \text{error} = \frac{h^{p+1}}{(p+1)!} \left(\frac{\partial^{p+1} f}{\partial x^{p+1}} \right) \tag{4.8}$$

$$\therefore \text{the order of the rate of convergence} = O(h^{p+1}) \tag{4.9}$$

Taking logs on both the sides results in

$$\log_{10}(\text{error}) = (p+1)\log_{10}(h) \quad \Rightarrow (p+1) = \frac{\log_{10}(\text{error})}{\log_{10}(h)} \quad (4.10)$$

where $(p+1)$ is the rate of the convergence or the rate at which the numerical solution converges to the corresponding analytical solution. This rate of the convergence is the slope of the graph $\log_{10}(\text{error})$ versus $\log_{10}(h)$. The error in Equation (4.10) is computed by Equation (4.6). The field nodes are created either in a uniform or random manner during the convergence analysis, and the virtual nodes are created by the cosine distribution as

$$x_i = x_0 + \frac{L}{2}\left[1 - \cos\left(\frac{i-1}{N-1}\pi\right)\right], \quad \text{for } i = 1,2,...,N \text{ virtual nodes} \quad (4.11)$$

where x_0 and L are the starting coordinate and domain length, respectively. According to Runge's phenomenon (Dahlquist and Björck, 2003; Mulay et al., 2009), with an increase in the order of the field variable interpolation by the field nodes distributed uniformly, the numerical solution gets unstable near the boundaries of the domain. This can be avoided if the nodes are distributed densely near the boundaries and are uniform within the domain. This requirement is fulfilled by Chebyshev nodes. The Chebyshev polynomials are classical orthogonal polynomials. The Chebyshev polynomials of the first kind $[T_n(x)]$ are cited here as orthogonal polynomials of the first kind (Christoph, 1997; Sarra, 2006), where n is the order of polynomial.

$[T_n(x)]$ represents the roots of the second order Chebyshev differential equation $(1 - x^2)y'' - xy' + n^2 y = 0$, and they are represented by trigonometric identity as $T_n(x) = \cos[n \cos^{-1}(x)]$, where $x \in [-1, 1]$. The roots or zeros of $[T_n(x)]$ are referred to as Chebyshev–Gauss (CG) points; they are evaluated as $x_k = \cos\{[\pi(2k+1)]/[2(n+1)]\}$, where $k = 0, 1, 2,..., n$. The CG points are generally not used to solve any PDE due to the difficulty while imposing the boundary conditions, as they do not include the values of domain boundaries $[-1, 1]$.

To overcome this restriction, Chebyshev–Gauss–Lobatto (CGL) points are constructed that are also considered the roots of $[T_n(x)]$. The CGL points are evaluated as $x_k = -\cos\{[(k-1)\pi]/(n-1)\}$, where $k = 1, 2,..., n$, and it is easy to impose the boundary conditions as well, as they cover the domain boundaries ± 1. Equation (4.11) is actually the equation of the computation of CGL points, but applied over the domain $[x_0,(x_0 + L)]$.

The approximate derivatives of function at the field nodes are computed by the improved weighted derivative approach (Mulay et al., 2009), as explained in Chapter 3. It is observed that the percentage relative numerical error in the computed derivatives is more than that in the values of function. To obtain a good convergence, it is essential to ensure that the error norm of derivative reduces as the field nodes are increased. Therefore, the complete or Sobolev error norm is computed as

$$(E)_0 = \sqrt{\sum_{I=1}^{N_r} (f_I^e - f_I^n)^2}, \quad \text{Sobolev norm of the order 0} \tag{4.12}$$

$$(E)_1 = (E)_0 + \sqrt{\sum_{I=1,\,j=1}^{N_r,\,3} \left[\left(\frac{\partial f}{\partial x_j} \right)_I^e - \left(\frac{\partial f}{\partial x_j} \right)_I^n \right]^2}, \quad \text{Sobolev norm of the order 1}$$

$$\tag{4.13}$$

where $(E)_0$ and $(E)_1$ are the Sobolev error norms of the zeroth and first-order, respectively, and e and n are the exact and numerical values, respectively. It is seen from Equation (4.13) that as the square of error in the values of function and derivative is added at each field node, the contribution of the term $(E)_0$ to the term $(E)_1$ becomes less as compared with the term of derivative error. Therefore, if $(E)_1$ reduces by successively increasing the field nodes, the error term of the derivative reduces. As a result, the Sobolev error norm computed by Equation (4.13) is fairly taken as an indicator of the reduction in the values of derivative error at the field nodes.

4.3.1 Computation of convergence rate for distribution of random field nodes

It is relatively easy to compute the global error and convergence rates for uniformly distributed field nodes as the nodal spacing h is constant, but it is not easy for randomly distributed field nodes. In order to compute the convergence rates for the randomly distributed field nodes, the values of function at the random field nodes are first evaluated by the RDQ method. The solution is then approximated over an equal number of uniformly distributed points by the values of function at the random field nodes (Aluru, 2000). The global error for the uniformly distributed points is computed and the convergence curves are plotted with it. For example, to compute the global error and the convergence rates for the 144 randomly distributed field nodes, 12 × 12 uniform points are created and the solution is approximated over them by the values of function at the 144 field nodes distributed randomly. The convergence curves are then plotted by the function values at the 12 × 12 uniform points.

4.3.2 Remarks about effects of random nodes on convergence rate

The random nodes are generated by a standard function provided by the C++ language. This function of random number generator is based on a pseudo-random number generator algorithm called the linear congruential generator (LCG; Scheinerman, 2006). There is an equal probability of generating any specific number between some lower and maximum values as per the LCG. Therefore, in general, the LCG can be called to be based on the uniform density. It is quite common to use the LCG algorithm to generate random numbers in computer simulation programs. A typical way to generate a pseudo-random number with uniform probability using the LCG in a known range is to use the modulo (%) of returned value by the range span and add the initial value of the range. For example, $number = 1 + [(value\,by\,LCG)\% \, 100]$ generates a number between 1 and 100.

If a current time is given as a reference value to generate the random numbers in the function of a random number generator, a new set of random numbers is generated each time. If no reference value is given, the same set of random numbers is always generated. To maintain the reproducibility of solution, all the results presented here are based on the random numbers generated without giving any starting reference value to the function of random number generator; therefore the same numerical results are always obtained. During the convergence analysis of the RDQ method, some problems are solved by different sets of random numbers to observe its effect on convergence rate, and it is observed that the convergence rate is affected but only slightly. Also, the convergence rate obtained by the random nodes generated without giving any starting reference value to the random number generator program are precise measures of the performance of the RDQ method for random field nodes. This remark is made after comparing the values of convergence rates obtained via the random nodes generated with and without giving a starting reference value to the random number generator program. This observation will be verified in the next subsection by solving some test problems by the different sets of random field nodes.

It is observed by conducting an extensive numerical analysis that the accuracy of a solution is affected by the random distribution of field nodes but only for low numbers of random field nodes. As the total number of random field nodes is increased, there is a sharp decline in the global error, as compared with the result for an equal number of uniform field nodes. This behaviour is verified from the convergence curves given in the next sections. Nevertheless, to make the comparison between the convergence rates obtained by the random distribution of field nodes with rates for an equal number of total uniform field nodes, all the convergence results for the random distribution of field nodes are computed with equal numbers of uniform points, as explained earlier. As a result, the convergence rates for equal numbers of uniform and random field nodes are directly comparable.

4.3.3 One-dimensional test problems

A number of 1-D test problems are solved in this section and their convergence curves are plotted to calculate the convergence rates. The first 1-D problem is a Poisson equation with a constant force term. The governing equation and boundary conditions are given as

$$\frac{d^2 f}{dx^2} = 2, \quad (0 < x < 8), \text{ and } f(x = 0) = 0, f(x = 8) = 64 \tag{4.14}$$

The exact solution is given as $f(x) = x^2$. This problem is solved by including up to second order monomials in the polynomial basis of function approximation and with 6, 21, 161, 321 field and 460 cosine virtual nodes. The convergence curves are plotted in Figure 4.1, and it is seen that the convergence rates obtained by the random field nodes are equally good as uniform field nodes. Also, reasonably good rates of the derivative convergence are obtained by both the uniform and random field nodes. The comparison between the analytical and numerical values of function is given in Figure 4.2.

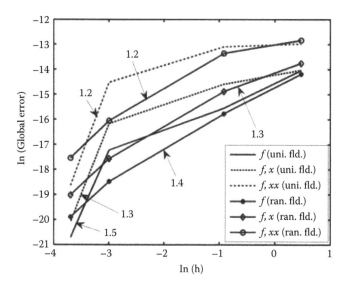

Figure 4.1 Convergence plots of uniform and random distributions of field nodes combined with cosine virtual nodes for the first 1-D problem of Poisson equation. (From H. Li, S. Mulay, and S. See. (2009b). *Computer Modeling in Engineering and Sciences*, 48, 43–82. With permission.)

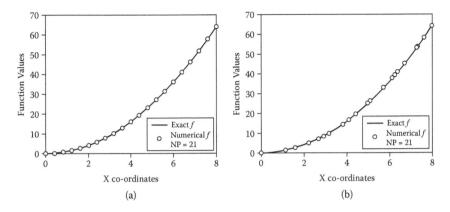

Figure 4.2 Comparison of numerical and analytical values of function by uniform (a) and random (b) field nodes for first 1-D problem of Poisson equation. (From H. Li, S. Mulay, and S. See. (2009b). *Computer Modeling in Engineering and Sciences*, 48, 43–82. With permission.)

The second 1-D problem is a mixed boundary value problem with the analytical solution containing a fourth order monomial. It is interesting to see how the RDQ method converges by including up to the second order monomials. The governing equation and boundary conditions are

$$\frac{d^2f}{dx^2} = \frac{105}{2}x^2 - \frac{15}{2}, (-1 < x < 1), \text{ and } f(x = 1) = -1, \frac{df}{dx}(x = 1) = 10 \quad (4.15)$$

The analytical solution is given as $f(x) = (35/8)\, x^4 - (15/4)\, x^2 + (3/8)$. This problem is solved by the superconvergence condition with $m_1 = 3$, $N_v = 641$, $h_{r1} = 0.05$, and $\lambda_1 = -1.2$. The successive values of λ are obtained by dividing the value of λ_1 by 3, and by Equation (4.5) to compute the corresponding values of h_{r2}. $N_r = 58$, 145, 299, 464, 569, and 632 are obtained. To compare the results, this problem is also solved by the second set of 21, 41, 81, 161, 321, 641 field and 641 cosine virtual nodes. The convergence curves obtained by both sets of field nodes are plotted in Figure 4.3 and the corresponding convergence rates are given in Table 4.2.

It is seen from Table 4.2 that all the convergence rates obtained by the field nodes computed with Equation (4.5) are improved, the rates of derivative convergence are superconvergent, and the convergence of function value by the random field nodes is also superconvergent. Figure 4.4a shows the reduction in the complete Sobolev norm with the number of field nodes obtained by the superconvergence condition.

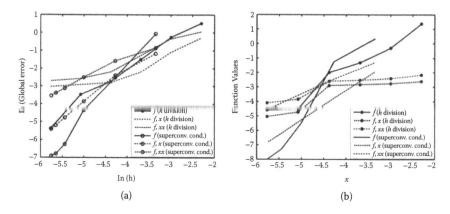

Figure 4.3 Convergence curves by uniform (a) and random (b) field nodes for second 1-D problem of Poisson equation. The curves are plotted with the field nodes obtained by uniformly decreasing *h* and the superconvergence condition. (From H. Li, S. Mulay, and S. See. (2009b). *Computer Modeling in Engineering and Sciences,* 48, 43–82. With permission.)

This problem is also solved by fourth order monomials, and the convergence curves are given in Figure 4.4b. When Table 4.2 and Figure 4.4b are compared, note that the rates of derivative convergence are considerably improved and the convergence rates for the random field nodes are almost equal to or better than those of the uniform field nodes. This indicates that the RDQ method is capable of equally handling the distributions of uniform and random nodes, which is one of the objectives behind the development of the RDQ method. A comparison of the analytical and numerical values of function is given in Figure 4.5. It is seen from all the results that reasonably good convergence rates of the function and the first and second order derivatives are achieved by the RDQ method with the domain discretised by either the uniform or random field nodes. Based on Figure 4.4b, the RDQ method converges at a faster rate with an increase in the monomial order of the function approximation.

Table 4.2 Convergence rates for second 1-D problem by second order monomials

Function	For Second Set (uniform nodes)	Superconvergence (uniform nodes)	For second set (random nodes)	Superconvergence (random nodes)
f	1.7	3.0	1.8	3.8
$f_{,x}$	0.9	1.8	0.8	2.1
$f_{,xx}$	0.9	1.2	0.7	1.4

Source: H. Li, S. Mulay, and S. See. (2009b). *Computer Modeling in Engineering and Sciences,* 48, 43–82. With permission.

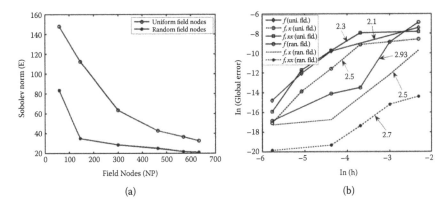

Figure 4.4 Reduction in complete error norm with field nodes obtained by superconvergence condition (a) and convergence plots by uniform and random field nodes with fourth order monomials (b) for second 1-D problem. (From H. Li, S. Mulay, and S. See. (2009b). *Computer Modeling in Engineering and Sciences*, 48, 43–82. With permission.)

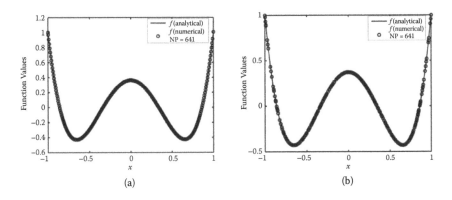

Figure 4.5 Comparison of numerical and analytical values of function by uniform (a) and random (b) field nodes for second 1-D problem. (From H. Li, S. Mulay, and S. See. (2009b). *Computer Modeling in Engineering and Sciences*, 48, 43–82. With permission.)

The third 1-D problem is with a high local gradient solved by second order monomials. The governing equation and boundary conditions are given as

$$\frac{d^2 f}{dx^2} = -6x - \left[\left(\frac{2}{\alpha^2} \right) - 4 \left(\frac{x-\beta}{\alpha^2} \right)^2 \right] \exp \left[-\left(\frac{x-\beta}{\alpha} \right)^2 \right] \quad (0 < x < 1) \quad (4.16)$$

$$f(x=0) = \exp \left[-\left(\frac{\beta^2}{\alpha^2} \right) \right], \frac{df(x=1)}{dx} = -3 - 2 \left(\frac{1-\beta}{\alpha^2} \right) \exp \left[-\left(\frac{1-\beta}{\alpha} \right)^2 \right]$$

$$(4.17)$$

The analytical solution is given as $f(x) = -x^3 + \exp\{-[(x-\beta)/\alpha]^2\}$. This problem is solved with the superconvergence condition by fixing $m_1 = 2$, $N_v = 641$, $h_{r1} = 0.05$ and $\lambda_1 = -1.2$. The successive values of λ are obtained by dividing the value of λ_1 by 2, and via Equation (4.5) to compute the corresponding values of h_{r2}. As a result, 21, 40, 75, 133, 219, 327, 433, 517 and 572 number of total field nodes are obtained. The convergence plots are given in Figure 4.6. The convergence rate of function by the uniform field nodes is $O(h^{p+2.1})$ and all the convergence rates of the derivatives are also superconvergent. The complete or Sobolev error norm is plotted in Figure 4.6c.

4.3.4 Two-dimensional test problems

The 2-D test problems are solved in this section and their convergence curves are plotted to find the convergence rates. The first 2-D problem is a Laplace equation with Dirichlet boundary conditions as

$$\nabla^2 f = 0, \qquad (0 < x < 1) \text{ and } (0 < y < 1) \tag{4.18}$$

$$f(x = 0, y) = -y^3, f(x = 1, y) = -1 - y^3 + 3y^2 + 3y \tag{4.19}$$

$$f(x, y = 0) = x^3, f(x, y = 1) = -1 - x^3 + 3x^2 + 3x \tag{4.20}$$

The analytical solution is given as $f(x, y) = -x^3 - y^3 + 3xy^2 + 3x^2y$. This problem is solved with the second order monomials, and 5×5, 9×9, 17×17, 33×33, 44×44 field, and 44×44 cosine virtual nodes. The convergence curves are plotted in Figure 4.7. A better convergence rate of the function value is achieved by the random field nodes, but the convergence rates of the derivatives remain unchanged. The comparison of

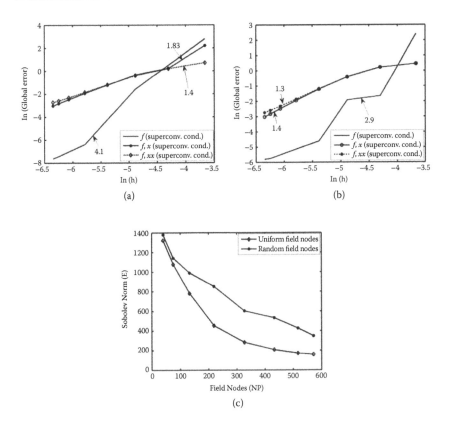

Figure 4.6 Convergence curves by uniform (a) and random (b) field nodes and reduction in Sobolev error norm (c) for third I-D problem of local high gradient. (From H. Li, S. Mulay, and S. See. (2009b). *Computer Modeling in Engineering and Sciences*, 48, 43–82. With permission.)

the analytical and numerical values of the function by the uniform and random field nodes is shown in Figure 4.8.

The second 2-D problem is also a Laplace equation with mixed boundary conditions as

$$\nabla^2 f = 0, \qquad (0 < x < 1) \text{ and } (0 < y < 1) \tag{4.21}$$

$$f(x = 0, y) = -y^3, \quad f(x = 1, y) = -1 - y^3 + 3y^2 + 3y \tag{4.22}$$

$$\frac{df}{dy}(x, y = 0) = 3x^2, \quad \frac{df}{dy}(x, y = 1) = -3 + 6x + 3x^2 \tag{4.23}$$

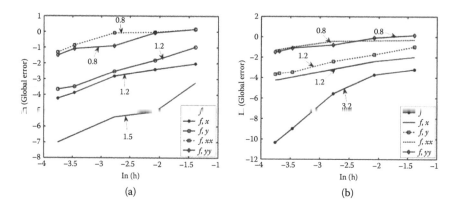

Figure 4.7 Convergence curves by uniform (a) and random (b) field nodes by including up to second order monomials for first 2-D problem of Laplace equation. (From H. Li, S. Mulay, and S. See. (2009b). *Computer Modeling in Engineering and Sciences*, 48, 43–82. With permission.)

The analytical solution is given as $f(x,y) = -x^3 - y^3 + 3xy^2 + 3x^2y$. This problem is solved with second order monomials, and 5×5, 9×9, 17×17, 33×33, 41×41, 44×44 field. and 44×44 cosine virtual nodes. The convergence curves are plotted in Figure 4.9, and it is observed that better convergence rate of the function is achieved by the random field nodes.

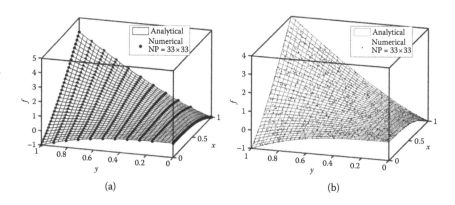

Figure 4.8 Comparison of numerical and analytical values of function by 33×33 uniform (a) and random (b) field nodes for first 2-D problem of Laplace equation. (From H. Li, S. Mulay, and S. See. (2009b). *Computer Modeling in Engineering and Sciences*, 48, 43–82. With permission.)

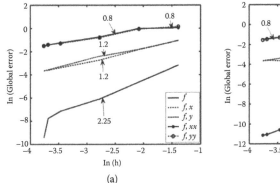
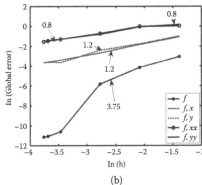

(a)   (b)

Figure 4.9 Convergence plots by uniform (a) and random (b) field nodes for second 2-D problem of Laplace equation. (From H. Li, S. Mulay, and S. See. (2009b). *Computer Modeling in Engineering and Sciences*, 48, 43–82. With permission.)

The third 2-D problem solved is of local high gradient value at the node (0.5, 0.5). The governing equation and boundary conditions are given as

$$\nabla^2 f = -6x - 6y - \left[\left(\frac{4}{\alpha^2}\right) - 4\left(\frac{x-\beta}{\alpha^2}\right) - 4\left(\frac{y-\beta}{\alpha^2}\right)\right]$$

$$\times \exp\left[-\left(\frac{x-\beta}{\alpha}\right)^2 - \left(\frac{y-\beta}{\alpha}\right)^2\right], \quad (0 < x < 1) \text{ and } (0 < y < 1) \quad (4.24)$$

$$f(x=0,y) = -y^3 + \exp\left[-\left(\frac{\beta}{\alpha}\right)^2 - \left(\frac{(y-\beta)}{\alpha}\right)^2\right] \quad (4.25)$$

$$f(x=1,y) = -1 - y^3 + \exp\left[-\left(\frac{(1-\beta)}{\alpha}\right)^2 - \left(\frac{(y-\beta)}{\alpha}\right)^2\right] \quad (4.26)$$

$$\frac{df}{dy}(x,y=0) = \frac{2\beta}{\alpha^2}\exp\left[-\left(\frac{\beta}{\alpha}\right)^2 - \left(\frac{(x-\beta)}{\alpha}\right)^2\right] \quad (4.27)$$

$$\frac{df}{dy}(x,y=1) = -3 - 2\left(\frac{(1-\beta)}{\alpha^2}\right)\exp\left[-\left(\frac{(x-\beta)}{\alpha}\right)^2 - \left(\frac{(1-\beta)}{\alpha}\right)^2\right] \quad (4.28)$$

The analytical solution is given as $f(x, y) = -x^3 - y^3 + \exp\{-[(x - \beta)/\alpha]^2 - [(y - \beta)/\alpha]^2\}$. This problem is solved with the second order monomials and the superconvergence condition by fixing $m_1 = 2$, $N_v = 44 \times 44$, $h_{r1} = 0.25$, and $\lambda_1 = -1.2$. The successive values of λ are obtained by dividing the value

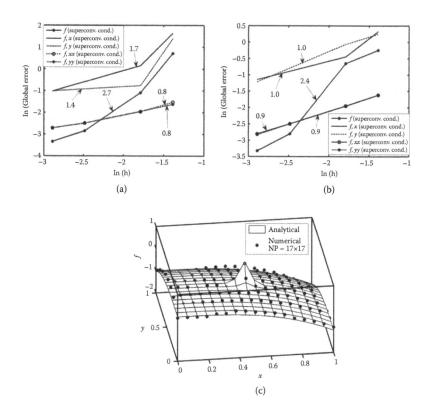

Figure 4.10 Convergence plots by uniform (a) and random (b) field nodes and numerical and analytical distributions of field variable by uniform field nodes (c) for third 2-D problem. (From H. Li, S. Mulay, and S. See. (2009b). *Computer Modeling in Engineering and Sciences*, 48, 43–82. With permission.)

of λ_1 by 2, and with Equation (4.5) to compute the values of h_{r2}. As a result, $N_r = 5 \times 5, 9 \times 9, 13 \times 13$, and 19×19 are obtained. The convergence curves are plotted in Figure 4.10a and Figure 4.10b and the convergence rates of the derivative are superconvergent. Figure 4.10c shows a comparison between the numerical and analytical values of the distributed field variable.

The fourth 2-D problem is of steady-state heat conduction in a rectangular plate domain with a heat source. The governing equation and boundary conditions are given as

$$\nabla^2 T = -2\, s^2 \sec h^2[s(y-0.5)]\tanh[s(y-0.5)], (0 < x < 0.5), \quad (0 < y < 1) \quad (4.29)$$

$$\frac{\partial f}{\partial n} = 0 \text{ along } x = 0 \text{ and } 0.5,\ T(y = 0) = -\tanh\left(\frac{s}{2}\right),\ T(y = 1) = \tanh\left(\frac{s}{2}\right)$$

$$(4.30)$$

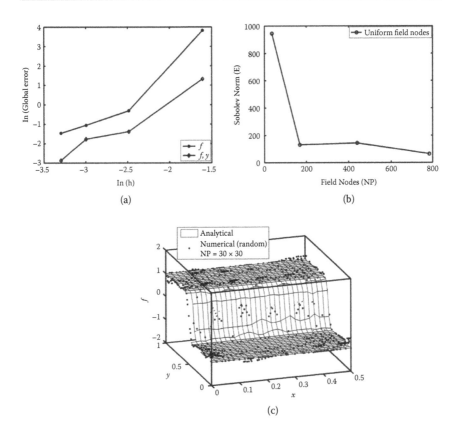

Figure 4.11 Convergence plots by uniform field nodes (a), reduction in Sobolev norm (b), and numerical distribution of temperature within domain (c) for fourth 2-D problem of steady-state heat conduction. (From H. Li, S. Mulay, and S. See. (2009b). *Computer Modeling in Engineering and Sciences*, 48, 43–82. With permission.)

The successive values of λ are obtained by dividing the value of λ_1 by 3, and with Equation (4.5) to compute the corresponding values of h_{r2}. As a result, $N_r = 6 \times 6$, 13×13, and 21×21 28×28 are obtained. The convergence plots of the temperature and its gradient are given in Figure 4.11a and the corresponding convergence rates are 3.2 and 2.4. The reduction in the Sobolev norm with increasing the field nodes and the comparison between the numerical and analytical values of temperature are given in Figure 4.11b and Figure 4.11c, respectively. According to Figure 4.11, good rates of the function and derivative convergences are achieved with field nodes obtained by the superconvergence condition. The superconvergent rates of derivative confirm that the improved weighted derivative approach gives reasonably good rates of convergence.

4.3.5 Elasticity problems

All the problems presented in this section are based on the plane stress condition with the mechanical equilibrium equation expressed as

$$\sigma_{ij,j} + B_i = 0 \tag{4.31}$$

where σ_{ij} and B_i are the Cauchy stress tensor and body force, respectively.
The body force is assumed zero for the present results. The equations of
2-D strains in the plane-stress condition are given as

$$\varepsilon_{xx} = \left(\frac{1}{E}\right)[\sigma_{xx} - (v_0\, \sigma_{yy})], \varepsilon_{yy} = \left(\frac{1}{E}\right)[\sigma_{yy} - (v_0\sigma_{xx})], \text{ and } \varepsilon_{xy} = \frac{2(1+v_0)}{E}\sigma_{xy}$$

$$\tag{4.32}$$

where ε_{xx} and ε_{yy}, and σ_{xx} and σ_{yy} are the normal strains and stresses, respectively, in the x and y directions, respectively, ε_{xy} and σ_{xy} are the engineering
shear strain and stress, respectively, and E and v_0 are the Young's modulus
and Poisson ratio, respectively. Equation (4.32) is known as the 2-D Hook's
law. The geometrical equations for the plane-stress condition are given as

$$\varepsilon_{xx} = \frac{\partial u}{\partial x}, \quad \varepsilon_{yy} = \frac{\partial v}{\partial y}, \quad \text{and} \quad \gamma_{xy} = \left[\frac{\partial u}{\partial y} + \frac{\partial v}{\partial x}\right] \tag{4.33}$$

where u and v are the displacements in the x and y directions, respectively, and γ_{xy} is shear strain. Substituting terms of σ_{ij} in Equation (4.31) by
Equations (4.32) and (4.33), and simplifying results in

$$\frac{E}{\left(1-v_0^2\right)}\left[\frac{\partial^2 u}{\partial x^2} + \left(\frac{1-v_0}{2}\right)\frac{\partial^2 u}{\partial y^2} + \left(\frac{1+v_0}{2}\right)\frac{\partial^2 v}{\partial x \partial y}\right] + B_x = 0 \tag{4.34}$$

$$\frac{E}{\left(1-v_0^2\right)}\left[\frac{\partial^2 v}{\partial y^2} + \left(\frac{1-v_0}{2}\right)\frac{\partial^2 v}{\partial x^2} + \left(\frac{1+v_0}{2}\right)\frac{\partial^2 u}{\partial x \partial y}\right] + B_y = 0 \tag{4.35}$$

Equations (4.34) and (4.35) are governing equations in the displacement
form for 2-D plane stress. The Neumann boundary conditions are given as

$$\left(\frac{E}{1-v_0^2}\right)\left[l\left(\frac{\partial u}{\partial x}+v_0\frac{\partial v}{\partial y}\right)_s + m\left(\frac{1-v_0}{2}\right)\left(\frac{\partial u}{\partial y}+\frac{\partial v}{\partial x}\right)_s\right] = \overline{X} \tag{4.36}$$

$$\left(\frac{E}{1-v_0^2}\right)\left[m\left(\frac{\partial v}{\partial y}+v_0\frac{\partial u}{\partial x}\right)_s + l\left(\frac{1-v_0}{2}\right)\left(\frac{\partial v}{\partial x}+\frac{\partial u}{\partial y}\right)_s\right] = \overline{Y} \tag{4.37}$$

where l and m are the direction cosines of surface normal, and $\bar{X}$ and $\bar{Y}$ are the prescribed boundary values. The values of $l = 0$ and $m = \pm 1$ are for a boundary parallel to the x axis, and $l = \pm 1$ and $m = 0$ are for a boundary parallel to the y axis. The surface normals for the curved shape boundary are computed by the curve equation. In order to use Equations (4.34) to (4.37) for plane strain conditions, replace $E = E / (1 - v_0^2)$ and $v_0 = v_0 / (1 - v_0)$, keeping the other factors the same.

4.3.5.1 Cantilever beam under pure bending

A cantilever beam under a pure bending load is shown in Figure 4.12a. The analytical solutions are derived as (Timoshenko and Goodier, 1970; Xu, 1990)

$$u = \frac{M}{E\,I}\, x\, y, v = \left(\frac{-v_0\, M}{2\,E\,I} \right) y^2 - \left(\frac{M}{2\,E\,I} \right) x^2 \text{ and } \sigma_{xx} = \frac{M\, y}{I}, \sigma_{yy} = 0, \sigma_{xy} = 0$$

$$(4.38)$$

The coordinate system for Figure 4.12a is used as shown in Popov (1990). The transverse and longitudinal directions are designated by the y and x axes, respectively, and the depth by the z axis. The bending moment is applied about the z axis. This problem is solved to test the ability of the RDQ method to capture the second order continuity of field variable distribution in the x and y directions, as seen in Equation (4.42).

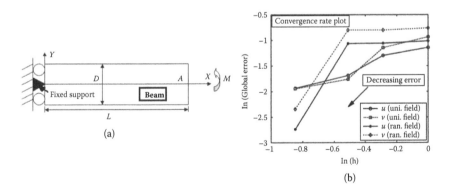

(a)

(b)

Figure 4.12 Cantilever beam under pure bending load (a) and convergence plots of displacements u and v by uniform and random field nodes coupled with cosine virtual nodes (b) (From H. Li, S. Mulay, and S. See. (2009b). *Computer Modeling in Engineering and Sciences*, 48, 43–82. With permission; modified from S. Mulay, H. Li, and S. See. (2009). *Computational Mechanics*, 44, 563–590.)

A negative bending moment M is applied at the free end of the cantilever beam, which will generate pure normal stress. It was assumed while deriving the analytical equations of the displacements that the stresses $\sigma_{yy} = 0$, $\sigma_{xy} = 0$, and σ_{xx} vary linearly with respect to x. When the analytical equation $\sigma_{xx} = (M\,y)/I$ is derived with these constraints, it coincides with the solution given by mechanics of materials (Xu, 1992). That means that when normal stress caused by M and proportional to y is applied at the free end, the exact solution in Equation (4.38) should obtain. But, if normal stress is applied by any other manner, Equation (4.38) will not represent the exact solution.

However, the numerical solution will approach the exact solution away from the boundary, as per Saint Venant's principle. The intention here is to solve the problem of a cantilever beam under pure bending for 2-D plane stress, rather than only the transverse deflection of a beam. In transverse deflection, the longitudinal strain ε_x is non-zero, the transverse strains ε_y and ε_z are equal to zero, and the change in the length of the beam is negligible as the beam deflects transversely (Popov, 1990). Due to the 2-D plane stress, both ε_x and ε_y are non-zero. Therefore, both nodal displacements u and v exist in the x and y directions, respectively, (Aluru, 2001; Liu and Gu, 2001; Aluru and Li, 2001).

As per the boundary conditions, the displacements u and v are set to zero at a node $(0, 0)$ due to hinge support, and $u = 0$ at all the nodes along $(0, y)$. The stresses σ_{xy}, σ_{yy} and σ_{xx} are imposed as zero along all the boundaries except at $x = 0$ and $x = L$, where σ_{xx} is imposed as a normal stress caused by the bending moment M, as per Equation (4.38). The condition $dv/dx = 0$ is imposed at the node $(0,0)$ to avoid the rotation of beam.

This problem is solved for $L = 48$, $D = 12$, $M = -24000$, $v_0 = 0.3$, $E = 3.0 \times 10^7$, 13×13, 17×17, 21×21, and 29×29 field and 41×41 cosine virtual nodes with up to the first-order monomials included in the approximation of function. The convergence rate of the displacements u and v, by separately discretizing the domain with the uniform and random field nodes, are given as 1.0 and 1.3, and 1.96 and 2.0, respectively; the corresponding convergence plots are given in Figure 4.12b.

When this problem is solved by 9×9 uniform field and 41×41 cosine virtual nodes with the second order monomials in the approximation of function, the analytical solutions are almost exactly reproduced. The global error values for the displacements u and v are 3.53×10^{-13} and 2.54×10^{-12}, respectively, as shown in Figure 4.13. It is also observed from Figure 4.13 that if the complete required order of the monomials is included in the polynomial basis of the function approximation, the field variables are almost exactly reproduced. Based on Equation (4.38) and Figure 4.12a, the lower half of the beam is expected to have +ve displacement u due to tension and the upper half is expected to have −ve displacement v due to compression. Overall, the distribution of u should be like a parabola (as seen from analytical equation), and the displacement v should have the maximum value near the free end and zero toward the fixed end. The displacement v is also

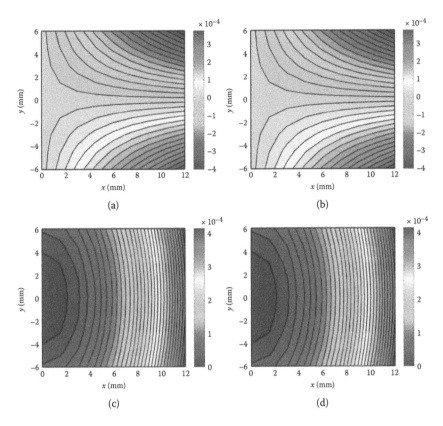

Figure 4.13 Analytical displacement u (a), numerical displacement u (b), analytical displacement v (c), and numerical displacement v (d) where domain is discretised by 9 × 9 uniform field and 41 × 41 cosine virtual nodes and the functions are approximated by second order monomials for cantilever beam under pure bending.

a function of the coordinate y at a specific value of the coordinate x. As a result, the vertical distribution of v is expected. These expected behaviours in the displacements u and v are verified in Figure 4.13.

4.3.5.2 Cantilever beam under pure shear

A cantilever beam is loaded under a pure shear, as shown in Figure 4.14a. The displacements u and v are set to zero at the node $(0,0)$ due to the hinge support. The stresses σ_{xx} and σ_{xy} are imposed along the boundaries $x = 0$ and $x = L$ according to

$$\sigma_{xx} = \frac{P\,(L-x)\,y}{I}, \sigma_{xy} = \frac{-P}{2\,I}\left(\frac{D^2}{4} - y^2\right) \text{ and } \sigma_{yy} = 0 \qquad (4.39)$$

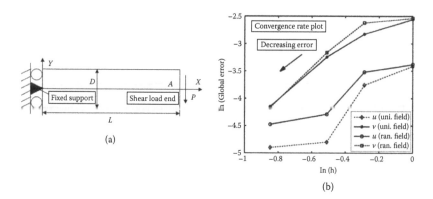

Figure 4.14 Cantilever beam under pure shear load (a) and convergence plots of displacements u and v by uniform and random field nodes coupled with cosine virtual nodes (b). (From H. Li, S. Mulay, and S. See. (2009b). *Computer Modeling in Engineering and Sciences*, 48, 43–82. With permission; modified from S. Mulay, H. Li, and S. See. (2009). *Computational Mechanics*, 44, 563–590.)

The stresses $\sigma_{xy} = 0$ and $\sigma_{yy} = 0$ are imposed along the boundary $y = \pm (D/2)$. The condition $dv/dx = 0$ is imposed at the node $(0,0)$ to avoid beam rotation. The analytical equations are derived below for the coordinate system shown in Figure 4.14a. The distribution of stresses given in Equation (4.39) is initially assumed, and substituting it in Equation (4.33) results in

$$\varepsilon_{xx} = \frac{\partial u}{\partial x} = \frac{\sigma_{xx}}{E} = \frac{P\, y\, (L-x)}{E\, I}, \varepsilon_{yy} = \frac{-v_0\, \varepsilon_{xx}}{E} = \frac{-v_0\, P\, y\, (L-x)}{E\, I}, \text{ and}$$

(4.40)

$$\gamma_{xy} = \left[\frac{\partial u}{\partial y} + \frac{\partial v}{\partial x} \right] = \frac{-P}{2\, G\, I} \left(\frac{D^2}{4} - y^2 \right)$$

(4.41)

Integrating Equation (4.40) results in

$$u = \frac{P\, y\, x}{E\, I} \left(L - \frac{x}{2} \right) + f(y) \text{ and } v = \frac{-v_0\, P\, y^2\, (L-x)}{2\, E\, I} + f_1(x)$$

(4.42)

where $f(y)$ and $f_1(x)$ are the integral constants. Differentiate the displacements u and v in Equation (4.42) with respect to y and x, respectively, and substitute in Equation (4.41) to get

$$f_1(x) = dx - \frac{P}{EI} \left(\frac{L\, x^2}{2} - \frac{x^3}{6} \right) + h \text{ and } f(y) = ey - \frac{v_0\, P y^3}{6\, EI} + \frac{P y^3}{6\, GI} + g$$

(4.43)

where d and e are the functions of x and y, respectively. Substituting Equation (4.43) into (4.42) gives

$$u = \frac{Pyx}{EI}\left(L - \frac{x}{2}\right) + ey - \frac{v_0 Py^3}{6EI} + \frac{Py^3}{6GI} + g \qquad (4.44)$$

$$v = \frac{-v_0 Py^2 (L-x)}{2EI} + dx - \frac{P}{EI}\left(\frac{Lx^2}{2} - \frac{x^3}{6}\right) + h \qquad (4.45)$$

Imposing the boundary conditions in Equations (4.44) and (4.45) as $u(0,0)=0$, $v(0,0) = 0$ and $(\partial v/\partial x)_{(0,\,0)} = 0$ results in $g = h = d = 0$ and $e = (-PD^2)/(8\,GI)$. Therefore, Equations (4.44) and (4.45) are simplified as

$$u = \frac{Py\,x}{EI}\left(L - \frac{x}{2}\right) - \frac{v_0 Py^3}{6EI} + \frac{Py^3}{6GI} - \frac{PD^2 y}{8GI} \qquad (4.46)$$

$$v = \frac{-v_0\,Py^2\,(L-x)}{2EI} - \frac{Px^2}{EI}\left(\frac{L}{2} - \frac{x}{6}\right) \qquad (4.47)$$

where $G = E/[2(1+v_0)]$ and $I = (D^3/12)$; beam has unit thickness. As a result, the analytical expressions in Equations (4.46) and (4.47) correspond to the distributions of the stresses given in Equation (4.39), and the displacements u and v are third order continuous in the space domain.

This problem is solved for $L = 48$, $D = 12$, $P = -1000$, $v_0 = 0.3$, $E = 3.0 \times 10^7$, 13×13, 17×17, 21×21, and 29×29 field and 41×41 cosine virtual nodes with the second order monomials included in the monomial basis of shape function while approximating the function. The convergence rates of the displacements u and v by the uniform and random field nodes are 1.94 and 1.9, and 1.5 and 2.0, respectively, and the corresponding convergence curves are plotted in Figure 4.14b. When this problem is solved by 9×9 field and 41×41 virtual nodes with first, second, and third order monomials, the numerical results are steadily improved, as shown in Table 4.3. The plots in Figure 4.15 and Figure 4.16 show a comparison of the numerical and analytical values of the displacement and stresses, respectively, when

Table 4.3 Decreases in values of global error with increase in monomial order

Monomial order	Global error in u	Global error in v
1	1.38×10^{-4}	1.2
2	3.78×10^{-5}	5.03×10^{-2}
3	3.26×10^{-9}	2.43×10^{-9}

Source: H. Li, S. Mulay, and S. See. (2009b). *Computer Modeling in Engineering and Sciences*, 48, 43–82. With permission.

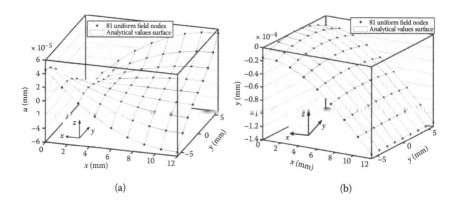

(a)   (b)

Figure 4.15 Comparison of displacements *u* (a) and *v* (b) by third order monomials in the polynomial basis of function approximation for cantilever beam under pure shear load. (Modified from S. Mulay, H. Li, and S. See. (2009). *Computational Mechanics*, 44, 563–590.)

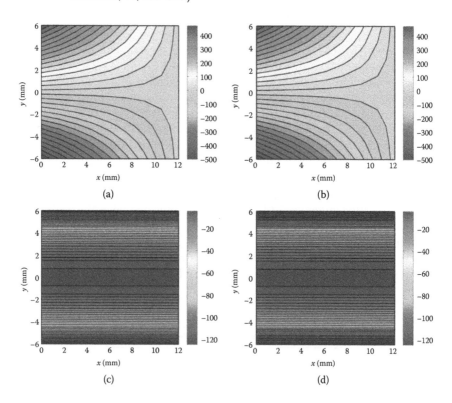

Figure 4.16 Analytical σ_{xx} (a), numerical σ_{xx} (b), analytical σ_{xx} (c) and numerical σ_{xy} (d) stresses where domain is discretised by 9 × 9 uniform field and 41 × 41 cosine virtual nodes for cantilever beam under pure shear load.

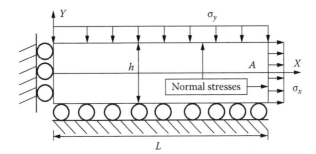

Figure 4.17 Beam under loading of constant normal stresses. (Modified from S. Mulay, H. Li, and S. See. (2009). *Computational Mechanics*, 44, 563–590.)

the numerical results are obtained by including up to third order monomials in the polynomial basis of shape function.

4.3.5.3 Beam under tensile and compressive normal stresses in x and y directions

A beam is loaded with normal stresses in the x and y directions, as shown in Figure 4.17. The displacements u and v are set to zero along the edge $x = 0$ and $y = 0$, respectively. The values of stresses σ_{xx} and σ_{xy} are imposed as 1.0 and 0 along the edge $x = L$, respectively, and the values of stresses σ_{xy} and σ_{yy} are imposed as 0 and -0.5 along the edge $y = L$, respectively. The analytical expressions of the displacements u and v are given as

$$u = \frac{9}{8}x \quad \text{and} \quad v = \frac{-3}{4}\left[y + \frac{h}{2}\right] \tag{4.48}$$

As the displacements u and v are first-order continuous, the problem is treated as a test case to study the reproducibility of the first-order continuous field variable by the RDQ method. This problem is solved for $L = 9$ mm and $h = 3$ mm, and by 9×9 uniform field nodes coupled with 41×41 cosine distributed virtual nodes. The global errors in the displacements u and v are obtained as 3.66×10^{-11} and 4.53×10^{-11}, respectively, and the displacements and stresses are shown in Figure 4.18. When this problem is solved by 9×9 randomly distributed field nodes coupled with 41×41 cosine distributed virtual nodes, the global errors in the displacements u and v are obtained as 1.53×10^{-11} and 4.64×10^{-12}, respectively. The displacements and stresses are shown in Figure 4.19.

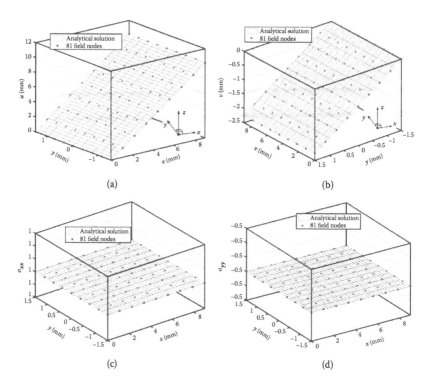

(a)

(b)

(c)

(d)

Figure 4.18 The displacements u (a) and v (b), and stresses σ_{xx} (c) and σ_{yy} (d) when the beam under normal stresses is solved by uniform field nodes coupled with cosine virtual nodes. (Modified from S. Mulay, H. Li, and S. See. (2009). *Computational Mechanics*, 44, 563–590.)

4.3.5.4 Semi-infinite plate with a central hole

The RDQ method is applied to solve the problem of a semi-infinite plate with a central hole, as shown in Figure 4.20a. Only one quarter of the plate is used as the computation domain due to the axis-symmetric nature, as shown in Figure 4.20b. The analytical solutions in a Cartesian coordinate system are given as

$$u = \left(\frac{1+v_0}{E}\right) P \left[\frac{r}{1+v_0}\cos(\theta) + \left(\frac{2}{1+v_0}\right)\frac{b^2}{r}\cos(\theta) + \frac{b^2}{2\,r}\cos(3\theta) - \frac{b^4}{2\,r^3}\cos(3\theta)\right]$$

$$(4.49)$$

$$v = \left(\frac{1+v_0}{E}\right) P \left[\left(\frac{-v_0\,r}{1+v_0}\right)\sin(\theta) - \left(\frac{1-v_0}{1+v_0}\right)\frac{b^2\sin(\theta)}{r} + \frac{b^2\sin(3\theta)}{2r} - \frac{b^4\sin(3\theta)}{2r^3}\right]$$

$$(4.50)$$

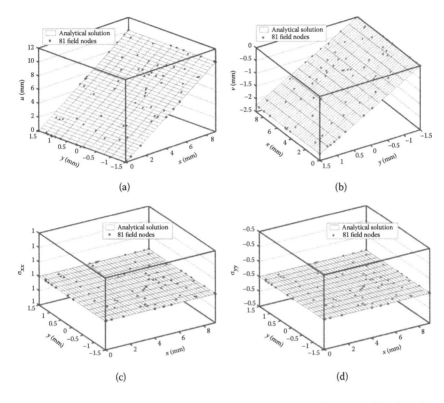

Figure 4.19 The displacements u (a) and v (b), and stresses σ_{xx} (c) and σ_{yy} (d) when the beam under normal stresses is solved by random field nodes coupled with cosine virtual nodes.

$$\sigma_{xx} = P\left[1 - \frac{a^2}{r^2}\left(\frac{3}{2}\cos(2\theta) + \cos(4\theta)\right) + \frac{3}{2}\frac{a^4}{r^4}\cos(4\theta)\right] \qquad (4.51)$$

$$\sigma_{yy} = -P\left[\frac{a^2}{r^2}\left(\frac{1}{2}\cos(2\theta) - \cos(4\theta)\right) + \frac{3}{2}\frac{a^4}{r^4}\cos(4\theta)\right] \qquad (4.52)$$

$$\sigma_{xy} = -P\left[\frac{a^2}{r^2}\left(\frac{1}{2}\sin(2\theta) + \sin(4\theta)\right) - \frac{3}{2}\frac{a^4}{r^4}\sin(4\theta)\right] \qquad (4.53)$$

where r and θ are the local polar coordinates. The conditions, $u = 0$ and $\sigma_{xy} = 0$, are imposed along edge 1 due to the symmetric boundary condition, and all the traction components are equal to zero along edge 2; therefore

$$\sigma'_{xx} = n_x\, t_x + n_y\, t_y = 0 \quad \text{and} \quad \sigma'_{xy} = -n_y\, t_x + n_x\, t_y = 0 \qquad (4.54)$$

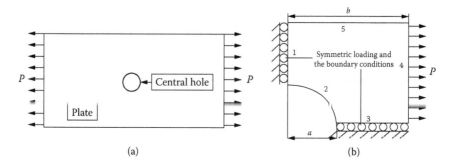

Figure 4.20 Semi-infinite plate with central hole (a) and actual computational domain with boundary and loading conditions (b). (Modified from S. Mulay, H. Li, and S. See. (2009). *Computational Mechanics*, 44, 563–590.)

where $t_x = n_x \sigma_{xx} + n_y \sigma_{xy}$ and $t_y = n_x \sigma_{yx} + n_y \sigma_{yy}$, n_x and n_y are the direction cosines in the x and y directions, respectively. The conditions, $v = 0$ and $\sigma_{xy} = 0$, are imposed along edge 3 due to the symmetric boundary condition. The stresses σ_{xx} and σ_{xy} are imposed along edge 4 by Equation (4.51) and Equation (4.53), respectively. The stresses σ_{yy} and σ_{yx} are imposed along edge 5 by Equation (4.52) and Equation (4.53), respectively.

The problem is solved for $a = 1$, $b = 5$, $P = 1$, $v_0 = 0.3$, $E = 1000$, and 6 × 6, 11 × 11, 21 × 21, and 31 × 31 uniform field and 34 × 34 cosine virtual nodes with second order monomials in the polynomial basis of shape function computation. The convergence rates of the displacements u and v by the uniform field nodes are obtained as 0.3 and 0.3, respectively, and the convergence curves are plotted in Figure 4.21a. The numerical and analytical values of σ_{xx} along boundary 1 are given in Figure 4.21b.

Based on all the test problems, the RDQ method gives convincingly good rates of convergence for the function and derivatives. The convergence rates obtained by discretizing the domain with either uniform or random field nodes are almost equal. In fact, sometimes the results obtained by the random field nodes are better than those for uniform field nodes. If the number of total field nodes is computed by the superconvergence condition, the function and its approximate derivatives converge at the rate of $O(h^{p+\alpha})$, where $\alpha \geq 1$ for the approximated function and $\alpha \approx 0.7$ to 1 for the approximated derivative. The convergence analysis also shows that if the complete and consistent order of the monomials is included in the approximation of function, the function and derivatives are almost exactly reproduced.

In the next section, the RDQ method is applied to analyze the MEMS devices of fixed–fixed and cantilever microswitches for pull-in instability caused by nonlinear electrostatic force between the fixed bottom plate and beam.

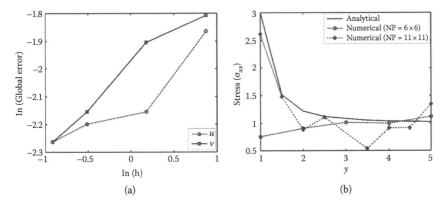

Figure 4.21 Convergence plots of displacements u and v by uniform field nodes (a) and comparison of numerical and analytical values of σ_{xx} along the boundary I (b) for problem of semi-infinite plate with central hole. (From H. Li, S. Mulay, and S. See. (2009b). *Computer Modeling in Engineering and Sciences*, 48, 43–82. With permission.)

4.4 APPLICATION OF RDQ METHOD FOR SOLVING FIXED–FIXED AND CANTILEVER MICROSWITCHES UNDER NONLINEAR ELECTROSTATIC LOADING

Several types of MEMS devices were developed in the past decade for various applications. They can be broadly categorised as actuation systems, i.e., elastic, thermal-elastic, electrostatic-elastic, magnetic-elastic, and microfluidics devices. At early stages, the MEMS devices chiefly employed mechanical deformation and electrical potential as actuation mechanisms. Today's MEMS devices, especially bioMEMS, employ various actuating mechanisms mostly from transport phenomena such as differences in ionic concentration, pH, and temperature, between the environmental solution and MEMS, resulting in ionic diffusion from the solution to MEMS.

Using the FEM to simulate the diffusion phenomenon coupled with large deformation may lead to distorted elements, as the diffusion of particles may result in large deformation. However, it is relatively easy for the particle-based meshless methods to simulate this diffusion phenomenon, as they do not involve meshes and no connectivity information between the particles is required, such that the free movement of particles is allowed.

Currently most MEMS problems are solved as multiphysics problems involving the governing differential equations from various engineering areas, such as chemical, mechanical, and electrical. If a meshless method is employed to solve these multi-physics problems, a user does not need

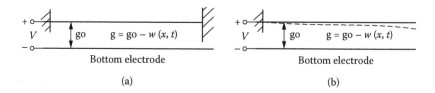

Figure 4.22 Fixed–fixed (a) and cantilever (b) microswitches. (From S. Mulay and H. Li. (2009). *Advanced Materials Research*, 74, 29–32. With permission.)

to worry about mesh quality and correctness that may require considerable attention. A user can better focus on the physics of the problem. Therefore, as the actuation mechanism becomes complicated, it becomes exceedingly important to go beyond the traditional techniques of analysis such as the FEM and FDM and develop more sophisticated methods of simulation.

Out of this motivation, the RDQ method was applied to analyze pull-in instability in two typical configurations of a microswitch, i.e., fixed–fixed and cantilever beam, as shown in Figure 4.22. The beam in the microswitch deflects toward the fixed electrode due to the applied voltage V, whereas the deflection of the beam can be controlled by the applied voltage. For a certain value of voltage V, the peak deflection w of the beam is equal to the initial gap g_0. This behaviour is called the pull-in instability, and the corresponding value of voltage is defined as the pull-in voltage.

The RDQ method is tested in this section for the reproducibility of the fourth order field variable by solving the microswitches of the fixed–fixed and cantilever beam types, as shown in Figure 4.22a and Figure 4.22b, respectively. The beam configurations are first solved for the values of slope and deflection under the uniformly distributed loads (UDLs) by the thin beam theory (Popov, 1990). The same beam configurations are then solved by applying the nonlinear electrostatic force field. The governing equation of the thin beam based on the thin beam theory (Timoshenko and Goodier, 1970; Popov, 1990) is given as

$$E\,I\frac{d^4w(x)}{dx^4} = q(x) \tag{4.55}$$

where $q(x)$ is an applied load, $E\,I$ is flexural rigidity, and w is beam deflection. The fixed–fixed beam under the UDL is solved with the boundary conditions as

$$w(x=0) = 0, \;\; w(x=L) = 0, \frac{dw(x=0)}{dx} = 0, \;\; \text{and} \;\; \frac{dw(x=L)}{dx} = 0 \tag{4.56}$$

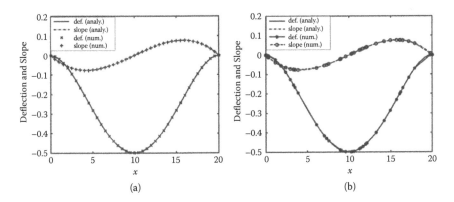

Figure 4.23 Comparison of numerical and analytical values of deflection and slope by uniform (a) and random (b) field nodes for fixed–fixed beam under UDL. (From H. Li, S. Mulay, and S. See. (2009b). *Computer Modeling in Engineering and Sciences*, 48, 43–82. With permission.)

The beam parameters are $q = -3.0$ N, $E = 3 \times 10^7$ Pa, $v_0 = 0.3$, length $(L) = 20$ m, thickness $(t) = 0.1$ m, and width $(D) = 1.0$ m. The analytical solutions of the deflection and slope are, respectively,

$$w(x) = \frac{q \, x^2 \, (L-x)^2}{24 \, EI}, \quad \text{and} \quad \frac{dw(x)}{dx} = \left(\frac{q}{EI}\right)\left[\frac{x \, L^2}{12} - \frac{L \, x^2}{4} + \frac{x^3}{6}\right] \quad (4.57)$$

The problem is solved by the 41 uniform and random field nodes separately combined with the 41 cosine virtual nodes. The numerical solutions are plotted in Figure 4.23, and the numerical results match almost exactly with the corresponding analytical values. The nonlinear electrostatic force field is then applied on the fixed–fixed beam as

$$q(x) = \left(\frac{\varepsilon_0 \, \tilde{v}^2 \, \tilde{w}}{2 \, g^2}\right)\left[1 + 0.65 \left(\frac{g}{\tilde{w}}\right)\right] \quad (4.58)$$

where ε_0, $\tilde{v}$, and $\tilde{w}$ are the vacuum permittivity ($8.8541878176 \times 10^{-12}$ F/m), applied voltage, and beam width, respectively, and $g = g_0 - w(x)$, where g_0 is an initial gap between the beam and bottom fixed plate. The parameters of beam are $E = 169$ GPa, $v_0 = 0.3$, $L = 80$ μm, $t = 0.5$ μm and $\tilde{w} = 10$ μm. As the beam starts deflecting due to the applied electrostatic force given by Equation (4.58), the load becomes increasingly nonlinear. As a result, it becomes an implicit problem that needs to be solved by inner iterations with a relaxation technique.

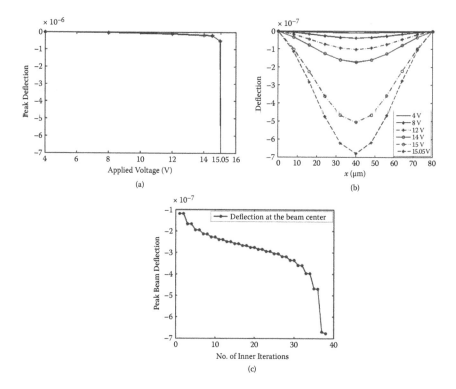

Figure 4.24 Peak deflection of beam for different applied voltages (a), full deflection of beam for different applied voltages (b), and convergence of peak deflection with inner iterations at $V = 15.05$ volts (c) for fixed–fixed microswitch under influence of nonlinear electrostatic force. (From H. Li, S. Mulay, and S. See. (2009b). *Computer Modeling in Engineering and Sciences*, 48, 43–82; S. Mulay and H. Li. (2009). *Advanced Materials Research*, 74, 29–32. With permission.)

Different values of the voltage are applied to study the pull-in behaviour and the results are plotted in Figure 4.24. It is seen from Figure 4.24b that the beam touches the bottom electrode for $V = 15.05$ volts. This value is closely matching with the literature value of 15.07 volts (Wang et al., 2007; Li et al., 2004d). The convergence of the peak deflection value during inner iterations by $V = 15.05$ volts is shown in Figure 4.24c. Figure 4.24b indicates that the values of the field node deflection are smoothly captured, demonstrating that the RDQ method can effectively handle the problems of nonlinear deformation. The inner iterations are performed by the fixed point and Newton methods to compare the results, and both are found to give similar results.

A cantilever beam microswitch, as shown in Figure 4.22b under the UDL is solved with the boundary conditions as

$$w(x = 0) = 0, \frac{dw(x = 0)}{dx} = 0, \quad \frac{d^2w(x = L)}{dx^2} = 0, \quad \text{and} \quad \frac{d^3w(x = L)}{dx^3} = 0$$

(4.59)

All parameters of the beam are same as the fixed–fixed beam with the analytical solutions as

$$w(x) = \frac{q}{24\,EI}\left[x^4 - 4x^3L + 6x^2L^2\right], \quad \text{and} \quad \frac{dw(x)}{dx} = \frac{q}{EI}\left[\frac{x^3}{6} - \frac{L\,x^2}{2} + \frac{x^2L}{2}\right]$$

(4.60)

The problem is solved by the 41 uniform and random field nodes separately combined with the 41 cosine virtual nodes, and the numerical solutions are plotted in Figure 4.25. The nonlinear electrostatic force is then applied on the cantilever beam, as given in Equation (4.58). The beam parameters are $E = 169$ GPa, $v_0 = 0.3$, $L = 80$ μm, $t = 0.5$ μm, and $\tilde{w} = 10$ μm. The deflections of the beam by the different values of applied voltages are plotted in Figure 4.26a. The pull-in instability occurs at $V = 2.5286$ volts, which closely matches with the value 2.33 volts found (Wang et al., 2007; Li et al., 2004d).

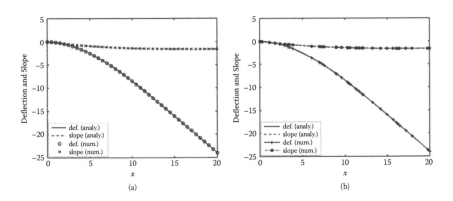

Figure 4.25 Comparison of numerical and analytical values of deflection and slope by uniform (a) and random (b) field nodes for cantilever beam under UDL. (From H. Li, S. Mulay, and S. See. (2009b). *Computer Modeling in Engineering and Sciences*, 48, 43–82. With permission.)

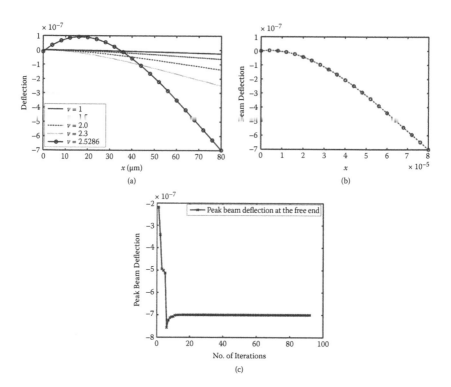

Figure 4.26 Deflection of beam under different applied voltages (a), deflection of beam at $\bar{v} = 2.64453$ volt by modified boundary conditions (b) and convergence of peak deflection during inner iterations by Newton method (c). (From H. Li, S. Mulay, and S. See. (2009b). *Computer Modeling in Engineering and Sciences*, 48, 43–82; S. Mulay and H. Li. (2009). *Advanced Materials Research*, 74, 29–32. With permission.)

There should be only one equation of discretisation per node for the technique of Newton iteration, but it is seen from Equation (4.59) that the two boundary conditions are imposed at each of the nodes located at the leftmost and rightmost boundaries of the domain. Therefore, this difficulty is overcome by transferring one of the boundary conditions to their neighbouring nodes as

$$\left(\frac{dw}{dx}\right)_1 = \frac{(w_2 - w_1)}{(x_2 - x_1)} = 0, \quad \text{as } w_1 = 0, \text{ so } w_2 = 0, \text{ and}$$

$$w_{,xxx}\big|_N = \frac{w_{,xx}\big|_N - w_{,xx}\big|_{N-1}}{(x_N - x_{N-1})} = 0, \text{ as } w_{,xx}\big|_N = 0 \therefore w_{,xx}\big|_{N-1} = 0 \qquad (4.61)$$

where w_1 and w_2, and x_1 and x_2 are the values of deflection and location at the first and second domain nodes, respectively, and N is the number of total field nodes. Therefore, the modified boundary conditions are given as

$$w_1 = 0, \quad w_2 = 0, \quad \left(\frac{d^2w}{dx^2}\right)_N = 0, \quad \text{and} \quad \left(\frac{d^2w}{dx^2}\right)_{N-1} = 0 \qquad (4.62)$$

The Newton method is employed by Equation (4.62) with Equations (4.55) and (4.58) imposed at the virtual nodes with the residual $R = KU-F$ as a function. The values of the virtual node deflection at the nth iteration where $n > 1$ are computed by the values of the deflections at the zeroth and first iterations as

$$J\left[\frac{\partial R(w^1_{vir})}{\partial w}\right]\left[w^2_{vir} - w^1_{vir}\right] = -R\left(w^1_{vir}\right) \quad \text{applied at the } j^{\text{th}} \text{ virtual node}$$

$$(4.63)$$

$$J\left[\frac{\partial R(w^1_{vir})}{\partial w}\right]_j = \frac{k_j U^1 - k_j U^0}{w^1_j - w^0_j} - \left.\frac{\partial F(w^1)}{\partial w}\right|_j, \quad \left.\frac{\partial F(w^1)}{\partial w}\right|_j = \frac{F(w^1_j) - F(w^0_j)}{w^1_j - w^0_j}, \quad \text{and}$$

$$R(w^1)_j = K_j U^1 - F(w^1_j)$$

$$(4.64)$$

where $[\partial R(w^1_{vir})/\partial w]_j$ is the Jacobean matrix computed at the jth virtual node by the residual function R. Equations (4.63) and (4.64) are together iteratively solved till the peak deflection value of the field node located at the free end converge. The beam deflection is computed by Newton method for the given beam parameters and $\tilde{v} = 2.64453$ volt as plotted in Figure 4.26b. Figure 4.26c shows the convergence of the peak deflection value during the inner iterations. Note that in Figure 4.26a for the pull-in instability voltage $V = 2.5286$ volts, the beam becomes unstable and the Neumann boundary conditions are not exactly imposed. But this behaviour is rectified and the Neumann boundary conditions are exactly imposed by the modified boundary conditions in Equation (4.60) for $\tilde{v} = 2.64453$ volt as evident from Figure 4.26b. It is seen in Figure 4.26c that the values of the peak beam deflection are smoothly converged, therefore it can be said that the modified boundary conditions in Equation (4.62) lead to the stable solution.

In summary, the RDQ method is applied in this section to study the phenomenon of pull-in instability in fixed–fixed and cantilever microswitches.

Based on the simulation results, the pull-in voltages obtained by the RDQ method closely match the literature values based on the experiments. It can also be said that the RDQ method handles the nonlinear electrostatic force field well.

It is seen from Figure 4.24b and Figure 4.26b that the peak deflection equal to 7 μm of the beam is about 9% of the total length equal to 80 μm of the beam, which is a case of large deflection. Therefore, the linear structural model, as given by Equation (4.55), may not be sufficient to solve this problem correctly. However, it is adequate to capture the pull-in voltages correctly because no transient analysis is performed. Therefore, the remeshing of the geometry is not required. A nonlinear structural model is necessary for the transient analysis of the same problems. The main objective here is to test the ability of the RDQ method in correctly capturing the pull-in voltages in spite of the linear structural model, which is fairly achieved.

4.5 INTRODUCTION TO CONSISTENCY ANALYSIS OF RDQ METHOD

The objective of the presented work is to perform the detailed consistency analysis of the locally applied DQ method and verify the results by convergence analysis of the RDQ method. The performance of the RDQ method is systematically studied by discretizing the domain with uniformly or randomly distributed field nodes coupled with the cosine or uniform distribution of the virtual nodes. The convergence analysis of RDQ method was performed in the previous section by considering only the cosine distribution of virtual nodes. It is performed in the present section by considering the cosine as well as the uniform distributions of the virtual nodes, and their performance is compared by solving several 1-D, 2-D, and elasticity problems.

The main difference between the DQ and RDQ methods is that the derivative at a specific node in the former method is approximated by all the nodes in a domain, whereas it is approximated in the latter case by the surrounding nodes in a local domain associated with the concerned node. Therefore, the consistency equations obtained after the derivative discretisation will be different for both methods. Another important difference is that in the RDQ method, the terms of the function values at the virtual nodes in the expression of derivative approximation are replaced by their expression of interpolation correlating with the values of nodal parameter at the field nodes obtained by the function of fixed RKPM interpolation. As a result, it is vital to investigate the effect of this substitution on the overall consistency of the RDQ method.

The terms of the virtual node function values from the discretised governing equation are replaced by Taylor series expansion in the present consistency analysis to achieve a reasonable judgment of an error in the analytical discretisation. Therefore, with reference to the final consistency equations obtained after using Taylor series expansion, the present analysis represents the consistency analysis for the DQ method applied over a local domain or the RDQ method.

A problem of a semi-infinite plate with a central hole is solved here to demonstrate that the RDQ method can handle the problems of irregular boundaries. The intention of this demonstration is to ascertain the applicability of RDQ method to solve such problems. The only factor that changes with the different types of irregular boundaries is how the virtual nodes are created within domains. The RDQ method is then directly applied as it is independent of the type or domain of the problem.

4.6 CONSISTENCY ANALYSIS OF LOCALLY APPLIED DQ METHOD

Given a PDE $P u = f$ and a scheme of derivative discretisation $P(h_x, h_y) v = f$, the scheme of derivative discretisation becomes consistent with the PDE if any smooth function $\phi(x,y)$ satisfies the following condition

$$P\phi - P(h_x, h_y)\, \phi \to 0, \quad \text{as } h_x, h_y \to 0 \tag{4.65}$$

where P is a differential operator, $h_x = L_x / (N_x - 1)$ and $h_y = L_y / (N_y - 1)$ are the nodal spacings, L_x and L_y are the domain lengths, and N_x and N_y are the numbers of total field nodes in the x and y directions, respectively. The $h_x = h$ and $h_y = t$ are the grid spacings in the space (x) and time (T) domains, respectively, for a hyperbolic differential equation. In a nutshell, the consistency analysis determines how closely the continuous form of the governing differential equation is represented by the corresponding discretised form.

The consistency studies of the locally applied DQ method are performed in this section, based on the theory (Strikwerda, 1989), and by the virtual nodes with different configurations of the DQ local domains. The 1-D wave and Laplace equations are discretised to perform the consistency analysis.

There are two important differences between the DQ and RDQ methods. The first is that the RDQ method employs the fixed RKPM interpolation function to approximate the value of function at the concerned virtual node in the form of the values of nodal parameter at the field nodes that surround the concerned virtual node. The second is that the DQ method approximates the derivative at a specific node by including all the nodes in a domain, but is performed in the RDQ method by the virtual nodes located in the surrounding region of the concerned virtual node. Therefore, the final consistency equations for both the cases would be different due to these differences.

In the present work of consistency analysis, the governing equation is discretised at the concerned virtual node by the virtual nodes located in the local domains of the concerned virtual node. The terms of function values at the virtual nodes resulting from the discretised form of the governing equation are replaced by Taylor series expansion. The important task is to numerically verify the conclusions drawn from the consistency analysis by the RDQ method. This is done while performing convergence analysis of the RDQ method by the cosine and uniform distributions of the virtual nodes.

The weighting coefficients in this section are computed by Quan and Chang (1989a and b) and Shu's general approach (Shu et al., 1994; Shu, 2000). The governing equation is discretised at a virtual node (x_m, t_m) by the uniform and cosine distributions of the virtual nodes with three different possibilities of the location of the concerned virtual node, i.e., between, at the left end, and at the right end of the computational domain. For certain configurations of the DQ local domains, Shu's approach gives consistent discretisation of the governing equation. Also, the governing equation discretised by the cosine distribution of virtual nodes converges to that of the uniform distribution of virtual nodes. All the configurations of the local domains of the DQ method are presented here to illustrate the consistency of the method only; the stability aspects of the corresponding schemes are not considered.

4.6.1 Consistency analysis of one-dimensional wave equation by uniform distribution of virtual nodes

4.6.1.1 Shu's general approach

The DQ weighting coefficients in this section are computed by Shu's approach (Shu et al., 1994; Shu, 2000) with the uniform distribution of virtual nodes. A virtual node (x_m, t_m) is initially assumed as an internal domain node around which the DQ local domain is shown in Figure 4.27a. The 1-D wave equation is given as

$$\phi_{,t} + a\phi_{,x} = 0, \quad P\phi = 0 \tag{4.66}$$

where $P = (\partial/\partial t) + a(\partial/\partial x)$ is a derivative operator, and m and n are the nodal indices for the space and time domains, respectively. Equation (4.66) is discretised at the virtual node (x_m, t_n) to get

$$P(h, t)\,\phi = \frac{\partial\phi(x_m, t_n)}{\partial t} + a\,\frac{\partial\phi(x_m, t_n)}{\partial x} = 0 \tag{4.67}$$

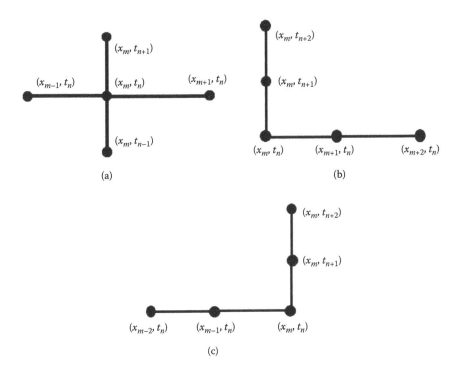

Figure 4.27 DQ local domains at concerned virtual node when virtual node is assumed in internal domain (a), at left end (b), and right end (c) of domain for 1-D wave equation. (Modified from S. Mulay, H. Li, and S. See. (2009). *Computational Mechanics, 44,* 563–590.)

where h and t are the nodal spacings along the space (x) and time (T) axes, respectively. The term of space derivative in Equation (4.67) is discretised by the DQ method to give

$$\frac{\partial \phi(x_m, t_n)}{\partial x} = \left\{ a_{m,m-1} \quad a_{m,m} \quad a_{m,m+1} \right\} \left\{ \begin{array}{c} \phi(x_{m-1}, t_n) \\ \phi(x_m, t_n) \\ \phi(x_{m+1}, t_n) \end{array} \right\} \tag{4.68}$$

The weighting coefficients by Shu's approach are computed as

$$a_{i,j} = \frac{1}{x_i - x_j} \prod_{k=1; k \neq i,j}^{N_x} \frac{(x_i - x_k)}{(x_j - x_k)}, \quad a_{ii} = -\sum_{j=1; j \neq i}^{N_x} a_{ij} \tag{4.69}$$

The weighting coefficients from Equation (4.68) are computed by Equation (4.69) as

$$a_{m,m-1} = \frac{-1}{2h}, a_{m,m+1} = \frac{1}{2h}, a_{m,m} = 0 \tag{4.70}$$

As per Taylor series expansion

$$\phi(x_{m-1}, t_n) = \phi(x_m - h) = \phi - h\phi_{,x} + \frac{h^2}{2}\phi_{,xx} - O(h)^3 \tag{4.71}$$

$$\phi(x_{m+1}, t_n) = \phi(x_m + h) = \phi + h\phi_{,x} + \frac{h^2}{2}\phi_{,xx} + O(h)^3 \tag{4.72}$$

where $\phi = \phi(x_m, t_n)$. Substituting Equations (4.70) to (4.72) into (4.68) results in

$$\frac{\partial \phi(x_m, t_n)}{\partial x} = \frac{-1}{2h}\left[\phi - h\phi_{,x} + \frac{h^2}{2}\phi_{,xx} - O(h)^3\right]$$

$$+ 0[\phi] + \frac{1}{2h}\left[\phi + h\phi_{,x} + \frac{h^2}{2}\phi_{,xx} + O(h)^3\right] \tag{4.73}$$

Simplifying Equation (4.73) results in

$$\frac{\partial \phi(x_m, t_n)}{\partial x} = \phi_{,x} + O(h)^2 \tag{4.74}$$

Equation (4.74) is a final expression for the space derivative term in Equation (4.67). The time derivative term is also similarly determined as

$$\frac{\partial \phi(x_m, t_n)}{\partial t} = \left\{a_{n,n-1} \quad a_{n,n} \quad a_{n,n+1}\right\} \begin{Bmatrix} \phi(x_m, t_{n+1}) \\ \phi(x_m, t_n) \\ \phi(x_m, t_{n-1}) \end{Bmatrix} \tag{4.75}$$

$$a_{n,n-1} = \frac{-1}{2t}, a_{n,n+1} = \frac{1}{2t}, a_{n,n} = 0 \tag{4.76}$$

As per Taylor series expansion

$$\phi(x_m, t_{n-1}) = \phi(t_m - t) = \phi - t\,\phi_{,t} + \frac{t^2}{2}\phi_{,tt} - O(t)^3 \tag{4.77}$$

$$\phi(x_m, t_{n+1}) = \phi(t_n + t) = \phi + t\,\phi_{,t} + \frac{t^2}{2}\phi_{,tt} + O(t)^3 \tag{4.78}$$

where $\phi = \phi(x_m, t_n)$. We then substitute Equations (4.76) to (4.78) into Equation (4.75) and simplify it to

$$\frac{\partial\phi(x_m, t_n)}{\partial t} = \phi_{,t} + O(t)^2 \tag{4.79}$$

Substituting Equation (4.74) and Equation (4.79) into Equation (4.67) results in

$$P(h,t)\phi = \phi_{,t} + O(t)^2 + a\left[\phi_{,x} + O(h)^2\right] \tag{4.80}$$

Subtracting Equation (4.80) from Equation (4.66) gives

$$P\phi - P(h,t)\phi = \phi_{,t} + a\phi_{,x} - \phi_{,t} - O(t)^2 - a\phi_{,x} - a\,O(h)^2 \tag{4.81}$$

Equation (4.81) tends to zero as h and $t \to 0$, which means that the presented discretisation scheme is consistent.

The virtual node (x_m, t_n) is assumed at the left end of a domain with the DQ local domain around it as shown in Figure 4.27b. The term of space derivative in Equation (4.67) is discretised as per the DQ method as

$$\frac{\partial\phi(x_m, t_n)}{\partial x} = \left\{a_{m,m} \quad a_{m,m+1} \quad a_{m,m+2}\right\}\left\{\begin{array}{c} \phi(x_m, t_n) \\ \phi(x_{m+1}, t_n) \\ \phi(x_{m+2}, t_n) \end{array}\right\} \tag{4.82}$$

The weighting coefficients are computed by Equation (4.69) as

$$a_{m,m+1} = \frac{2}{h}, \quad a_{m,m+2} = \frac{-1}{2h}, \quad a_{m,m} = -\frac{3}{2h} \tag{4.83}$$

As per Taylor series

$$\phi(x_{m+1},t_n) = \phi(x_m + h) = \phi + h\phi_{,x} + \frac{h^2}{2}\phi_{,xx} + O(h)^3 \qquad (4.84)$$

$$\phi(x_{m+2},t_n) = \phi(x_m + 2h) = \phi + (2h)\phi_{,x} + \frac{(2h)^2}{2}\phi_{,xx} + O(2h)^3 \qquad (4.85)$$

where $\phi = \phi(x_m, t_n)$. Substitute Equations (4.83) to (4.85) into (4.82) to give

$$\frac{\partial\phi(x_m,t_n)}{\partial x} = \frac{-3}{2h}[\phi] + \frac{2}{h}\left[\phi + h\phi_{,x} + \frac{h^2}{2}\phi_{,xx} + O(h)^3\right]$$

$$- \frac{1}{2h}\left[\phi + (2h)\phi_{,x} + \frac{(2h)^2}{2}\phi_{,xx} + O(2h)^3\right] \qquad (4.86)$$

Simplify Equation (4.86) to obtain

$$\frac{\partial\phi(x_m, t_n)}{\partial x} = \phi_{,x} + 2\,O(h)^2 - O(2h)^2 \qquad (4.87)$$

Equation (4.87) is a final expression for the term of space derivative in Equation (4.67). The term of time derivative is discretised in a similar manner as

$$\frac{\partial\phi(x_m, t_n)}{\partial t} = \left\{a_{n,n} \quad a_{n,n+1} \quad a_{n,n+2}\right\}\left\{\begin{array}{c}\phi(x_m, t_n) \\ \phi(x_m, t_{n+1}) \\ \phi(x_m, t_{n+2})\end{array}\right\} \qquad (4.88)$$

The discretisation of the time derivative term similar to space derivative is obtained as

$$\frac{\partial\phi(x_m, t_n)}{\partial t} = \phi_{,t} + 2\,O(t)^2 - O(2t)^2 \qquad (4.89)$$

Substitute Equations (4.87) and (4.89) into (4.67), and subtract it from Equation (4.66) to get

$$P\phi - P(h,t)\phi = \phi_{,t} + a\phi_{,x} - \phi_{,t} - 2\,O(t)^2 + O(2t)^2$$

$$- a\,\phi_{,x} - 2a\,O(h)^2 + a\,O(2h)^2 \qquad (4.90)$$

Equation (4.90) tends to zero as h and $t \to 0$, which indicates that the presented discretisation scheme is consistent.

The virtual node (x_m, t_n) is assumed at the right end of the domain with the DQ local domain, as shown in Figure 4.27c. The space derivative term in Equation (4.67) is discretised as

$$\frac{\partial \phi(x_m, t_n)}{\partial x} = \left\{ a_{m,m-2} \quad a_{m,m-1} \quad a_{m,m} \right\} \left\{ \begin{array}{c} \phi(x_{m-2}, t_n) \\ \phi(x_{m-1}, t_n) \\ \phi(x_m, t_n) \end{array} \right\} \tag{4.91}$$

The weighting coefficients are determined by Equation (4.69) as

$$a_{m,m-2} = \frac{1}{2h}, \ a_{m,m-1} = \frac{-2}{h}, \ a_{m,m} = \frac{3}{2h} \tag{4.92}$$

As per Taylor series

$$\phi(x_{m-1}, t_n) = \phi(x_m - h) = \phi - h\phi_{,x} + \frac{h^2}{2}\phi_{,xx} - O(h)^3 \tag{4.93}$$

$$\phi(x_{m-2}, t_n) = \phi(x_m - 2h) = \phi - (2h)\phi_{,x} + \frac{(2h)^2}{2}\phi_{,xx} - O(2h)^3 \tag{4.94}$$

where $\phi = \phi(x_m, t_n)$. Substitute Equations (4.92) to (4.94) into Equation (4.91) and simplify it to obtain

$$\frac{\partial \phi(x_m, t_n)}{\partial x} = \phi_{,x} - O(2h)^2 + 2 \, O(h)^2 \tag{4.95}$$

Equation (4.95) is a final expression for the space derivative term in Equation (4.67). The discretisation of time derivative term is obtained by following the similar procedure as

$$\frac{\partial \phi(x_m, t_n)}{\partial t} = \left\{ a_{n,n} \quad a_{n,n+1} \quad a_{n,n+2} \right\} \left\{ \begin{array}{c} \phi(x_m, t_n) \\ \phi(x_m, t_{n+1}) \\ \phi(x_m, t_{n+2}) \end{array} \right\} \tag{4.96}$$

Equation (4.96) is given by weighting coefficients and Taylor series expansions as

$$\frac{\partial \phi(x_m, t_n)}{\partial t} = \phi_{,t} + 2 \, O(t)^2 - \left(\frac{1}{2}\right) O(2t)^2 \tag{4.97}$$

We then substitute Equations (4.95) and (4.97) into Equation (4.67) and subtract the result from Equation (4.66) to get

$$P\phi - P(h,t)\phi = \phi_{,t} + a\phi_{,x} - \phi_{,t} - 2\ O(t)^2 + (0.5)\ O(2t)^2$$
$$- a\ \phi_{,x} + a\ O(2h)^2 - (2a)\ O(h)^2$$

(4.98)

Equation (4.98) tends to zero as h and $t \to 0$, which shows that the presented discretisation scheme is consistent. From Equations (4.81), (4.90), and (4.98), we can see that the weighting coefficients computed by Shu's general approach (Shu et al., 1994; Shu, 2000) give consistent discretisation of the governing equation for the locally applied DQ method with uniform virtual nodes.

4.6.1.2 Quan and Chang approach

The weighting coefficients are computed in this section by Quan and Chang approach (Quan and Chang, 1989a and b) with the uniform distribution of virtual nodes. The virtual node (x_m, t_n) is assumed an internal domain node with the DQ local domain, as shown in Figure 4.27a. The space term derivative in Equation (4.67) is discretised as

$$\frac{\partial\phi(x_m,\ t_n)}{\partial x} = \left\{ a_{m,m-1} \quad a_{m,m} \quad a_{m,m+1} \right\} \begin{Bmatrix} \phi(x_{m-1},\ t_n) \\ \phi(x_m,\ t_n) \\ \phi(x_{m+1},\ t_n) \end{Bmatrix}$$

(4.99)

According to the Quan and Chang approach (1989a and b), the weighting coefficients are given as

$$a_{ij} = \frac{1}{x_j - x_i} \prod_{k=1, k\neq i, j}^{N_x} \frac{(x_i - x_k)}{(x_j - x_k)}, \quad a_{ii} = \sum_{j=1, j\neq i}^{N_x} \frac{1}{(x_i - x_j)}$$

(4.100)

The weighting coefficients in Equation (4.99) are computed by Equation (4.100) and given as

$$a_{m,m-1} = \frac{1}{2h}, \quad a_{m,m+1} = \frac{-1}{2h}, \quad a_{m,m} = 0$$

(4.101)

Substitute Equations (4.71), (4.72), and (4.101) into Equation (4.99) and simplify to

$$\frac{\partial\phi(x_m,\ t_n)}{\partial x} = -\phi_{,x} - \ O(h)^2$$

(4.102)

Equation (4.102) is a final equation for the space derivative term in Equation (4.67). The term of time derivative in Equation (4.67) is discretised as given in Equation (4.75), and the corresponding weighting coefficients are computed by Equation (4.100) as

$$a_{n,n-1} = \frac{1}{2t}, \quad a_{n,n+1} = \frac{-1}{2t}, \quad a_{n,n} = 0 \qquad (4.103)$$

Substituting Equations (4.77), (4.78), and (4.103) into Equation (4.75), and simplifying it results in

$$\frac{\partial \phi(x_m, t_n)}{\partial t} = -\phi_t - O(t)^2 \qquad (4.104)$$

We then substitute Equations (4.102) and (4.104) into Equation (4.67) and subtract result from Equation (4.66) to get

$$P\phi - P(h, t)\phi = \phi_{,t} + a\phi_{,x} + \phi_t + O(t)^2 + a\phi_{,x} + a\ O(h)^2 \qquad (4.105)$$

Equation (4.105) tends to $2P\phi$ as h and $t \to 0$. As a result, the presented discretisation scheme is not consistent. As the governing equation discretised at the virtual node in the internal domain with the weighting coefficients computed by the Quan and Chang approach is not consistent, it is unnecessary to consider the cases of the virtual nodes located at the left and right ends of the domain. Also, the consistent discretisation of the governing equation of the Quan and Chang approach (1989a and b) is highly dependent on the pattern of distribution of virtual nodes, as also observed by Quan and Chang, (1989a).

4.6.2 Consistency analysis of one-dimensional wave equation by cosine distribution of virtual nodes

In this section, the virtual nodes are distributed by cosine law, resulting in nodal coordinates similar to the roots of the Nth order orthogonal polynomial of the first kind as

$$x_i = x_0 + \frac{L}{2}\left[1 - \cos\left(\frac{i-1}{N-1}\pi\right)\right], \quad \text{for } i = 1, 2, ..., N \qquad (4.106)$$

where x_0 is a starting coordinate of the domain, L is the domain length, and N is the number of total virtual nodes in a domain. The index i is replaced by $i - 1$ for node x_{i-1}

$$x_{i-1} = x_0 + \frac{L}{2}\left[1 - \cos\left(\frac{i-2}{N-1}\pi\right)\right] \qquad (4.107)$$

The following notation is used to simplify the equation when the difference between the two cosine distributed virtual nodes is computed as

$$x_i - x_{i-1} = \frac{L}{2}\cos\left(\frac{i-2}{N-1}\pi\right) - \frac{L}{2}\cos\left(\frac{i-1}{N-1}\pi\right) = \Delta x(i-2, \ i-1)\frac{L}{2} \quad (4.108)$$

The virtual node (x_m, t_n) is assumed an internal domain node with the DQ local domain shown in Figure 4.27a. The term of space derivative in Equation (4.67) is discretised by the DQ method, as given in Equation (4.68), and the weighting coefficients are computed by Equation (4.69)

$$a_{m,m-1} = \frac{\Delta x(m, \ m-1)}{\Delta x(m-1, \ m-2)\,\Delta x(m, \ m-2)}\left(\frac{2}{L}\right),$$

$$a_{m,m+1} = \frac{\Delta x(m-2, \ m-1)}{\Delta x(m-2, \ m)\,\Delta x(m-1, \ m)}\left(\frac{2}{L}\right) \quad a_{m,m} = -(a_{m,m-1} + a_{m,m+1})$$

$$(4.109)$$

As per Taylor series

$$\phi(x_{m-1}, t_n) = \phi(x_m - h_1) = \phi - h_1\,\phi_{,x} + \frac{h_1^2}{2}\phi_{,xx} - O(h_1)^3 \quad (4.110)$$

where

$$h_1 = x_m - x_{m-1} = \Delta x(m-2, \ m-1)\frac{L}{2}$$

$$(4.111)$$

$$\phi(x_{m+1}, t_n) = \phi(x_m + h_2) = \phi + h_2\,\phi_{,x} + \frac{h_2^2}{2}\phi_{,xx} + O(h_2)^3 \quad (4.112)$$

where

$$h_2 = x_{m+1} - x_m = \Delta x(m-1, \ m)\frac{L}{2} \quad (4.113)$$

where $\phi = \phi(x_m, t_n)$. Substituting Equations (4.110) and (4.112) in Equation (4.68) results in

$$\frac{\partial\phi(x_m, t_n)}{\partial x} = \phi_{,x}\left[a_{m,m-1}\,(-h_1) + a_{m,m+1}h_2\right] + \phi_{,xx}\left[a_{m,m-1}\frac{h_1^2}{2} + a_{m,m+1}\frac{h_2^2}{2}\right]$$

$$-a_{m,m-1}\,O(h_1)^3 + a_{m,m+1}\,O(h_2)^3 \quad (4.114)$$

Simplify Equation (4.114) by Equations (4.109), (4.111), and (4.113) to obtain

$$\frac{\partial \phi(x_m,\, t_n)}{\partial x} = \phi_{,x} - a_{m,m-1}\, O(h_1)^3 + a_{m,m+1}\, O(h_2)^3 \tag{4.115}$$

Equation (4.115) is a final expression for the discretisation of space derivative term in Equation (4.67). Similarly, the term of time derivative in Equation (4.67) is discretised and simplified as

$$\frac{\partial \phi(x_m,\, t_n)}{\partial t} = \phi_{,t} - a_{n,n-1}\, O(t_1)^3 + a_{n,n+1}\, O(t_2)^3 \tag{4.116}$$

Substituting Equations (4.115) and (4.116) into Equation (4.67) and then subtracting the result from Equation (4.66) results in

$$P\phi - P(h,\, t)\phi = \phi_{,t} + a\phi_{,x} - \phi_{,t} + a_{n,n-1}\, O(t_1)^3 - a_{n,n+1}\, O(t_2)^3 - a\phi_{,x}$$
$$+ a\, a_{m,m-1}\, O(h_1)^3 - a\, a_{m,m+1}\, O(h_2)^3 \tag{4.117}$$

Equation (4.117) is a consistency equation obtained by the cosine distribution of virtual nodes. As h and $t \to 0$, $(N_x - 1) \to \infty$ and $(N_t - 1) \to \infty$, and $a_{m,m-1}, a_{m,m+1}, a_{n,n-1}, a_{n,n+1} \to 0$, Equation (4.117) tends to zero. Therefore, the presented discretisation scheme is consistent by the cosine distribution of virtual nodes. If the cosine distribution of virtual nodes is changed to the uniform one, the new weighting coefficients are given as

$$a_{m,m-1} = \frac{\Delta x(m,\, m-1)}{\Delta x(m-1,\, m-2)\,\Delta x(m,\, m-2)} \frac{2}{L} = \frac{-1}{2h},$$

$$a_{m,m+1} = \frac{\Delta x(m-2,\, m-1)}{\Delta x(m-2,\, m)\,\Delta x(m-1,\, m)} \frac{2}{L} = \frac{1}{2h} \quad a_{m,m} = -(a_{m,m-1} + a_{m,m+1}) = 0$$
$$\tag{4.118}$$

$$h_1 = \Delta x(m-2,\, m-1) = x_m - x_{m-1} = h \tag{4.119}$$

$$h_2 = \Delta x(m-1,\, m) = x_{m+1} - x_m = h \tag{4.120}$$

$$t_1 = \Delta x(n-2,\, n-1) = t_n - t_{n-1} = t \tag{4.121}$$

$$t_2 = \Delta x(n-1,\, n) = t_{n+1} - t_n = t \tag{4.122}$$

Substituting Equations (4.118) to (4.122) into Equation (4.117) results in

$$
P\phi - P(h,t)\phi = \phi_{,t} + a\phi_{,x} - \phi_{,t} + \left(\frac{-1}{2t}\right)O(t)^3 - \left(\frac{1}{2t}\right)O(t)^3
$$

$$
- a\phi_{,x} + a\left(\frac{-1}{2t}\right)O(h)^3 - a\left(\frac{1}{2t}\right)O(h)^3
$$

(4.123)

Simplify Equation (4.123) to get

$$
P\phi - P(h,\ t)\phi = \phi_{,t} + a\phi_{,x} - \phi_{,t} - O(t)^2 - a\phi_{,x} - a\ O(h)^2
$$

(4.124)

Equation (4.124) is the same as Equation (4.81). Therefore, the consistency equation obtained by the cosine distribution of virtual nodes converges to that obtained by the uniform distribution of virtual nodes. Similar results are obtained for the other configurations of the DQ local domains as well.

4.6.3 Consistency analysis of one-dimensional Laplace equation by uniform distribution of virtual nodes

The consistency analysis performed in this section is for the 1-D Laplace equation by the uniform virtual nodes. The weighting coefficients are computed by Equation (4.69). The virtual node (x_m) is assumed an internal domain node with the DQ local domain shown in Figure 4.28a. The 1-D Laplace equation is given as

$$
\phi_{,xx} = 0, \quad P\phi = 0
$$

(4.125)

where $P = (\partial^2 / \partial x^2)$. Equation (4.125) is discretised at the virtual node (x_m). Therefore

$$
P(h)\phi = \frac{\partial^2 \phi(x_m)}{\partial x^2}
$$

(4.126)

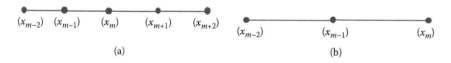

(a)   (b)

Figure 4.28 DQ local domains at virtual node when virtual node is assumed inside internal domain (a) and right end of domain (b) for 1-D Laplace equation. (Modified from S. Mulay, H. Li, and S. See. (2009). *Computational Mechanics*, 44, 563–590.)

where h is a virtual node spacing along the x axis. The term of space derivative in Equation (4.126) is approximated by the locally applied DQ method as

$$P(h)\phi = \left\{ \begin{array}{ccccc} b_{m,m-2} & b_{m,m-1} & b_{m,m} & b_{m,m+1} & b_{m,m+2} \end{array} \right\} \begin{bmatrix} \phi(x_{m-2}) \\ \phi(x_{m-1}) \\ \phi(x_m) \\ \phi(x_{m+1}) \\ \phi(x_{m+2}) \end{bmatrix}$$

(4.127)

where $b_{m,k}$, for $k = m - 2, \ldots, m + 2$, represents the DQ weighting coefficients for the second order derivative as given

$$b_{ij} = 2\, a_{ij} \left[a_{ii} - \frac{1}{x_i - x_j} \right]$$

(4.128)

where a_{ij} and a_{ii} are the DQ weighting coefficients for the first-order derivative. The weighting coefficients in Equation (4.127) are computed by Equations (4.69) and (4.128), and given in Equation (4.130) as

$$a_{m,m-2} = \frac{1}{12h}, \; a_{m,m-1} = \frac{-2}{3h}, \; a_{m,m+2} = \frac{-1}{12h}, \; a_{m,m+1} = \frac{2}{3h}, \; a_{m,m} = 0 \quad (4.129)$$

$$b_{m,m-2} = \frac{-1}{12h^2}, \; b_{m,m-1} = \frac{4}{3h^2}, \; b_{m,m+1} = \frac{4}{3h^2}, \; b_{m,m+2} = \frac{-1}{12h^2}, \; b_{m,m} = \frac{-5}{2h^2}$$

(4.130)

As per Taylor series

$$\phi(x_{m-2}) = \phi(x_m - 2h) = \phi - (2h)\phi_{,x} + \frac{(2h)^2}{2}\phi_{,xx} - O(2h)^3$$

(4.131)

$$\phi(x_{m-1}) = \phi(x_m - h) = \phi - h\phi_{,x} + \frac{h^2}{2}\phi_{,xx} - O(h)^3$$

(4.132)

$$\phi(x_{m+2}) = \phi(x_m + 2h) = \phi + (2h)\phi_{,x} + \frac{(2h)^2}{2}\phi_{,xx} + O(2h)^3$$

(4.133)

$$\phi(x_{m+1}) = \phi(x_m + h) = \phi + h\phi_{,x} + \frac{h^2}{2}\phi_{,xx} + O(h)^3$$

(4.134)

Substituting Equations (4.130) through Equation (4.134) into Equation (4.127) with simplification results in

$$P(h)\phi = \phi_{,xx} \tag{4.135}$$

Subtracting Equation (4.135) from Equation (4.125) yields

$$\Gamma\psi - \Gamma(h)\psi = \psi_{,xx} - \psi_{,xx} = 0 \tag{4.136}$$

The presented discretisation scheme is consistent since Equation (4.136) is equal to zero.

The virtual node (x_m) is assumed at the right end of the domain with the DQ local domain shown in Figure 4.28b. The term of space derivative in Equation (4.126) is discretised by the locally applied DQ method as

$$P(h)\phi = \left\{ \begin{array}{ccc} b_{m,m-2} & b_{m,m-1} & b_{m,m} \end{array} \right\} \left[\begin{array}{c} \phi(x_{m,m-2}) \\ \phi(x_{m,m-1}) \\ \phi(x_{m,m}) \end{array} \right] \tag{4.137}$$

The weighting coefficients are computed by Equations (4.69) and (4.128)

$$b_{m,m-2} = \frac{1}{h^2}, \ b_{m,m-1} = \frac{-2}{h^2}, \ b_{m,m} = \frac{1}{h^2} \tag{4.138}$$

As per Taylor series

$$\phi(x_{m-2}) = \phi(x_m - 2h) = \phi - (2h)\phi_{,x} + \frac{(2h)^2}{2}\phi_{,xx} - O(2h)^3 \tag{4.139}$$

$$\phi(x_{m-1}) = \phi(x_m - h) = \phi - h\phi_{,x} + \frac{h^2}{2}\phi_{,xx} - O(h)^3 \tag{4.140}$$

Substituting Equations (4.138) to (4.140) into Equation (4.137), we simplify the result and subtract it from Equation (4.125) to obtain

$$P\phi - P(h)\phi = \phi_{,xx} - \phi_{,xx} + O(8h) + 2O(h) \tag{4.141}$$

Equation (4.141) tends to zero as $h \to 0$. Therefore the current scheme of discretisation is consistent. It is shown from Equations (4.136) and (4.141) that the discretised form of the 1-D Laplace equation by locally applied DQ method is consistent with the continuous form in Equation (4.125). We can observe from the consistency analysis that the discretisation error by uniform distribution of virtual nodes is much smaller than by cosine distribution of virtual nodes.

4.7 EFFECT OF UNIFORM AND COSINE DISTRIBUTIONS OF VIRTUAL NODES ON CONVERGENCE OF RDQ METHOD

Several 1-D, 2-D, and elasticity problems are solved in this section by the RDQ method. Their convergence analyses are performed by considering the observations made from the consistency analysis. All the given problems are solved by the uniform and cosine distributions of virtual nodes, and their rates of function convergence are compared. The distribution pattern of the field and virtual nodes is identified, which gives better rates of convergence. For all the problems solved here, the global error is computed by Equation (4.6), which is same as the L^2 error norm but averaged over the total field nodes and normalized by an absolute maximum value of the analytical function.

4.7.1 One-dimensional test problems

The first 1-D problem is a Poisson equation with a variable force term of the function x and mixed boundary conditions. The governing equation and boundary conditions are given as

$$\frac{d^2 f}{dx^2} = -x, \quad (0 < x < 1) \tag{4.142}$$

$$f(x = 0) = 0, \quad \frac{df}{dx}(x = 1) = 0 \tag{4.143}$$

The analytical solution with third order continuity is given as $f(x) = -(x^3)/6 + (x/2)$. This problem is solved by including up to second order monomial terms in the basis of shape function, and with 21, 41, 81, 161, 321, 641 field and 641 virtual nodes, respectively. The uniform and randomly distributed field nodes are separately combined with the cosine and uniformly distributed virtual nodes. All the errors are computed by Equation (4.6).

The convergence rates for both types of virtual nodes are given in Table 4.4 and Table 4.5, and the corresponding convergence curves are plotted in Figure 4.29. We can see from Table 4.4 and Table 4.5 that the convergence rate of the function f by the uniform distribution of virtual nodes is improved as compared with that by the cosine distribution of virtual nodes. A good convergence rate of the function f is obtained by combining the uniformly distributed virtual nodes with the randomly distributed field nodes, whereas a good convergence rate of the derivative of function is obtained by combining the uniformly distributed virtual nodes with the uniformly distributed field nodes.

Table 4.4 Convergence rates for first 1-D problem of
          Poisson equation when cosine distributed
          virtual nodes are combined with uniform
          and randomly distributed field nodes

Function	Convergence rates (uniform field nodes)	Convergence rates (random field nodes)
f	1.8	1.73
$f_{,x}$	1.97	1.0
$f_{,xx}$	2.1	0.8

Source: Modified from S. Mulay, H. Li, and S. See. (2009). *Computational Mechanics*, 44, 563–590.

This problem is additionally solved by five different sets of the random nodes to study the effects of random field nodes on the rate of convergence. The corresponding convergence plots of the function value are shown in Figure 4.30. When the convergence rates shown in Figure 4.30 are compared with the value 1.73 given in Table 4.4 obtained by the random nodes generated without giving any starting reference value, as explained earlier, 1.73 is a good measure of the performance of RDQ method. It is also observed from Figure 4.30 that the global error varies significantly for low numbers of random field nodes, but converges to a fixed value as the number of random field nodes is increased.

The results obtained by solving this problem with five different sets of the random field nodes indicate that even if different sets of random field nodes are used, the values of the global error may vary a little, but the convergence rate of the function will be almost the same. This is an important result. It signifies that the global error is almost constant for any set of sufficiently high numbers of random field nodes. As a result, the computed values of function at the random field nodes have equal numerical accuracy.

The second 1-D problem is also a Poisson equation with a second order continuous force term and mixed boundary conditions. The fourth order

Table 4.5 Convergence rates for first 1-D problem of
          Poisson equation when uniformly distributed
          virtual nodes are combined with uniform and
          randomly distributed field nodes

Function	Convergence rates (uniform field nodes)	Convergence rates (random field nodes)
f	2.8	3.2
$f_{,x}$	3.1	1.4
$f_{,xx}$	2.9	1.3

Source: Modified from S. Mulay, H. Li, and S. See. (2009). *Computational Mechanics*, 44, 563–590.

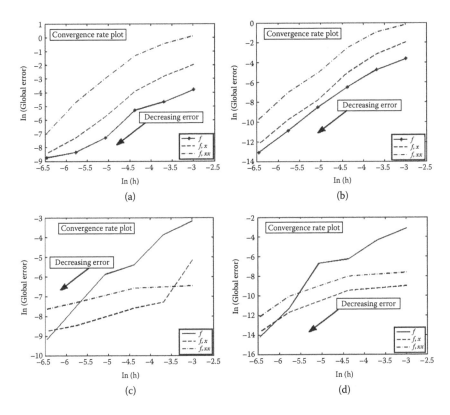

Figure 4.29 Convergence curves when uniform field nodes are combined with cosine (a) and uniform (b) virtual nodes and the random field nodes are combined with cosine (c) and uniform (d) virtual nodes for first 1-D problem. (Modified from S. Mulay, H. Li, and S. See. (2009). *Computational Mechanics*, 44, 563–590.)

continuous analytical solution is evaluated by including up to second order monomial terms in the polynomial basis of a shape function. The governing equation and boundary conditions are given as

$$\frac{d^2 f}{dx^2} = \frac{105}{2} x^2 - \frac{15}{2}, \quad (-1 < x < 1) \tag{4.144}$$

$$f(x = -1) = 1, \quad \frac{df}{dx}(x = 1) = 10 \tag{4.145}$$

The analytical solution is given as $f(x) = (35/8)x^4 - (15/4)x^2 + (3/8)$. This problem is solved by 21, 41, 81, 161, 641 field and 641 virtual nodes, respectively. The uniform and randomly distributed field nodes

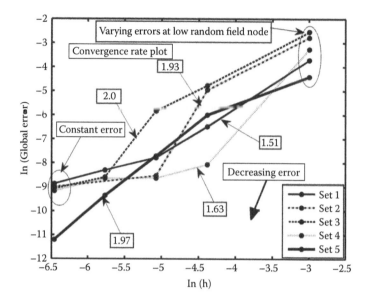

Figure 4.30 Convergence curves of the function by five different sets of the random field nodes for the 1ˢᵗ 1-D problem, where the global error varies for the low number of field nodes but converges to almost the constant value as the total random field nodes are increased. (Modified from S. Mulay, H. Li, and S. See. (2009). *Computational Mechanics*, 44, 563–590.)

are separately combined with the cosine and uniformly distributed virtual nodes. The convergence rates for both types of virtual nodes are given in Table 4.6 and Table 4.7 and corresponding convergence curves are plotted in Figure 4.31. Note from the tables that the convergence rate of the function by the uniform distribution of virtual nodes is improved, as compared with the cosine distribution of virtual nodes.

Table 4.6 Convergence rates for second 1-d problem of Poisson equation when cosine distributed virtual nodes are combined with uniform and randomly distributed field nodes

Function	Convergence rates (uniform field nodes)	Convergence rates (random field nodes)
f	1.7	1.8
$f_{,x}$	0.9	0.8
$f_{,xx}$	0.9	0.7

Source: Modified from S. Mulay, H. Li, and S. See. (2009). *Computational Mechanics*, 44, 563–590.

Table 4.7 Convergence rates for second 1-d problem of Poisson equation by combining uniform distribution of virtual nodes with uniform and random field nodes

Function	Convergence rates (uniform field nodes)	Convergence rates (random field nodes)
f	2.1	2.7
$f_{,x}$	1.3	0.8
$f_{,xx}$	1.4	0.7

Source: Modified from S. Mulay, H. Li, and S. See. (2009). Computational Mechanics, 44, 563–590.

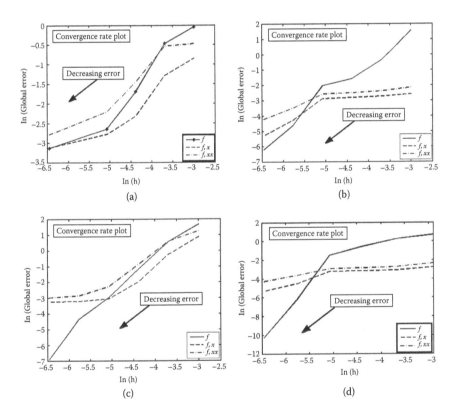

Figure 4.31 Convergence curves when uniform field nodes are combined with cosine virtual nodes (a), random field nodes are combined with cosine virtual nodes (b), uniform field nodes are combined with uniform virtual nodes (c), and random field nodes are combined with uniform virtual nodes (d) for second 1-D problem. (Modified from S. Mulay, H. Li, and S. See. (2009). Computational Mechanics, 44, 563–590.)

This problem is also solved in the earlier section of convergence analysis, but only for the cosine distribution of virtual nodes combined with the uniform and random distributions of field nodes. It is additionally solved here for the uniform virtual nodes combined with the uniform and random field nodes, and their convergence rates are compared.

The third 1-D example is a local high gradient problem with a force term containing an exponential expression. Exponential function can be approximated by an infinite series, such as the power series by Taylor expansion. It is solved here by including up to second order monomial terms in the polynomial basis of a shape function computation. The governing equation and boundary conditions are given as

$$\frac{d^2 f}{dx^2} = -6x - \left[\left(\frac{2}{\alpha^2} \right) - 4 \left(\frac{x-\beta}{\alpha^2} \right)^2 \right] \exp\left[-\left(\frac{x-\beta}{\alpha} \right)^2 \right], \qquad (0 < x < 1) \quad (4.146)$$

$$f(x=0) = \exp\left[-\left(\frac{\beta^2}{\alpha^2} \right) \right], \quad \frac{df(x=1)}{dx} = -3 - 2\left(\frac{1-\beta}{\alpha^2} \right) \exp\left[-\left(\frac{1-\beta}{\alpha} \right)^2 \right]$$

$$(4.147)$$

The analytical solution is given as $f(x) = -x^3 + \exp\{-[(x-\beta)/\alpha]^2\}$. This problem is solved by 21, 41, 81, 161, 321, 641 field and 641 virtual nodes, respectively. The uniform and randomly distributed field nodes are separately combined with the cosine and uniformly distributed virtual nodes. The convergence rates for both types of virtual nodes are given in Table 4.8 and Table 4.9 and corresponding convergence curves are plotted in Figure 4.32. We can again observed from the tables that very good convergence rates of the function and its derivative are achieved by combining the uniform distribution of virtual nodes with the random and uniform distributions of field nodes, respectively. Figure 4.33 shows a comparison of the numerical

Table 4.8 Convergence rates for third 1-D problem by uniform and random distributions of field notes combined with cosine distribution of virtual nodes

Function	Convergence rates (uniform field nodes)	Convergence rates (random field nodes)
f	3.3	2.5
$f_{,x}$	2.2	1.6
$f_{,xx}$	1.8	1.3

Source: Modified from S. Mulay, H. Li, and S. See. (2009). Computational Mechanics, 44, 563–590.

Table 4.9 Convergence rates for third 1-D problem by uniform and random distributions of field nodes combined with uniform distribution of virtual nodes

Function	Convergence rates (uniform field nodes)	Convergence rates (random field nodes)
f	3.0	3.3
$f_{,x}$	1.9	1.2
$f_{,xx}$	1.6	1.2

Source: Modified from S. Mulay, H. Li, and S. See. (2009). *Computational Mechanics*, 44, 563–590.

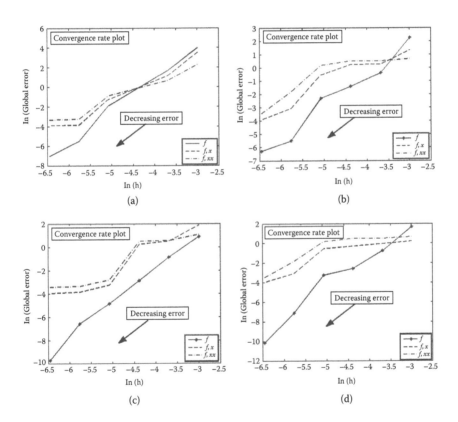

Figure 4.32 Convergence curves when uniform field nodes are combined with cosine virtual nodes (a), random field nodes are combined with cosine virtual nodes (b), uniform field nodes are combined with uniform virtual nodes (c), and random field nodes are combined with uniform virtual nodes (d) for third 1-D problem. (Modified from S. Mulay, H. Li, and S. See. (2009). *Computational Mechanics*, 44, 563–590.)

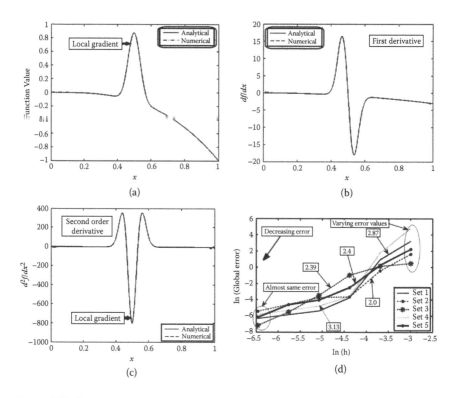

Figure 4.33 Comparison of analytical and numerical values of function f (a), the first derivative (b), and the second derivative (c) for third 1-D problem by 641 uniform field nodes combined with 641 cosine virtual nodes; plots of function value convergence by five sets of random field nodes combined with cosine distribution of virtual nodes (d) where the global error varies significantly for lower number of field nodes but their difference reduces with increase in number of random field nodes. (Modified from S. Mulay, H. Li, and S. See. (2009). *Computational Mechanics*, 44, 563–590.)

and analytical values of the function and its first and second order derivatives, respectively, by 641 uniform field and cosine distributed virtual nodes. The locally high gradient function value at $x = 0.5$ is captured very well, and the first and second order derivatives are also accurately captured.

This problem is additionally solved to study the effects of different sets of random nodes on the rate of convergence. The corresponding convergence plots of the function values are shown in Figure 4.33d by five sets of random field nodes. If the convergence rates of the function in the figure are compared with the rate 2.5 in Table 4.8 obtained by random nodes generated without giving a starting reference value as explained earlier, 2.5 is a good measure of the performance of RDQ method in a sense that it averagely represents the rates of convergence shown in Figure 4.33d. Also, the global error varies for the low number of field nodes, but their difference reduces with increases in the random field nodes.

It is seen from all the results that the weighted derivative approach to compute the approximate derivatives gives good rates of derivative convergence. Based on consistency analysis, the uniform distribution of virtual nodes yields smaller error in the discretisation of governing equations as compared with cosine distribution. This leads to better rates of convergence by the uniform distribution of virtual nodes. This observation is verified in this section by comparing the convergence results obtained by the uniform and cosine distributions of virtual nodes. For all the problems, much better convergence values of the function are achieved by the uniform distribution of virtual nodes than the cosine one. It is also observed from all the results that the higher convergence rates of the function values are achieved by the random distribution of field nodes. This is a very important result since it highlights one of the objectives of the development of RDQ method. All the convergence curves show that the global error by the random distributed field nodes reduces very fast with a sufficiently high number of random field nodes.

4.7.2 Two-dimensional test problems

The first 2-D problem is a Laplace equation with mixed boundary conditions. The governing equation and boundary conditions are given as

$$\frac{d^2 f(x,y)}{dx^2} + \frac{d^2 f(x,y)}{dy^2} = 0 , \qquad (0 < x < 1) \quad \text{and} \quad (0 < y < 1) \qquad (4.148)$$

$$f(x = 0, y) = -y^3, \quad f(x = 1, y) = -1 - y^3 + 3y^2 + 3y \qquad (4.149)$$

$$\frac{df}{dy}(x, y = 0) = 3x^2, \quad \frac{df}{dy}(x, y = 1) = -3 + 6x + 3x^2 \qquad (4.150)$$

The analytical solution is given as $f(x,y) = -x^3 - y^3 + 3xy^2 + 3x^2 y$. This problem is solved by including up to second order monomials in the shape function basis, and with 5×5, 17×17, 33×33, 41×41, and 44×44 field and 44×44 virtual nodes respectively. The uniform and randomly distributed field nodes are separately combined with the cosine and uniformly distributed virtual nodes. The convergence rates for both types of virtual nodes are given in Table 4.10 and Table 4.11, and the corresponding convergence curves are plotted in Figure 4.34.

Comparing Table 4.10 and Table 4.11 indicates that the convergence rate of the function by the uniform distribution of virtual nodes is better than that by the cosine distribution of virtual nodes, notably with random field nodes. The convergence rates of the derivatives are not much affected by the different distributions of the field and virtual nodes.

Table 4.10 Convergence rates for first 2-D problem by cosine distribution of virtual nodes combined with uniform and random distributions of field nodes

Function	Convergence rates (uniform field nodes)	Convergence rates (random field nodes)
f	2.23	3.75
$f_{,x}$	1.2	1.2
$f_{,y}$	1.2	1.2
$f_{,xx}$	0.8	0.8
$f_{,yy}$	0.8	0.8

Source: Modified from S. Mulay, H. Li, and S. See. (2009). *Computational Mechanics*, 44, 563–590.

This problem is also solved during the convergence analysis, but only by the cosine distribution of virtual nodes combined with the uniform and random distributions of field nodes. It is additionally solved here by the uniform distribution of virtual nodes separately combined with the uniform and random field nodes. Nevertheless, all the plots of the convergence are given here for an easy comparison.

The second 2-D problem concerns steady-state heat conduction without an internal heat generation in a slab with no temperature gradient in the z direction, as shown in Figure 4.35a. The governing equation and boundary conditions are given as

$$\frac{d^2T}{dx^2} + \frac{d^2T}{dy^2} = 0, \qquad (0 < x < a) \text{ and } (0 < y < b) \tag{4.151}$$

$$T(x = 0) = 0, \ T(x = a) = 0, \ T(y = 0) = 0, \ T(y = b) = 100 \tag{4.152}$$

Table 4.11 Convergence rates for first 2-D problem by combining uniformly distributed virtual nodes with uniformly and randomly distributed field nodes

Function	Convergence rates (uniform field nodes)	Convergence rates (random field nodes)
f	1.95	4.73
$f_{,x}$	1.2	1.2
$f_{,y}$	1.2	1.2
$f_{,xx}$	0.8	0.8
$f_{,yy}$	0.8	0.8

Source: Modified from S. Mulay, H. Li, and S. See. (2009). *Computational Mechanics*, 44, 563–590.

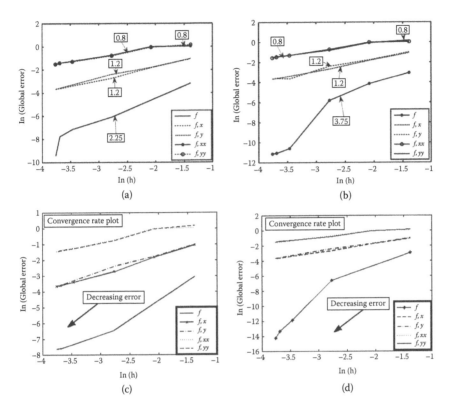

Figure 4.34 Convergence curves when uniform field nodes are combined with cosine virtual nodes (a), random field nodes are combined with cosine virtual nodes (b), uniform field nodes are combined with uniform virtual nodes (c), and random field nodes are combined with uniform virtual nodes (d) for the first 2-D problem. (From H. Li, S. Mulay, and S. See. (2009b). *Computer Modeling in Engineering and Sciences*, 48, 43–82. With permission; modified from S. Mulay, H. Li, and S. See. (2009). *Computational Mechanics*, 44, 563–590.)

The analytical solution derived using the Fourier series is given as $T(x,y) = T_s \sum_{n=1}^{\infty} [n\pi \sinh(n\pi b/a)]^{-1} 2[1-(-1)^n] \sin(n\pi x/a) \sinh(n\pi y/a)$. This problem is solved, by including up to second order monomials in the shape function basis and with 5×5, 9×9, 17×17, 21×21, 26×26, and 33×33 field and 44×44 virtual nodes, respectively. The uniform and randomly distributed field nodes are combined with the cosine and uniformly distributed virtual nodes. The convergence rates of both types of virtual nodes are given in Table 4.12 and Table 4.13 and corresponding convergence curves are plotted in Figure 4.35. The tables indicate that the convergence rate of function is improved by the uniform distribution of virtual nodes as compared with the cosine one. The comparison between the numerical and analytical values of temperature along

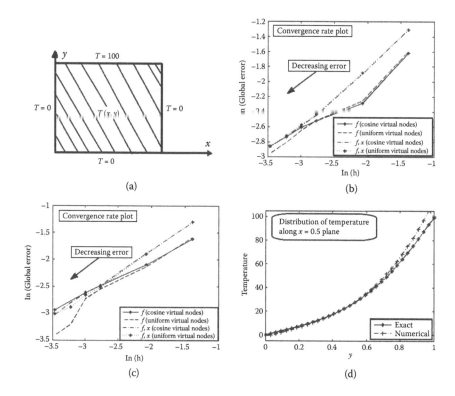

(a)   (b)

(c)   (d)

Figure 4.35 Computational domain and boundary conditions for steady-state heat conduction problem with heat source at $y = b$ (a), convergence curves for uniform distribution of field nodes combined with uniform and cosine distributions of virtual nodes (b), convergence curves for random distribution of field nodes combined with uniform and cosine distributions of virtual nodes (c), and comparison of numerical and analytical values of temperature along central x-axis plane (d). (Modified from S. Mulay, H. Li, and S. See. (2009). *Computational Mechanics*, 44, 563–590.)

Table 4.12 Convergence rates for second 2-d problem by cosine distribution of virtual nodes combined with uniform and random distributions of field nodes

Function	Convergence Rates (uniform field nodes)	Convergence Rates (random field nodes)
f	0.6	0.65
$f_{,x}$	0.8	0.9

Source: Modified from S. Mulay, H. Li, and S. See. (2009). *Computational Mechanics*, 44, 563–590.

Table 4.13 Convergence rates for second 2-d problem by combining uniform distribution of virtual nodes with uniform and random distributions of field nodes

Function	Convergence Rates (uniform field nodes)	Convergence Rates (random field nodes)
f	0.7	0.9
$f_{,x}$	0.8	0.9

Source: Modified from S. Mulay, H. Li, and S. See. (2009). Computational Mechanics, 44, 563–590.

the central x-axis plane is given in Figure 4.35d by the uniform distribution of 33 × 33 field and the cosine distribution of 44 × 44 virtual nodes, and it is noted that the boundary conditions are exactly imposed.

4.7.3 Elasticity problems

All the problems presented in this section are based on the plane stress condition with the conventional equilibrium equation expressed in the form of the displacements, as explained earlier.

4.7.3.1 Cantilever beam under pure bending load

A cantilever beam is loaded with a bending load, as shown in Figure 4.36a. This problem was solved earlier by combining the cosine virtual nodes with the uniform and random field nodes. Here it is solved again by combining the uniformly distributed virtual nodes with the uniform and randomly distributed field nodes for comparison.

This problem is solved for $L = 48$, $D = 12$, $M = -24000$, $v_0 = 0.3$, $E = 3.0 \times 10^7$, and 13 × 13, 17 × 17, 21 × 21, and 29 × 29 field and 41 × 41 virtual nodes respectively, by including up to first-order monomial terms in the shape function basis. The convergence rates in the displacements u and v by the cosine distribution of virtual nodes combined with the uniform and random distributions of field nodes, respectively, are 1.0, 1.3 and 1.96, 2.0, respectively, and the corresponding convergence curves are given in Figure 4.36b. Similarly, the convergence rates in the displacements u and v by combining the uniform distribution of virtual nodes with the uniform and random distributions of the field nodes, respectively, are 0.2, 0.2 and 0.4, 0.4, respectively, and the corresponding curves are plotted in Figure 4.36. The convergence rates by the cosine distribution of virtual nodes are better than that of the uniform distribution.

It is observed from Figure 4.36c that the value of global error in the displacement v decreases slowly after 21 ×2 1 field nodes, as compared

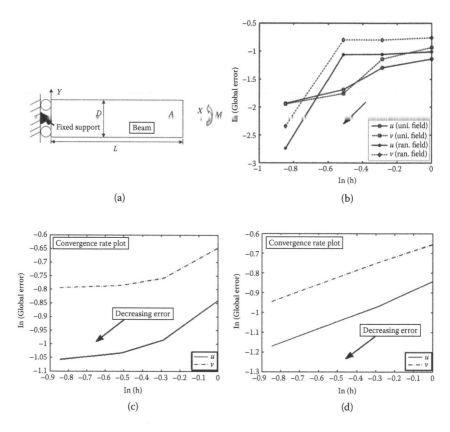

Figure 4.36 Cantilever beam under pure bending load (a), convergence plots obtained by combining cosine virtual nodes with uniform and random field nodes (b), and convergence plots obtained by combining uniform virtual nodes with uniform (c) and random (d) field nodes. (From H. Li, S. Mulay, and S. See. (2009b). *Computer Modeling in Engineering and Sciences*, 48, 43–82. With permission; modified from S. Mulay, H. Li, and S. See. (2009). *Computational Mechanics*, 44, 563–590.)

with that in the displacement u. With reference to the convergence curve of displacement u, the convergence curve of displacement v is shifted in the y direction because the absolute values of the global error in the displacement u are lower as compared with those in the displacement v for the equal number of field nodes. By comparing Figure 4.36b, c, and d, we note that the global error by the cosine virtual nodes is higher at lower field nodes, as seen in Figure 4.36b, as compared with that by the uniform virtual nodes, as seen in c and d. Therefore naturally the rate of decrease of error using the cosine virtual nodes is better as compared with the uniform virtual nodes.

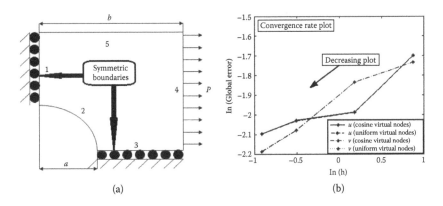

Figure 4.37 One-quarter computational domain of semi-infinite plate (a) and convergence plots of displacements *u* and *v* by uniform field nodes combined with cosine and uniformly distributed virtual nodes (b) for problem of semi-infinite plate with central hole. (Modified from S. Mulay, H. Li, and S. See. (2009). *Computational Mechanics*, 44, 563–590.)

4.7.3.2 Semi-infinite plate with central hole

A semi-infinite plate with a central hole along with the boundary conditions is shown in Figure 4.37a. This problem was solved earlier by combining the cosine virtual nodes with the uniform and random field nodes, respectively. Here it is solved again by combining the uniform distribution of virtual nodes with the uniform and random distributions of field nodes for comparison.

This problem is solved for $a = 1$, $b = 5$, $p = 1$, $v_0 = 0.3$, and $E = 1000$ with 6×6, 11×11, 21×21, and 31×31 uniform field nodes and 34×34 cosine and uniformly distributed virtual nodes, respectively, by including up to second order monomial terms in the polynomial basis of a function approximation. The rates of convergence in the displacements *u* and *v* by the uniform field nodes combined with the cosine and uniform virtual nodes, respectively, are 0.3, 0.3, and 0.3, 0.3 respectively, and the corresponding convergence plots are given in Figure 4.37b.

4.8 SUMMARY

Initially, the convergence analysis of RDQ method is performed by solving several 1-D and 2-D test problems. The superconvergence condition is derived at the beginning for the RDQ method. If the field nodes are computed by the superconvergence condition given in Equation (4.5), the function and its approximate derivatives converge at the rate of $O(h^{p+\alpha})$, where $\alpha \geq 1$ for the approximated function and $\alpha \approx 0.7$ to 1 for the approximated derivative. It is noted from the test problems that the RDQ method can work well for the domain discretised by either the uniform or random field nodes.

Next, the RDQ method is applied to solve several elasticity problems with the first, second, and third order distributions of the field variables in the x and y directions. When the complete order of monomials is added in the approximation of function, the numerical results closely match the corresponding analytical solution.

The RDQ method is applied to study the pull-in instability in the microswitches of the fixed–fixed and cantilever types under the influence of a nonlinear electrostatic force field. The simulation results show that the pull-in voltages obtained by the RDQ method closely match the literature values. A novel approach of transferring the multiple boundary conditions from the boundary nodes to their neighbouring nodes is used to get the modified boundary conditions. It is shown by the cantilever microswitch that the simulation results obtained by implementing the modified boundary conditions are more stable than those obtained by original boundary conditions.

Finally, it is shown that the RDQ method successfully handles uniformly and randomly distributed field nodes using the fixed RKPM interpolation function. It is seen from the convergence rates of the derivatives that the new weighted derivative approach gives very good convergence of the approximate derivatives. It is observed from Figure 4.33a through c that the first and second order derivatives are well captured even though the function has a high local gradient. To perform the consistency analysis, the terms of function values at the virtual nodes from the discretised governing equation are replaced by Taylor series expansion to get a reasonable judgment of an analytical error in the discretisation. As a result, with reference to the final consistency equations obtained by Taylor series, the same analysis can be applied to both the locally applied DQ and RDQ methods.

It is concluded from the consistency analysis that the DQ method is consistent for certain configurations of DQ local domains if the weighting coefficients are computed by Shu's general approach instead of the Quan and Chang approach. The consistency of the discretised governing equation by the locally applied DQ method is independent of the distribution pattern used for the virtual nodes. For example, even if the governing equation is discretised by the cosine distribution of virtual nodes, it converges to the consistency equation discretised by the uniform distribution of virtual nodes when the nodal spacing is changed from the cosine to a uniform one. This observation indicates that the error in the discretisation of governing equation by the uniform distribution of virtual nodes should be less in comparison with the cosine one. Extensive numerical analysis is performed by the RDQ method, taking into consideration the observations made during the consistency analysis, by solving several 1-D, 2-D, and elasticity problems with different orders of field variable distributions.

When the convergence rates obtained by the cosine and uniform distributions of virtual nodes are compared, good convergence rates of the function are achieved by combining the uniform distribution of virtual nodes with the random distribution of field nodes, whereas good convergence rates of the function derivatives are achieved by combining the uniform distribution of virtual nodes with the uniform distribution of field nodes. This is a very important result as it highlights the applicability of the RDQ method.

Overall, it is concluded that the RDQ method is capable of effectively handling uniform and randomly distributed field nodes coupled with the uniform or cosine distributed virtual nodes over the regular or irregular domains. The weighted derivative approach also ensures the convergence of derivatives.

Chapter 5

Stability analyses

5.1 INTRODUCTION

A numerical method must have a stable behaviour to simulate time-dependent problems. Thus, a rigorous stability analysis of a numerical method is essential before it is relied upon to solve time-dependent problems; in particular, it is vital for strong form- based methods. This aspect of a meshless method is elaborated in this chapter by studying the stability characteristics of meshless RDQ method. This is accomplished by analyzing the locations of zeros or roots of its characteristic polynomials with respect to the unit circle in the complex plane, by discretizing the domain with either uniform or random field nodes.

This is illustrated by performing a stability analysis of the RDQ method by the first-order wave, transient heat conduction, and transverse beam deflection equations via both the analytical and numerical approaches. The stability analysis of the locally applied DQ and RDQ methods is performed by different single and multistep schemes using Von Neumann (VN) and Schur polynomials (Strikwerda, 1989).

The stable schemes are identified, and their consistency analysis is performed to obtain the additional constraints on the temporal spacing. The analytical results obtained by the stability and consistency analyses of the stable schemes are effectively verified by numerically implementing the RDQ method to solve the first-order wave, transient heat conduction, and transverse beam deflection equations with the domain discretised by either uniform or random field nodes. As a result, it is shown that the time-dependent problems can be solved successfully by the RDQ method via systematic stability analysis through the Neumann and Schur polynomials (Strikwerda, 1989), compared with the FEM and other meshless methods.

The stability and consistency analyses of the numerical schemes have been major topics of interest among the applied mathematics research community for a long time. A lot of studies have been performed in this area for schemes such as the FDM and FEM and well developed theories are available. The DQ and SPH methods are relatively new techniques as compared with these schemes. In general, stability analysis has two foci. The first is to study the ill-conditioning of the stiffness matrix, and the second is to study the propagation or magnification of numerical errors over time when different schemes of the discretisations in time and space are implemented. This chapter is involved in the second focus of the stability analysis.

In order to ensure the applicability of a numerical method in solving time-dependent problems, it becomes vital to know behaviours for different schemes of time and space discretisations. A computing error may be introduced in the numerical solution during an initial time increment for a specific scheme of temporal discretisation. Sometimes it may not be possible to completely eliminate the error, but it may be possible to restrict its growth or magnification as time passes. Therefore, the study of the numerical error propagation or magnification is a very important aspect of the strong form-based meshless method.

Since the RDQ method is a strong form meshless approach, the stability analysis is pursued in this chapter. As the first-order wave, transverse beam deflection and heat conduction equations are typical examples of hyperbolic and parabolic partial differential equations (PDEs), they are chosen for the present stability analysis.

The SPH method was proposed by Lucy (1977) and Gingold and Monaghan (1977). The function is approximated by its integral form via the window function. Balsara (1995) performed the VN stability analysis of the SPH method, and suggested the optimal ranges of the parameters used in the SPH. He performed the stability analysis by applying a constant velocity to an unperturbed state, verified the results using different kernels, and drew conclusions based on the extensive numerical analyses.

Børve et al. (2004) applied the SPH method to solve equations of magnetohydrodynamics, and performed their stability analysis by perturbation with small amplitudes. Swegle et al. (1995) obtained the stability criterion of SPH in terms of the stress state because it was observed that when the SPH method is used with the cubic B-spline kernel, it becomes unstable in the tensile region but remains stable in the compressive region. They explained that the instability is caused by a negative effective modulus resulting from the kernel function interaction with the constitutive equations. In other words, the instability is caused due to the effective stress, which amplifies rather than reduces the applied strain.

Sigalotti and López (2008) proposed an alternate method to reduce tensile instability by an adaptive density kernel estimation algorithm. As a result, the amount of smoothing applied to the data is controlled and the final smoothing length of the kernel is changed from point to point. Di Lisio et al. (1998) attempted the convergence of SPH method for a generic polytrophic fluid. A unified stability analysis was given by Belytschko et al. (2000) while addressing three instabilities (rank deficiency and tensile and material instability). They applied the perturbation method to momentum equations and studied Eigenvalues of the resulting matrix.

Fung (2002) used the DQ method to solve the second order ordinary differential equation in time. He studied the stability and accuracy properties of the DQ method by different spacings of the grid points. Tomasiello (2003) performed stability and accuracy analyses of the iterative-based DQ approach called the iterative differential quadrature method. Aceto and Trigiante (2007) reviewed the stability properties of linear multistep methods. Ata and Soulaïmani (2005) used the SPH method to solve shallow water equations and improved the dynamic stability of the method by introducing the Lax-Friedrich scheme. Different approaches were discussed in the past to impose the Dirichlet boundary conditions as the SPH method does not possess the property of delta function.

When a PDE contain terms of time and space derivatives, its stable and consistent (without oscillations or dispersion in the solution) discretisation depends on how accurately temporal marching is performed. The criterion of temporal marching can be determined by performing the stability analysis of a numerical method by the specific PDE. Therefore, it is crucial for the numerical method to appropriately carry out the stability analysis of a time-dependent problem. It is shown in this chapter that the RDQ method can be used very well for stability analysis, and provides stable results with the domain discretised by either the uniform or random field nodes. This capability of the RDQ method makes it suitable to solve the complex problems involving phenomena such as moving boundaries and hydrogel swelling or de-swelling, in which the solution is highly sensitive to the chosen increment in time.

The motivation behind the presented work is to comprehensively study the stability characteristics of the RDQ method by analytical and numerical approaches with the domains discretised by uniform or random field nodes. Most of the research on the stability analysis performed in the past is based on either Eigenvalue analysis or the perturbation theory, or the numerical results based on uniform spacing of the field nodes.

In the present novel approach, the stability analysis of the locally applied DQ and RDQ methods is demonstrated for several single and multistep schemes of temporal and spatial discretisation based on the VN and Schur polynomials (Miller, 1971; Strikwerda, 1989). The stable schemes are identified for the first-order wave equation and their consistency analysis is performed to identify the additional constraints on the temporal spacing.

The analytical results of both analyses are numerically verified by implementing the stable schemes via the RDQ method with the domains discretised by uniform or random field nodes. The stability analysis is further extended for the equations of transient heat conduction and transverse beam deflection. An analytical equation is developed for the problem of transverse beam deflection to compute the reduction in the successive amplitudes of beam deflection for both the implicit and explicit approaches. Also, it is shown by the RDQ method that the numerical reduction in the values of beam amplitude closely matches the analytically determined values.

For transverse beam deflection, it is also demonstrated by the RDQ method that an explicit approach involves the dispersion effect. Thus $(v_2/v_1) > 1.0$ and $(v_2/v_1) \to 1.0$ as $k \to 0$, and an implicit approach involves the dissipation effect. Therefore $(v_2/v_1) < 1.0$ and $(v_2/v_1) \to 1.0$ as $k \to 0$, where k is the temporal spacing, v_1 and v_2 are the reference amplitudes of beam deflection at the field nodes with respect to the time t_1 and $[t_1 + (2\pi / \omega_d)]$, respectively, and ω_d is the damped natural frequency of the system.

Therefore, the broad motivation of this chapter is to study analytically and numerically the propagation or magnification of numerical errors over time increments when different schemes of the space and time discretisations are implemented by the RDQ method.

The objective of this chapter is to study the analytical and numerical stability characteristics of the locally applied DQ and RDQ methods in accordance with the VN and Schur polynomials. Moreover, the governing equations of the first-order wave, transient heat conduction and the transverse beam deflection are solved with their stability criteria as an application of the stability analysis by the RDQ method coupled with the domains discretised by uniform or random field nodes.

It may not be always possible to explicitly evaluate the stability criteria for the complex PDE. This limitation is overcome here by plotting the roots ϕ of the characteristic polynomial as $|\phi|^2$ versus (θ/π), and identifying the stability criterion from $|\phi|^2$, namely the scheme is stable if $|\phi|^2 \leq 1$ and unstable if $|\phi|^2 > 1$. Therefore, a broad framework is developed in this chapter that facilitates the solution of transient PDE by the RDQ method via detailed stability analysis.

We show in this chapter that time-dependent problems can be very well solved by the RDQ method via stability analysis, as compared with the other existing meshless methods and the FEM, as no mesh is involved. As a result, the RDQ method can be effectively used for time-dependent problems (linear and nonlinear) such as hydrogel swelling and crack propagation.

The outline of the subsequent subsections of this chapter is as follows. The VN and Schur polynomials are explained in Section 5.2 where detailed stability analysis of the first-order wave equation is performed by the locally applied DQ and RDQ methods. The stable schemes are identified and their elaborate consistency analysis is performed to obtain the additional constraints on the temporal spacing. In Section 5.3, the stability analysis is extended to solve the transient heat conduction equation by the RDQ method with the spatial domain discretised by either uniform or random field nodes. The transverse beam deflection equation is solved by the RDQ method in Section 5.4 with the implicit and explicit approaches, and the numerical reduction of amplitude in the beam deflection is compared with the corresponding analytical values. Finally, a summary is given in Section 5.5.

5.2 STABILITY ANALYSIS OF FIRST-ORDER WAVE EQUATION BY RDQ METHOD

In this section, the stability analysis of the first-order wave equation $\phi_{,t} = a\phi_{,x}$ is performed by the RDQ method with several single step and multistep temporal discretisation schemes. The single step discretisation schemes are studied by the VN stability analysis (Strikwerda, 1989), and the multistep discretisation schemes by Schur and Neumann polynomials (Miller, 1971; Strikwerda, 1989).

The VN and Schur polynomials appear frequently in studies of PDEs. Let us consider a general polynomial represented as $\phi(z) = \sum_{l=0}^{d} a_l z^l$, where d is the order of the polynomial and a_1 are the coefficients. The polynomial $\phi(z)$ is called the Schur polynomial if all its roots are located within the circle of unit radius, namely $|\phi_i| < 1.0$, where $i = 1, 2,..., d$. The polynomial $\phi(z)$ is called the VN polynomial if some of the roots are located within the circle of unit radius and the remaining are on the circle but none is outside of it, namely $|\phi_i| \leq 1.0$, where $i = 1, 2,..., d$.

In the Neumann analysis, the amplification factor $g(h\xi)$ of the discretised wave equation is derived by taking the inverse Fourier transform and simplifying it to get $\hat{v}_n(\xi) = g(h\xi)\hat{v}_0(\xi)$, where h is the spatial spacing, $\xi \subset [-\pi, \pi]$ is the Fourier domain, and $\hat{v}_n$ and $\hat{v}_0$ are the values of function at the nth and 0th levels of time, respectively (Strikwerda, 1989). The amplification factor $g(h\xi)$ is further analyzed for the stability criterion. In Schur and Neumann polynomial analyses, the amplification or characteristic polynomial is derived by discretizing the wave equation, and the roots of this polynomial are examined to classify the polynomial as either Schur $P(n, 0, 0)$ or VN $P(k, n-k, 0)$, (Miller, 1971) where n represents total roots of polynomial of which k roots lie within the circle of unit radius on the complex plane.

5.2.1 Stability analysis of first-order wave equation by different schemes for discretisation of domains

Different schemes of the domain discretisation are studied in this section for the stability of the first-order wave equation. The characteristic polynomial is developed for all the discretisation schemes at first, and its roots are then analyzed to identify the stable discretisation schemes. The constraints or conditions obtained from the roots of the polynomial of stable schemes are thus the stability criteria of the wave equation.

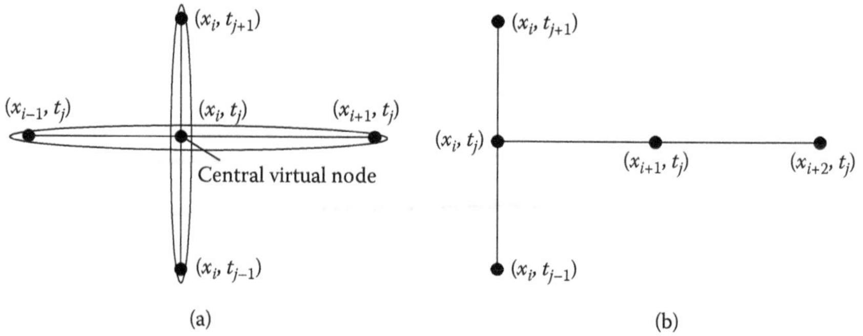

Figure 5.1 Local DQ domain around a central virtual node inside computational domain (a) and central time and forward space scheme of discretisation of time and space domains (b) for first -order wave equation. (From H. Li, S. Mulay, and S. See. (2009c). *Computer Modeling in Engineering and Sciences*, 54, 147–199. With permission.)

5.2.1.1 Central time and central space multistep scheme

In this scheme, three virtual nodes are considered over time and space domains, respectively, as shown in Figure 5.1a. The first-order wave equation $\phi_{,t} = a\phi_{,x}$ in the discretised form is given as

$$\{a_{j,j-1} \quad a_{j,j} \quad a_{j,j+1}\} \left\{ \begin{array}{c} \phi(x_i,t_{j-1}) \\ \phi(x_i,t_j) \\ \phi(x_i,t_{j+1}) \end{array} \right\} = a\ \{a_{i,i-1} \quad a_{i,i} \quad a_{i,i+1}\} \left\{ \begin{array}{c} \phi(x_{i-1},t_j) \\ \phi(x_i,t_j) \\ \phi(x_{i+1},t_j) \end{array} \right\}$$

(5.1)

The weighting coefficients in Equation (5.1) are computed by the locally applied DQ method, as explained in Chapter 3. The inverse Fourier transform is employed in the resulting equation to get the characteristic polynomial as (see Appendix A)

$$\phi(z) = z^2 - [2\ r\ i\ \sin(\theta)]\ z - 1$$

(5.2)

where $r = (at/h)$ is Courant number. Comparing Equation (5.2) with the generalized polynomial in Equation (5.3), the coefficients of Equation (5.2) are given in Equation (5.4) as

$$\phi(z) = \sum_{l=0}^{d} a_l z^l$$

(5.3)

$$a_0 = -1, \quad a_1 = 2\ r\ i\sin(\theta), \quad \text{and} \quad a_2 = 1$$

(5.4)

where d is the order of polynomial. The complex conjugate polynomial of Equation (5.2) is given in Equation (5.6) by Equation (5.5) as

$$\phi^*(z) = \sum_{l=0}^{d} \bar{a}_{d-l}\, z^l \tag{5.5}$$

$$\phi^*(z) = 1 + (2\, r\, i\, \sin\theta)\, z - z^2 \tag{5.6}$$

It is observed from Equations (5.2) and (5.6) that

$$|\phi^*(0)| = 1 \quad \text{and} \quad |\phi(0)| = 1, \text{ so } \phi_1(z) = \frac{\phi_0^*(0)\, \phi(z) - \phi(0)\, \phi_0^*(z)}{z} = 0 \tag{5.7}$$

where $\phi_1(z)$ is a reduced polynomial (see Appendix B). As $\phi_1(z) = 0$, $\phi(z)$ is a self-inversive polynomial (Miller, 1971). As per the theorem of Miller, a self-inversive polynomial $\phi(z)$ is the VN polynomial if $\phi^{(1)}(z)$ is the VN polynomial, where $\phi^{(1)}(z)$ is the first-order derivative of $\phi(z)$. Therefore, the first-order derivative of Equation (5.2) is written as

$$g(z) = \phi^{(1)}(z) = 2\, z - 2\, r\, i\, \sin(\theta) \tag{5.8}$$

The complex conjugate polynomial of Equation (5.8) is given as

$$g^*(z) = 2 + 2\, r\, i\, z\sin(\theta) \tag{5.9}$$

It is seen from Equations (5.8) and (5.9) that $|g^*(0)| = 2$ and $|g(0)| = 2\, r\, i\, \sin(\theta)$. The condition $|g^*(0)|^2 - |g(0)|^2 \geq 0$ must hold true for Equation (5.8) to be the VN polynomial, which results in

$$|r| \leq 1 \tag{5.10}$$

This shows that Equation (5.10) must be satisfied for Equation (5.8) to be the VN polynomial. Therefore, if Equation (5.10) is satisfied, $g(z)$ becomes the VN polynomial and consequently Equation (5.2) becomes the VN polynomial. As a result, it is concluded that Equation (5.10) is the stability condition for this scheme. As this scheme is multistep in the time domain, it needs another scheme as an initialization, such that the forward time (two nodes in the time domain) and central space (three nodes in the space domain) scheme can be used as an initialization scheme to begin the computation (Strikwerda, 1989). The initialization scheme does not affect the stability of multistep schemes as the amplification factor is controlled by the stability criterion. Even if instability is introduced in the numerical solution due to initialization, it does not propagate with time.

5.2.1.2 Central time and forward space multistep scheme

In this scheme, three virtual nodes are considered in the time and space domains, respectively, as shown in Figure 5.1b. The first-order wave equation in discretised form is given as

$$\phi_{,t} = a\phi_{,x} \Rightarrow$$

$$\{a_{j,j-1} \quad a_{j,j} \quad a_{j,j+1}\} \begin{Bmatrix} \phi(x_i,t_{j-1}) \\ \phi(x_i,t_j) \\ \phi(x_i,t_{j+1}) \end{Bmatrix} = a \{a_{i,i} \quad a_{i,i+1} \quad a_{i,i+2}\} \begin{Bmatrix} \phi(x_i,t_j) \\ \phi(x_{i+1},t_j) \\ \phi(x_{i+2},t_j) \end{Bmatrix}$$

$$(5.11)$$

The weighting coefficients are computed by the locally applied DQ method, and the characteristic polynomial is derived by applying the inverse Fourier transform (Miller, 1971; Appendix A) to obtain

$$\phi(z) = z^2 + (\alpha + i\beta) z - 1 \tag{5.12}$$

where $\gamma = \cos(\theta)-2$, $\alpha = 2r[1+\cos(\theta)\gamma]$, $\beta = 2r\sin(\theta)\gamma$. Comparing Equation (5.12) with Equation (5.3), the coefficients of Equation (5.12) are given as

$$a_0 = -1, \quad a_1 = \alpha + i\beta, \quad \text{and} \quad a_2 = 1 \tag{5.13}$$

The complex conjugate polynomial of Equation (5.12) is given as

$$\phi^*(z) = 1 + (\alpha - i\beta) z - z^2 \tag{5.14}$$

It is observed from Equations (5.12) and (5.14)

$$|\phi^*(0)| = 1 \quad \text{and} \quad |\phi(0)| = 1, \quad \therefore \phi_1(z) = 0 \tag{5.15}$$

where $\phi_1(z)$ is a reduced polynomial (Appendix B). $\phi(z)$ is a self-inversive polynomial, as $\phi_1(z) = 0$. Therefore, Equation (5.12) is the VN polynomial only if $\phi^{(1)}(z)$ is the VN polynomial (Miller, 1971), where $\phi^{(1)}(z)$ is the first-order derivative of $\phi(z)$ given as

$$g(z) = \phi^{(1)}(z) = 2\ z + (\alpha + i\beta) \tag{5.16}$$

$$g^*(z) = 2 + (\alpha - i\beta)\ z \tag{5.17}$$

Note from Equations (5.16) and (5.17) that $|g^*(0)| = 2$ and $|g(0)| = \alpha + i\beta$. The condition $|g^*(0)|^2 - |g(0)|^2 \geq 0$ must hold true for Equation (5.17) to be the VN polynomial; therefore this results in

$$r^2[1 + \gamma^2 + 2\gamma\cos(\theta)] \leq 1 \tag{5.18}$$

where $|1 + \gamma^2 + 2\gamma\cos(\theta)| \leq 1$ for $|r| \leq 1$. The maximum value of $[1 + \gamma^2 + 2\gamma\cos(\theta)]$ is at $\cos(\theta) = (4/3)$, which gives

$$|r| \leq 3i \tag{5.19}$$

Therefore, Equation (5.19) is the stability condition for this scheme and a complex number; however, it is impossible to have a complex number as the stability criterion. As a result, $g(z)$ is not the VN polynomial and consequently $\phi(z)$ is also not the VN polynomial. Therefore, it is concluded that this discretisation scheme is unstable.

Let us consider ϕ_+ and ϕ_- as the roots of Equation (5.12). A graph of $|\phi_+|^2$ versus (θ/π) is plotted and ϕ_+ is found to be unconditionally stable, namely $|\phi|^2 \leq 1$ even for $r > 1$, and a graph of $|\phi_-|^2$ versus (θ/π) is plotted and ϕ_- is found to be unstable, namely $|\phi|^2 > 1$ even for $r < 1$, as shown in Figure 5.2a and b, respectively, by considering $0 \leq r \leq 2$ and $0 \leq r \leq 1$, respectively.

5.2.1.3 Forward time and forward space multistep scheme

In this scheme, three virtual nodes are considered in the time and space domains, respectively, as shown in Figure 5.3a. The first-order wave equation $\phi_{,t} = a\phi_{,x}$ in the discretised form is given as

$$\{a_{j,j}\ \ a_{j,j+1}\ \ a_{j,j+2}\} \begin{Bmatrix} \phi(x_i, t_j) \\ \phi(x_i, t_{j+1}) \\ \phi(x_i, t_{j+2}) \end{Bmatrix} = a\ \{a_{i,i}\ \ a_{i,i+1}\ \ a_{i,i+2}\} \begin{Bmatrix} \phi(x_i, t_j) \\ \phi(x_{i+1}, t_j) \\ \phi(x_{i+2}, t_j) \end{Bmatrix}$$

$$\tag{5.20}$$

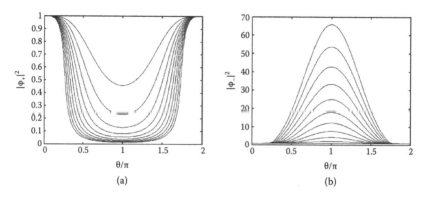

Figure 5.2 $|\phi_+|^2$ versus (θ/π) plot with r from 0 to 2 (a) and $|\phi_-|^2$ versus (θ/π) plot with r from 0 to 1 (b) by numerically implementing the second stability scheme. (From H. Li, S. Mulay, and S. See. (2009). *Computer Modeling in Engineering and Sciences*, 54, 147–199. With permission.)

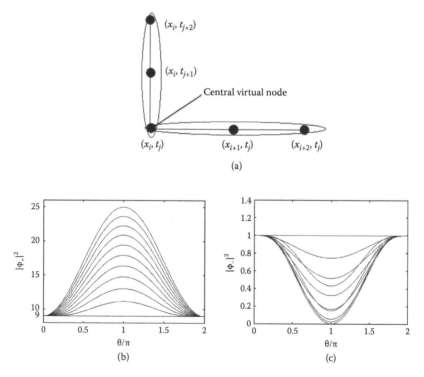

Figure 5.3 Multi-step discretisation scheme of forward time and space domains (a), plot of $|\phi_+|^2$ versus (θ/π) (b) and $|\phi_-|^2$ versus (θ/π) (c) by numerically implementing this scheme with r from 0 to 1 for first-order wave equation. (From H. Li, S. Mulay, and S. See. (2009). *Computer Modeling in Engineering and Sciences*, 54, 147–199. With permission.)

The weighting coefficients from Equation (5.20) are computed by the locally applied DQ method, and the characteristic polynomial (Appendix A) is given as

$$\phi(z) = z^2 - 4 z + 3 - (\alpha + i\beta) \tag{5.21}$$

where $\gamma = \cos(\theta) - 2$, $\alpha = 2 r [1 + \cos(\theta) \gamma]$, and $\beta = 2 r \sin(\theta) \gamma$. The complex conjugate polynomial of Equation (5.21) is given as

$$\phi^*(z) = 1 - 4 z + (3 - \alpha + i\beta) z^2 \tag{5.22}$$

The conditions $|\phi^*(0)| = 1$ and $|\phi(0)| = 3 - \alpha - i\beta$ are obtained from Equations (5.21) and (5.22). For Equation (5.21) to be the VN polynomial, $[|\phi^*(0)|^2 - |\phi(0)|^2] \geq 0$ results in (Miller, 1971)

$$1 - (3 - \alpha)^2 - \beta^2 \geq 0 \tag{5.23}$$

It is difficult to analytically solve Equation (5.23). Therefore the roots of $\phi(z)$ are studied by plotting the graph of $|\phi|^2$ versus (θ/π) by incrementing r from 0 to 1 and angle θ from 0 to 2π. The graph of $|\phi_+|^2$ versus (θ/π) shows that, $|\phi_+|^2$ has the lowest magnitude of 9, as shown in Figure 5.3b. The graph of $|\phi_-|^2$ versus (θ/π) shows that $|\phi_-|^2$ is stable for $|r| \leq 1$, as shown in Figure 5.3c. As a result, this scheme is unstable due to an instability in $|\phi_+|$, as $|\phi_+|^2 \geq 9$.

5.2.1.4 Forward time and forward space single step scheme

In this scheme, two and three virtual nodes are used in the time and space domains, respectively, as shown in Figure 5.4a. The first-order wave equation in the discretised form is given as

$$\phi_{,t} = a\phi_{,x} \Rightarrow$$

$$\{a_{j,j} \quad a_{j,j+1}\} \begin{Bmatrix} \phi(x_i, t_j) \\ \phi(x_i, t_{j+1}) \end{Bmatrix} = a \{a_{i,i} \quad a_{i,i+1} \quad a_{i,i+2}\} \begin{Bmatrix} \phi(x_i, t_j) \\ \phi(x_{i+1}, t_j) \\ \phi(x_{i+2}, t_j) \end{Bmatrix} \tag{5.24}$$

The characteristic polynomial of Equation (5.24), by computing the locally applied DQ weighting coefficients (Appendix A) is given as

$$\phi(z) = z - (1 + \alpha + i\beta) \tag{5.25}$$

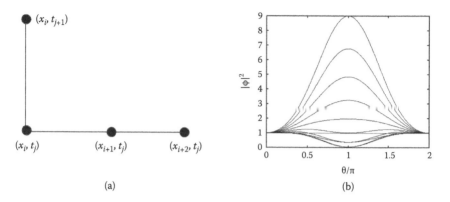

(x_i, t_{j+1})

(x_i, t_j) (x_{i+1}, t_j) (x_{i+2}, t_j)

(a)

$|\phi|^2$

θ/π

(b)

Figure 5.4 Single step scheme of forward time and forward space domains (a), and the plot of $|\phi|^2$ versus (θ/π) (b) with *r* from 0 to 1, in which the root is inside the unit circle when $|r| \leq 0.2$ for the discretisation of the first-order wave equation (From H. Li, S. Mulay, and S. See. (2009). *Computer Modeling in Engineering and Sciences*, 54, 147–199. With permission.)

where $\gamma = 2 - \cos(\theta)$, $\alpha = r\,[\cos(\theta)\,\gamma - 1]$ and $\beta = r\,sin(\theta)\,\gamma$. The complex conjugate polynomial of Equation (5.25) is given as

$$\phi^*(z) = -\,(1 + \alpha - i\beta)\,z + 1 \tag{5.26}$$

Note from Equations (5.25) and (5.26) that $|\phi^*(0)| = 1$ and $|\phi(0)| = -(1 + \alpha + i\beta)$. The condition $[\,|\phi^*(0)|^2 - |\phi(0)|^2\,] \geq 0$ must hold true for Equation (5.25) to be the VN polynomial, which results in

$$1 - [(1 + \alpha)^2 + \beta^2] \geq 0 \tag{5.27}$$

Simplifying Equation (5.27) results in

$$|r| \leq \frac{2}{7} \quad \text{and} \quad |r| > 0 \tag{5.28}$$

Equation (5.28) is the stability condition for this discretisation scheme. The plot of $|\phi|^2$ versus (θ/π) by incrementing *r* from 0 to 1 and angle θ from 0 to 2π also shows that the scheme is stable for $|r| \leq 0.2$ which is very close to Equation (5.28), as shown in Figure 5.4b. This scheme is nothing but an upwind technique, as the wave $\phi_{,t} = a\phi_{,x}$ is traveling toward the left side of the computational domain.

5.2.1.5 Forward time and central space single step scheme

In this scheme, two and three virtual nodes are used in the time and space domains, respectively, as shown in Figure 5.5a. The first-order wave equation in the discretised form is given as

$$\phi_{,t} = a\phi_{,x} \Rightarrow \{a_{j,j} \quad a_{j,j+1}\} \begin{Bmatrix} \phi(x_i,t_j) \\ \phi(x_i,t_{j+1}) \end{Bmatrix} = a \{a_{i,i-1} \quad a_{i,i} \quad a_{i,i+1}\} \begin{Bmatrix} \phi(x_{i-1},t_j) \\ \phi(x_i,t_j) \\ \phi(x_{i+1},t_j) \end{Bmatrix}$$

$$(5.29)$$

The amplification factor of Equation (5.29) after computing the locally applied DQ weighting coefficients is given as

$$g^{n+1} = \left\{ \frac{r}{2}[2\,i\,\sin(\theta)] + 1 \right\} g^n \tag{5.30}$$

$$\therefore\ g(\theta) = 1 + i\,r\sin(\theta) \tag{5.31}$$

Equation (5.31) is found unstable when it is plotted in the complex plane, as shown in Figure 5.5b. If $\phi^m = 0.5(\phi^n_{m-1} + \phi^n_{m+1})$ is substituted in Equation (5.29), similar to Lax-Friedrich scheme in the FDM (Strikwerda, 1989), it is rewritten as

$$\phi^{n+1}_m = \frac{r}{2}\left[\phi^n_{m+1} - \phi^n_{m-1}\right] + \left[\frac{\phi^n_{m-1} + \phi^n_{m+1}}{2}\right] \tag{5.32}$$

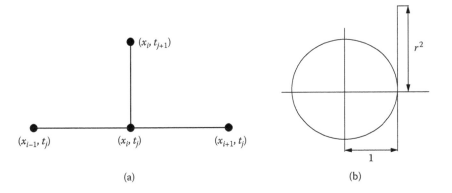

(a)

(b)

Figure 5.5 Single step scheme of forward time and central space domains (a) and plot of amplification factor in the complex plane with a circle having a unit radius (b), where the scheme is found unstable for the discretisation of first-order wave equation. (From H. Li, S. Mulay, and S. See. (2009c). *Computer Modeling in Engineering and Sciences*, 54, 147–199. With permission.)

The new amplification factor from Equation (5.32) is given as

$$g(\theta) = \cos\theta + i\, r\sin\theta, \;\therefore\, |g(\theta)| \Rightarrow |r| \le 1 \tag{5.33}$$

It is concluded from Equation (5.33) that the scheme is stable for $|r| \le 1$.

5.2.2 Consistency analysis of stable schemes and verification by numerically implementing first-order wave equation by locally applied DQ method

In order to obtain the stable numerical discretisation of the first-order wave equation, it is essential to ensure that the constraints on Courant number r obtained from stability analysis do not contradict results of consistency analysis. To fulfill this requirement, it is essential to perform the consistency analysis of the stable schemes cited in the previous section. As a result, the final condition of constraint is obtained by combining the results obtained from the stability and consistency analyses. The stable schemes are then numerically implemented to solve the first-order wave equation $\phi_{,t} = a\phi_{,x}$ with $\phi(x, 0) = \sin(2\pi x)$ and $\phi(0, t) = \phi(1, t) = 0$ when $x \in [0, 1]$ as the initial and periodic boundary conditions, respectively.

Let us consider the PDE in continuous and discretised forms as $P\,u = f$ and $P(h_x, h_y)\,u = f$, respectively, where $h_x = L_x/(NP_x - 1)$ and $h_y = L_y/(NP_y - 1)$ are the nodal spacings, L_x and L_y are the domain lengths, and NP_x and NP_y are the numbers of total field nodes in the x and y directions, respectively. The numerical discretisation scheme is considered consistent if the discretised form of the PDE closely approximates the continuous form as the nodal spacings h_x and $h_y \to 0$, namely $P\,u - P\,(h_x, h_y)\,u \to 0$ as $h_x, h_y \to 0$, such that there is no discretisation error involved in the numerical simulation and the continuous PDE is actually solved.

The discretised form of the first-order wave equation by the central time and space scheme (Section 5.2.1.1) given in Equation (5.1). The terms of function values in the equation are substituted by Taylor series expansion (Mulay et al., 2009) and written (Appendix C) as

$$\phi_{,t} - a\,\phi_{,x} = \phi_{,xxx}\left[\frac{a\,h^2}{6} - \frac{t^2\,a^3}{6}\right] \tag{5.34}$$

This scheme is consistent as $\phi_{,t} - a\,\phi_{,x} \to 0$ when h and $t \to 0$. Equation (5.34) is the exact or actual wave equation that is numerically solved. The term $\phi_{,xxx}$ in Equation (5.34) represents the numerical dispersion with respect to the term of space domain $\phi_{,x}$ from the wave equation. Therefore it is desirable to have a negative or zero value for $\phi_{,xxx}$ to reduce the oscillations in the solution as

$$\left[\frac{a\,h^2}{6} - \frac{t^2\,a^3}{6} \right] \leq 0 \tag{5.35}$$

Simplifying Equation (5.35) results in the consistency condition as

$$|r| \geq 1 \tag{5.36}$$

where $r = (a\,t/h)$. It is concluded from Equation (5.36) that the consistency condition matches the stability condition as given in Equation (5.10) for $|r| = 1$. This scheme is numerically implemented to solve the first-order wave equation. The wave is preserved during the total time of simulation for $r = 1$ and no instability is observed, as shown in Figure 5.6a. The initial wave is also preserved throughout the time of simulation for $r = 0.5$ and no dissipation or the damping is observed, as shown in Figure 5.6b. This is expected as no dissipation term is present in Equation (5.34). However, the instability is observed for $r = 2$ due to the violation of stability condition from Equation (5.10), as shown in Figure 5.6c.

The discretised form of the wave equation by the forward time and forward space scheme (Section 5.2.1.4) is given in Equation (5.24). The terms of the function values are substituted by Taylor series expansion (Mulay et al., 2009; Appendix C) and written as

$$\phi_{,t} - a\,\phi_{,x} = -\phi_{,xx} \left[\frac{a\,h\,r}{2} \right] \tag{5.37}$$

This scheme is consistent as $\phi_{,t} - a\,\phi_{,x} \to 0$ when h and $t \to 0$. Note from Equation (5.37) the negative dissipation effect (removal of dissipation from the system) for the positive wave speed a, as the term $\phi_{,xx}$ represents the dissipation or the damping corresponding to the term $\phi_{,x}$ from the wave equation. This is an interesting revelation. The dissipation may not be completely avoided irrespective of the values of h and r. In other words, even if this scheme is stable from the aspects of stability, there will always be some negative dissipation effect due to the +ve wave speed a from the view of consistency that will result in oscillations.

There will always be a positive dissipation effect due to the −ve wave speed a that will result in the damping of the wave. This observation is verified by numerically implementing this scheme to solve the first-order wave equation. The results are stable for $r = 0.2$ as the stability condition in Equation (5.28) is satisfied, but slight instability is added due to the negative dissipation effect, as given in Equation (5.37) and shown in Figure 5.7a. The wave becomes unstable for $r = 0.5$, as shown in Figure 5.7b due to

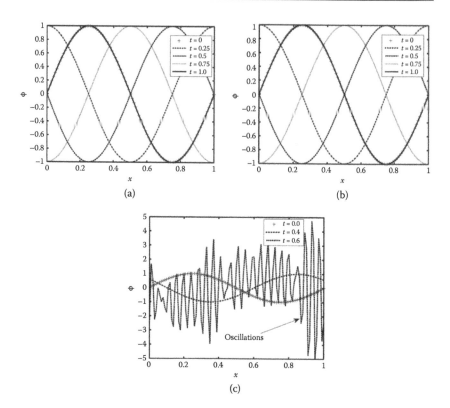

Figure 5.6 The wave is preserved for $r = 1$ (a) and $r = 0.5$ (b), but oscillations are observed for $r = 2$ (c) when the first-order wave equation is solved by locally applied DQ method with the first discretisation scheme. (From H. Li, S. Mulay, and S. See. (2009c). *Computer Modeling in Engineering and Sciences*, 54, 147–199. With permission.)

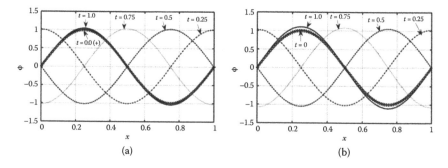

Figure 5.7 Small effect of dispersion is observed for $r = 0.2$ (a) and oscillations are observed for $r = 0.5$ (b) when the first-order wave equation is solved by the fourth scheme. (From H. Li, S. Mulay, and S. See. (2009c). *Computer Modeling in Engineering and Sciences*, 54, 147–199. With permission.)

the violation of stability condition. The instability in the wave is identified when the maximum value of function along the y axis exceeds 1.

The wave equation is discretised by the forward time and central space scheme (Section 5.2.1.5) given in Equation (5.29), and the terms of the function values are substituted by Taylor series expansion (Mulay et al., 2009; Appendix C) and written as

$$\phi_{,t} - a\,\phi_{,x} = \phi_{,xx}\left[\frac{h^2}{t} - \frac{a\,h\,r}{2}\right] \tag{5.38}$$

This scheme is consistent as $\phi_{,t} - a\,\phi_{,x} \to 0$ when h and $t \to 0$. As the term $\phi_{,xx}$ represents the dissipation or damping with respect to the term of space domain $\phi_{,x}$ from the governing wave equation, it is desirable to have a positive or zero value for the coefficient of $\phi_{,xx}$ as

$$\left[\frac{h^2}{t} - \frac{a\,h\,r}{2}\right] \geq 0 \tag{5.39}$$

Simplifying Equation (5.39) results in the consistency condition as

$$|r| \leq \sqrt{2} \tag{5.40}$$

Equation (5.40) gives the constraint on r by the approach of consistency. It is clear by comparing Equation (5.33) with Equation (5.40) that if the stability condition $|r| \leq 1$ is satisfied, automatically the consistency condition $|r| \leq \sqrt{2}$ is also satisfied, but the reverse is not true. As a result, the stability condition is taken as the final dominating criterion.

This scheme is implemented to numerically solve the first-order wave equation. The results are stable for $r = 1$ and the initial wave is preserved for all the time increment steps, as shown in Figure 5.8a. The dissipation effect is observed for $r = 0.5$, which is expected from the consistency analysis results in Equation (5.38), although the results are stable, as shown in Figure 5.8b. The instability is observed for $r = 2$, as shown in Figure 5.8c due to the violation of both the stability and consistency conditions.

In summary, all the schemes found stable in the previous section are analyzed for consistency in this section. The analytical results obtained from the stability and consistency analyses are successfully verified by numerically solving the first-order wave equation by the locally applied DQ method. The numerical behaviour of the method is similar to what was analytically predicted. The first-order wave equation is further solved by the RDQ method in the next section to test whether the RDQ method

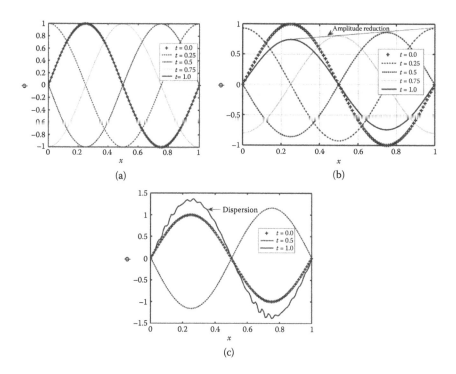

Figure 5.8 The wave is preserved for *r* = I (a), numerical damping is observed for *r* = 0.5 (b), and oscillations are observed for *r* = 2 (c) when the first-order wave equation is solved by the locally applied DQ method with the fifth scheme. (From H. Li, S. Mulay, and S. See. (2009c). *Computer Modeling in Engineering and Sciences*, 54, 147–199. With permission.)

satisfies the analytical conditions obtained so far from the stability and consistency analyses.

5.2.3 Implementation of RDQ method for first-order wave equation by forward time and central space scheme

The RDQ method is implemented in this section to solve the first-order wave equation by the forward time and central space scheme (Section 5.2.1.5). A local domain of interpolation is created around each virtual node, as shown in Figure 5.9, and the nearby field nodes falling in it are considered for the approximation of function value at the concerned central virtual node.

As the fixed RKPM interpolation function does not possess the property of delta function, a second level interpolation is performed to compute the values of approximate function at all the field nodes. A local domain

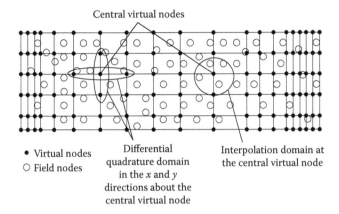

Central virtual nodes

• Virtual nodes

○ Field nodes

Differential
quadrature domain
in the x and y
directions about the
central virtual node

Interpolation domain at
the central virtual node

Figure 5.9 Local interpolation of function and creation of DQ local domains for discretisation of governing equation at all virtual nodes by RDQ method. (From H. Li, S. Mulay, and S. See. (2009c). *Computer Modeling in Engineering and Sciences*, 54, 147–199. With permission.)

is created around each field node, in which the nearby surrounding field nodes are considered for the approximation of function at the concerned central field node. The unknown values of nodal parameter at the field nodes are computed for the time $t = 0$ by the initial condition as

$$\{\phi\}^0_{M\times1} = [N]_{M\times M} \, \{U\}^0_{M\times1} \tag{5.41}$$

where 0 indicates $t = 0$, M is the number of total field nodes in whole computational domain, $\{\phi\}^0_{M\times1}$ is a known vector of function values at all the field nodes at time $t = 0$ (initial condition), $[N]_{M\times M}$ is a matrix of shape functions representing the interpolation of function values at the field nodes by the surrounding field nodes, and $\{U\}^0_{M\times1}$ is the unknown vector of nodal parameter values at all the field nodes at the time $t = 0$. The values of function at all the virtual nodes are then computed at the time $t = 0$ as

$$\{\phi\}^0_{N\times1} = [N]_{N\times M} \, \{U\}^0_{M\times1} \tag{5.42}$$

where N is the number of total virtual nodes in a whole computational domain, $\{\phi\}^0_{N\times1}$ is the unknown vector of function values for all the virtual nodes at $t = 0$, $[N]_{N\times M}$ is a matrix of shape functions interpolating the function values at all the virtual nodes by the function values at the surrounding field nodes, and $\{U\}^0_{M\times1}$ is the known vector of nodal parameter values at all the field nodes already computed by Equation (5.42). The values of function at all the virtual nodes for all the time increments are computed as

$$\{\phi\}^t_{N\times1} = [A]_{N\times N} \, \{\phi\}^{t-1}_{N\times1} \tag{5.43}$$

where $[A]_{N \times N}$ is a coefficient matrix of the discretisation scheme from Equation (5.32), $\{\phi\}_{N \times 1}^t$ and $\{\phi\}_{N \times 1}^{t-1}$ are the vectors of function values for all the virtual nodes at times t and $t - 1$, respectively. Equations (5.42) and (5.41) are used successively after computing the values of function at all the virtual nodes to compute the unknown values of nodal parameter and function at all the field nodes, respectively.

5.2.4 Remarks on solution of first-order wave equation by RDQ method

The first-order wave equation is solved by the RDQ method, as discussed earlier, with the cosine distribution of virtual nodes and the uniform distribution of field nodes in the computational domain. The numerical results are analyzed by the consistency condition of the wave equation. The wave equation in the discretised form by the RDQ method is given as

$$
\{a_{j,j} \quad a_{j,j+1}\}
\begin{Bmatrix} \phi(x_i, t_j) \\ \phi(x_i, t_{j+1}) \end{Bmatrix}
= a \{a_{i,i-1} \quad a_{i,i} \quad a_{i,i+1}\}
\begin{Bmatrix} \phi(x_{i-1}, t_j) \\ \phi(x_i, t_j) \\ \phi(x_{i+1}, t_j) \end{Bmatrix}
\qquad (5.44)
$$

If the time and space domains contain the uniform and cosine distributed virtual nodes, respectively, Equation (5.44) is simplified by computing the weighting coefficients for the time domain using Shu's general approach (Shu, 2000) as discussed earlier and given as

$$
\frac{1}{t}\{-1 \quad 1\}
\begin{Bmatrix} \phi(x_i, t_j) \\ \phi(x_i, t_{j+1}) \end{Bmatrix}
= a \{a_{i,i-1} \quad a_{i,i} \quad a_{i,i+1}\}
\begin{Bmatrix} \phi(x_{i-1}, t_j) \\ \phi(x_i, t_j) \\ \phi(x_{i+1}, t_j) \end{Bmatrix}
\qquad (5.45)
$$

where t is the time increment. Equation (5.45) is simplified by rearranging the terms and the final consistency equation $Pu - P(h_x, h_y)u$ by subtracting Equation (5.45) from the first-order wave equation for the scheme in Section 5.2.1.5 is given as

$$
\phi_{,t} - a\phi_{,x} - \phi_{,t} + a\phi_{,x}\left[\frac{Bh_2 - Ah_1}{a\,t}\right] = \phi\left[\frac{1 - A - B}{t}\right] + \phi_{,xx}\left[\frac{Ah_1^2 + Bh_2^2}{2t} + \frac{a^2\,t}{2}\right]
$$

$$
(5.46)
$$

where h_1 and h_2 are the spacings between the virtual nodes ϕ_m^n and ϕ_{m-1}^n, and ϕ_m^n and ϕ_{m+1}^n, respectively, t is an increment in time, and A and B are the constants in the above equation given as A = 0.5 + (0.5 $a\ t\ a_{i,i}$) + ($a\ t$ $a_{i,i-1}$) and B = 0.5 + (0.5 $a\ t\ a_{i,i}$) + ($a\ t\ a_{i,i+1}$), respectively, where $a_{i,i}$, $a_{i,i-1}$, and $a_{i,i+1}$ are the DQ weighting coefficients. Following conditions are thus derived for Equation (5.46) to be consistent, namely for $P\ u - P(h_x, h_y)\ u \rightarrow$ 0 to be true.

$$\frac{Bh_2 - Ah_1}{at} = 1, \quad \frac{1-A-B}{t} = 0, \quad \text{and} \quad \left[\frac{Ah_1^2 + Bh_2^2}{2t} + \frac{a^2t}{2}\right] \geq 0 \tag{5.47}$$

≥ 0 for + ve or zero dissipation

for +ve or zero dissipation must be satisfied. Equation (5.47) provides the constraints that, when satisfied, result in the consistent discretisation of the wave equation by the uniform and cosine distributions of the virtual nodes in the time and space domains, respectively.

To verify the satisfaction of Equation (5.47) by the uniform distribution of virtual nodes in the spatial domain, the wave equation $\phi_t = a\phi_x$ is numerically solved by the RDQ method with the spatial domain discretised separately by the uniform and random field nodes and the time domain discretised by uniform field nodes coupled with the uniform virtual nodes in the time and space domains, respectively. It is seen from Figure 5.10a and b that the constraints in Equation (5.47) are exactly satisfied, resulting in the consistent discretisation of the first-order wave equation.

When the first-order wave equation is solved by RDQ method with the spatial domain discretised by the uniform and random field nodes coupled with cosine virtual nodes and the temporal domain discretised by the uniform field nodes coupled with the uniform virtual nodes, the constraints in Equation (5.47) are not fully satisfied. This clearly shows that some amount of dissipation or damping is involved, as indicated by the equation when the spatial domain has cosine distributed virtual nodes. This is verified by tracking the numerical and analytical values of function for the field node located, say, at $x = 1$ with the time as shown in Figure 5.10c.

In summary, several single and multistep schemes of the discretisation of time and space domains are studied in this section by the VN and Schur polynomials to achieve a stable discretisation of the first-order wave equation and stable schemes are identified. The stable schemes are further studied for their consistency by discretizing the wave equation. The analytically predicted characteristics of the stable scheme are numerically verified by solving the first-order wave equation with the locally applied DQ and RDQ methods.

It is seen in the Section 5.2.1.5 scheme of discretisation that numerical dissipation is involved when the wave equation is solved by the RDQ method with the cosine distribution of virtual nodes in the space domain. But, the wave is preserved during all the time steps by the uniform distribution of

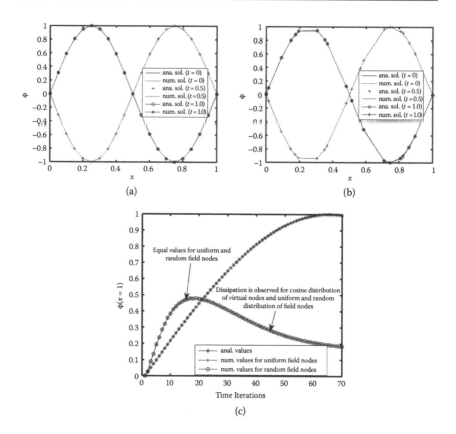

Figure 5.10 Comparison of numerical and analytical function values with spatial domain discretised by uniform (a) and random (b) field nodes coupled with uniform virtual nodes in both time and space domains and time history plot of comparison of numerical and analytical function values at field node (1, *t*) when the spatial domain is discretised by uniform and random field nodes coupled with cosine virtual nodes (c). (From H. Li, S. Mulay, and S. See. (2009c). *Computer Modeling in Engineering and Sciences*, 54, 147–199. With permission.)

virtual nodes in the space domain. This is a highly illuminating result that indicates the applicability of the RDQ method when the computational domain is discretised by random field nodes.

In order to accomplish stable and consistent discretisation of the governing PDE by the RDQ method, the spatial domain is discretised by either uniform or random field nodes coupled with uniform virtual nodes. The temporal domain is discretised by the uniform field nodes coupled with uniform virtual nodes.

5.3 STABILITY ANALYSIS OF TRANSIENT HEAT CONDUCTION EQUATION

The transient heat conduction equation $u_{,t} = \alpha u_{,xx}$ where α is the thermal conductivity of a material is analyzed for stability and consistency and then numerically solved by the RDQ method with the spatial domain discretised by uniform or random field nodes.

Several discretisation schemes of the time and space domains for the transient heat conduction equation are evaluated first for stability by the RDQ method and the stable schemes are identified. The consistency analysis of the stable schemes is then performed to ensure that the constraints obtained by consistency analysis do not conflict with those obtained by stability analysis. The transient heat conduction equation is finally solved by the RDQ method with the stable and consistent scheme.

5.3.1 Forward time and forward space scheme

In this scheme, three and two virtual nodes are used in the space and time domains, respectively, as shown in Figure 5.4a. The transient heat conduction equation is discretised here at the uniformly distributed virtual nodes in the time and space domains with the weighting coefficients computed by Shu's general approach as

$$\frac{\partial u(x_i, t_j)}{\partial t} = \alpha \frac{\partial^2 u(x_i, t_j)}{\partial x^2} \Rightarrow$$

$$[u(x_i, t_{j+1}) - u(x_i, t_j)] = \left(\frac{\alpha k}{h^2}\right) [u(x_i, t_j) - 2\, u(x_{i+1}, t_j) + u(x_{i+2}, t_j)] \tag{5.48}$$

where k and h are the spacings between the virtual nodes in the temporal and spatial domains, respectively. The u terms in Equation (5.48) are substituted by Taylor series expansion (Mulay et al., 2009) and simplified to give the discretised form (Appendix C) as

$$u_{,t} - \alpha\, u_{,xx} = \left(\frac{-k}{2}\right) u_{,tt} - (2\alpha\, h)\, u_{,xxx} \tag{5.49}$$

The terms $u_{,tt}$ and $u_{,xxx}$ represent the dissipation with respect to the time and space domains $u_{,t}$ and $u_{,xx}$, respectively. Both the dissipation terms have negative coefficients, which means that there is definitely going to be negative dissipation (dissipation removal from the system); therefore this scheme is not stable. As the equation of heat conduction is inconsistently discretised by this scheme, there is no need to perform the stability analysis.

The equation of heat conduction is diffusive in behavior; therefore the values of temperature at the field nodes are governed by the boundary conditions. This can be captured by including the virtual nodes from both the left and right sides of the central concerned virtual node in the space discretisation. Therefore, the spatial parameter is changed in the next subsection by considering the virtual nodes on both sides of the concerned virtual node.

5.3.2 Forward time and central space scheme

The forward time and central space discretisation scheme is used in this section to discretize the transient heat conduction equation by the uniform and cosine distributions of the virtual nodes with the spatial domain discretised by uniformly or randomly distributed field nodes. The stability analysis is first performed by the uniform distribution of virtual nodes in the space and time domains, then the consistency analysis is performed, after which the feasible working range of the temporal spacing k is obtained by combining both results.

The consistency analysis is further extended to the cosine distributed virtual nodes in the spatial domain, and the constraint on k is derived, which gives the stable discretisation. The transient heat conduction equation is finally numerically solved by the RDQ method with the space domain discretised by the uniform or random field nodes combined with the uniform or cosine virtual nodes, and the time domain discretised by the uniform field nodes combined with the uniform virtual nodes. The stable and accurate numerical solution of the transient heat conduction equation by the RDQ method follows the constraints on the temporal spacing k obtained from the consistency and stability analyses. The numerical results of the temperature distribution are compared with the corresponding analytical and FEM solutions. They almost exactly match the analytical solutions.

5.3.2.1 Stability analysis of forward time and central space scheme

In this scheme, three and two virtual nodes are considered for the discretisation of the space and time computation domains, respectively, as shown in Figure 5.5a. The discretised form of the heat conduction equation by the RDQ method is finally given as

$$[u(x_i, t_{j+1}) - u(x_i, t_j)] = \left(\frac{\alpha\, k}{h^2}\right)[u(x_{i-1}, t_j) - 2\, u(x_i, t_j) + u(x_{i+1}, t_j)] \quad (5.50)$$

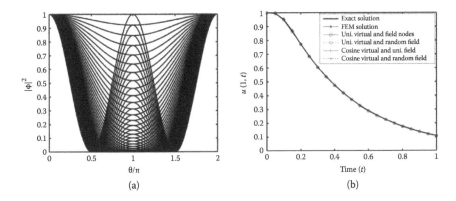

Figure 5.11 plot by $0 \le m \le 0.5$ with forward time and central space scheme (a) and time history plot of temperature at the field node (1, t) and its comparison with analytical and FEM solutions (b) for transient heat conduction. (From H. Li, S. Mulay, and S. See. (2009c). *Computer Modeling in Engineering and Sciences*, 54, 147–199. With permission.)

The inverse Fourier transform is performed on Equation (5.50) and the characteristic or amplification polynomial (Appendix A) is given as

$$\phi(\theta) = 1 - 4\ m\ \sin^2\left(\frac{\theta}{2}\right), \quad \text{where} \quad m = \left(\frac{\alpha\ k}{b^2}\right) \tag{5.51}$$

It is observed in Equation (5.51) that $|\phi| \le 1$ for $m \le 0.5$ (using MATLAB®), therefore

$$\frac{\alpha\ k}{b^2} \le \frac{1}{2} \Rightarrow k \le \frac{b^2}{2\alpha} \tag{5.52}$$

Equation (5.52) is the stability condition for the stable discretisation of the transient heat conduction equation by this scheme. The graph of $|\phi|^2$ versus (θ / π) is plotted in Figure 5.11a to study the value of Equation (5.51) by changing m as $0 \le m \le 1/2$, and it is observed that $\phi(\theta) \le 1.0$. As a result, this scheme is stable with Equation (5.52) as the stability criterion.

5.3.2.2 Consistency analysis of forward time and central space scheme

Let us begin the consistency analysis with Equation (5.50). The terms of function value u in Equation (5.50) are substituted by Taylor series expansion (Mulay et al., 2009) and simplifying it results (Appendix C) in

$$u_{,t} - \alpha\ u_{,xx} = \frac{\alpha\ b^2}{12}\ u_{,xxxx} - \frac{k}{2}\ u_{,tt} - \frac{k^2}{2}\ u_{,ttt} \tag{5.53}$$

where h and k are the spacings between the virtual nodes situated in the space and time computation domains, respectively. The terms $u_{,xxxx}$, $u_{,tt}$ and $u_{,ttt}$ in Equation (5.53) represent the dispersion corresponding to the space domain term $u_{,xx}$ and the dissipation and dispersion corresponding to the term $u_{,t}$ in the time domain, respectively. The dissipation or damping in the time domain term is replaced as $u_{,tt} = \alpha^2 u_{,xxxx}$ by $u_{,t} = \alpha u_{,xx}$. As a result, Equation (5.53) is modified as

$$u_{,t} - \alpha u_{,xx} = \left(\frac{\alpha h^2}{12} \right) u_{,xxxx} - \left(\frac{k \alpha^2}{2} \right) u_{,xxxx} - \left(\frac{k^2}{2} \right) u_{,ttt} \tag{5.54}$$

In order to make the dispersion or oscillation term $u_{,xxxx} \leq 0$

$$\left[\left(\frac{\alpha h^2}{12} \right) - \left(\frac{k \alpha^2}{2} \right) \right] \leq 0 \Rightarrow k \geq \frac{h^2}{6\alpha} \tag{5.55}$$

Equation (5.55) is the stability constraint on k from the consistency analysis. The final constraint on the temporal spacing k is obtained by combining Equations (5.52) and (5.55) as $(h^2/6\alpha) \leq k \leq (h^2/2\alpha)$.

The space domain in the RDQ method has either uniform or cosine distribution of virtual nodes and the constraints for the uniform distribution of virtual nodes are given by Equations (5.52) and (5.55). It is essential to study the consistency analysis of the transient heat conduction equation by the cosine distribution of virtual nodes as well, which is done in the next subsection.

5.3.2.3 Consistency analysis of forward time and central space scheme by cosine distribution of virtual nodes in space

The RDQ method uses the uniform or cosine distribution of virtual nodes to discretize the governing equation. Therefore it is crucial to understand the stability behaviour of the RDQ method by both distributions. The spatial nodal spacing h is not constant for the cosine virtual nodes but changes at each field node as

$$x_i = x_0 + \frac{L}{2} \left[1 - \cos \left(\frac{i-1}{NP-1} \pi \right) \right] \quad \text{for } i = 1, 2, ..., NP \tag{5.56}$$

where x_0 and x_i are the coordinates of starting and the ith virtual node, and L and NP are the length of computational domain and the total virtual nodes in the computation domain, respectively. The heat conduction equation in the discretised form by the cosine virtual nodes is given as

$$u_{,t} - \alpha\, u_{,xx} = \frac{\alpha}{3}\,(h_2 - h_1)\, u_{,xxx} + \alpha \left[\left(\frac{h_1^2 - h_1\, h_2 + h_2^2}{12} \right) - \left(\frac{k\,\alpha}{2} \right) \right] u_{,xxxx} \quad (5.57)$$

where the spatial node spacings h_1 and h_2 are given as $h_1 = x_i - x_{i-1}$ and $h_2 = x_{i+1} - x_i$, respectively. In order to keep the dispersion or oscillation term $u_{,xxxx} \leq 0$

$$k \geq \left[\frac{h_1^2 - h_1\, h_2 + h_2^2}{6\,\alpha} \right] \quad (5.58)$$

The nodal spacings h_1 and h_2 are different at each virtual node due to the cosine distribution. As a result, the temporal spacing k is computed at each virtual node by Equation (5.58), and the lowest value among them is chosen to get the stable solution. Therefore, Equation (5.58) is the stability criterion when the transient heat conduction equation is solved by discretizing the spatial domain with the cosine distribution of virtual nodes.

The transient heat conduction equation is numerically solved in the next subsection via the RDQ method to verify the results obtained by the stability analysis. The spatial domain is discretised by the uniform or random field nodes coupled with either uniform or cosine virtual nodes, and the time domain is discretised by the uniform field nodes coupled with the uniform virtual nodes.

5.3.2.4 Numerical solution of transient heat conduction equation by RDQ method

In this subsection, the RDQ method is implemented to solve the transient heat conduction equation. The numerical results are compared with the analytical and FEM solutions, and they are shown to match closely. The transient heat conduction equation is discretised at the virtual nodes located in the internal computational domain, with the space domain discretised by the uniform virtual nodes combined with the uniform or random field nodes as

$$u_i^{n+1} = m \left(u_{i-1}^n + u_{i+1}^n \right) + (1 - 2\,m)\, u_i^n, \quad \text{where } i = 1,\, 2,\, ...,\, NP \quad (5.59)$$

$$\frac{\partial u}{\partial t}(1,\, t) = 0 \quad \Rightarrow u_{NP}^n = \left(\frac{2}{3} \right) \left(2\, u_{NP-1}^n - \frac{u_{NP-2}^n}{2} \right) \quad (5.60)$$

where NP is the number of total virtual nodes in the domain, and n indicates the nth time step. Equation (5.60) is obtained by the locally applied DQ method, and is used to impose the Neumann boundary condition for the virtual node located at $x = 1$. Note that both sides of the equation are at the nth time level. The governing differential equation and the initial and boundary conditions are given as

$$\frac{\partial u}{\partial t} = \frac{\partial^2 u}{\partial x^2} \quad \text{for } 0 \le x \le 1, \quad \text{where } \alpha = 1 \tag{5.61}$$

$$u(x,0) = 1.0, \quad \text{and} \quad u(0, t) = 0.0 \text{ and } \frac{\partial u}{\partial t}(1, t) = 0.0 \tag{5.62}$$

At time $t = 0$, the values of temperature at all the field and virtual nodes are computed by Equation (5.62), and the values of boundary conditions on the left and right sides of the domain are computed by Equations (5.62) and (5.60), respectively. Let $total\_it = T/k$ be the total increments of time, where T and k are the total simulation time and the single time increment, respectively.

Equation (5.59) is applied for the increment number $n = 1$ to $total\_it$ to compute the values of temperature at all the virtual nodes. The unknown values of the nodal parameters at all the field nodes are computed by $[S]_{NP \times N} \{U\}_{N \times 1}^{n+1} = \{u\}_{NP \times 1}^{n+1} \Rightarrow \{U\}_{N \times 1}^{n+1}$, where $[S]_{NP \times N}$ is a shape function matrix approximating the values of function $\{u\}_{NP \times 1}^{n+1}$ at all the virtual nodes by their surrounding field nodes as per the RDQ method. $\{U\}_{N \times 1}^{n+1}$ is the unknown vector of nodal parameter values at all the field nodes at the $(n + 1)^{th}$ time level, and NP and N are the numbers of total virtual and field nodes in the computational domain, respectively. The values of temperature at all the field nodes are then evaluated by, $\{U_f\}_{N \times 1}^{n+1} = [S']_{N \times N} \{U\}_{N \times 1}^{n+1}$, where $[S']_{N \times N}$ is the matrix of shape function interpolating the values of function $\{U_f\}_{N \times 1}^{n+1}$ at all the field nodes by their surrounding field nodes, as per the RDQ method. Equations (5.55) and (5.58) are used to compute the temporal spacing k for the uniform and cosine distributions of the virtual nodes by the RDQ method, respectively.

The available FEM solution is computed with four 1-D quadratic elements (total nine field nodes; Reddy, 1993). Thus, for a comparison purpose, the results obtained by the RDQ method are also computed by the nine field nodes with either uniform or random distribution. The transient distribution of temperature for the field node located at $x = 1$ is evaluated by separately discretizing the spatial domain with the uniform and random field nodes coupled with the uniform and cosine virtual nodes, and compared with the analytical and FEM solutions, as shown in Figure 5.11b. Note that the solutions obtained by the RDQ method closely match the analytical and FEM solutions and accurate results are obtained by the randomly scattered field nodes in the domain. This problem is again solved by separately considering the 31 uniform and 21 random field nodes scattered in the domain, the diffusion from the time $t = 0$ to 1 sec. is shown in

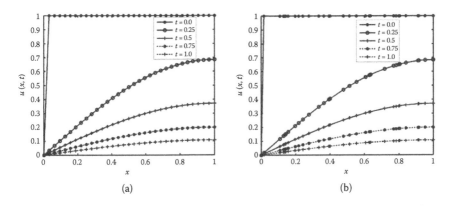

Figure 5.12 Diffusion of temperature from time $t = 0$ to 1 sec by uniform field and uniform virtual nodes (a) and random field and uniform virtual nodes (b). (From H. Li, S. Mulay, and S. See. (2009c). *Computer Modeling in Engineering and Sciences*, 54, 147–199. With permission.)

Figure 5.12 and Figure 5.13. It is seen from Figure 5.12b and Figure 5.13b that the diffusion phenomenon is well captured by the random field nodes.

In summary, two schemes of computation domain discretisation are studied in this section for the stability of transient heat conduction equation. The forward time and central space scheme is found to be stable. This stable scheme is further analyzed for the consistent discretisation of the governing equation with the spatial domain having either uniform or cosine virtual nodes. The stability conditions are derived in Equations (5.55) and

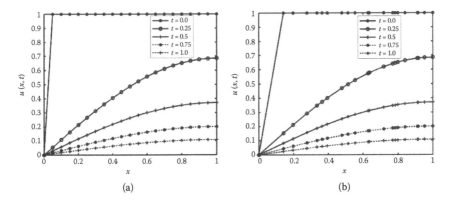

Figure 5.13 Diffusion of temperature from time $t = 0$ to 1 sec by uniform field and cosine virtual nodes (a) and random field and cosine virtual nodes (b). (From H. Li, S. Mulay, and S. See. (2009c). *Computer Modeling in Engineering and Sciences*, 54, 147–199. With permission.)

(5.58), respectively. The transient heat conduction problem is then numerically solved with the uniform and random field nodes scattered in the computation domain. The numerical results obtained by the RDQ method are closely match the corresponding analytical and FEM solutions, as seen in Figure 5.11b. The equation of transverse beam deflection is solved in the next section by the RDQ method with explicit and implicit approaches.

5.4 STABILITY ANALYSIS OF TRANSVERSE BEAM DEFLECTION EQUATION

The transverse motion of the beam is studied in this section for the fixed–fixed configuration by the Euler-Bernoulli beam equation $\rho A(\partial^2 v/\partial t^2) + EI(\partial^4 v/\partial x^4) = 0$, where ρ, A, E and I are the density of material, cross sectional area, modulus of elasticity, and moment of inertial of the beam, respectively. This is accomplished by performing stability and consistency analyses of the transverse beam deflection equation. The governing equation is then numerically solved for the free vibration case by the RDQ method with the feasible range of temporal spacing k obtained from the stability and consistency analyses. It is shown that the numerical results closely match the analytically predicted values.

The discretisation of domains is performed by both the explicit and implicit approaches. The explicit approach adds the numerical dispersion or oscillations, however small the temporal spacing k is. As a result, the ratio of the values of successive beam deflection is more than 1, which leads to instability. An implicit approach adds numerical dissipation or damping. As a result, the ratio of the values of successive beam deflection is less than 1, which leads to a stable numerical solution.

5.4.1 Explicit approach to solve the transverse beam deflection equation by the RDQ method

The terms of derivative from the governing differential equation are discretised in this section by the forward time and central space approach, as shown in Figure 5.14. The DQ weighting coefficients for a general m^{th} order derivative of function are given as

$$(w_{i,j})^m = m\left[a_{i,j}\,(w_{i,i})^{m-1} - \frac{(w_{i,j})^{m-1}}{x_i - x_j} \right], \text{ and } (w_{i,i})^m = -\sum_{k=1,k\neq i}^{N_p} (w_{i,k})^m \qquad (5.63)$$

where m is the order of the derivative, x_i and x_j are the nodal coordinates at the ith and j^{th} virtual nodes, respectively, N_p is the total virtual nodes in the local discretisation domain of the virtual node (x_i, t_j), and $a_{i,j}$ are

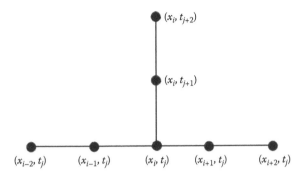

Figure 5.14 Discretisation of time and space domains for transverse beam deflection problem by explicit approach. (From H. Li, S. Mulay, and S. See. (2009c). *Computer Modeling in Engineering and Sciences*, 54, 147–199. With permission.)

the DQ weighting coefficients for the first-order derivative. The discretised beam equation at the virtual node (x_i, t_j) by the RDQ method is given as

$$
\{b_{j,j} \quad b_{j,j+1} \quad b_{j,j+2}\}
\begin{Bmatrix}
v_i^j \\
v_i^{j+1} \\
v_i^{j+2}
\end{Bmatrix}
+ \frac{EI}{\rho A}
\{(w_{i,i-2})^4 \quad (w_{i,i-1})^4 \quad (w_{i,i})^4
$$

$$
(w_{i,i+1})^4 \quad (w_{i,i+2})^4\} \times
\begin{Bmatrix}
v_{i-2}^j \\
v_{i-1}^j \\
v_i^j \\
v_{i+1}^j \\
v_{i+2}^j
\end{Bmatrix}
= 0 \tag{5.64}
$$

The values of the fourth order DQ weighting coefficients in Equation (5.64) are computed by recursively applying Equation (5.63) with $m = 4$. The final equation of transverse beam deflection in the discretised form is given as

$$
\{1 \quad -2 \quad 1\}
\begin{Bmatrix}
v_i^j \\
v_i^{j+1} \\
v_i^{j+2}
\end{Bmatrix}
+ \left(\frac{EI\,k^2}{\rho A\,h^4}\right)
\{1 \quad -4 \quad 6 \quad -4 \quad 1\}
\begin{Bmatrix}
v_{i-2}^j \\
v_{i-1}^j \\
v_i^j \\
v_{i+1}^j \\
v_{i+2}^j
\end{Bmatrix}
= 0 \tag{5.65}
$$

where k and h are the temporal and spatial spacings, respectively. Equation (5.65) is used for the stability and consistency analyses, and is similar to the FDM equation for beam deflection. The important difference between the FDM and RDQ techniques is that the domain is discretised by either the uniform or random nodes in the RDQ method, while it is discretised only by uniform nodes in the FDM.

5.4.1.1 Stability analysis of discretisation of transverse beam deflection equation by explicit scheme

The stability analysis of explicit approach is performed in this section by Equation (5.65) as

$$v_i^{j+2} = 2\,v_i^{j+1} - (1+6\,M)v_i^j + M\,(4v_{i-1}^j - v_{i-2}^j + 4v_{i+1}^j - v_{i+2}^j) \qquad (5.66)$$

where $M = (EI\,k^2)/(\rho A\,h^4)$. The characteristic polynomial obtained by the inverse Fourier transform of Equation (5.66), is given (Appendix A) as

$$\phi(z) = z^2 - 2\,z - \{4\,M\cos(\theta)\,[2 - \cos(\theta)] + 1 + 4\,M\} \qquad (5.67)$$

The roots, $\phi_+ = 1 + 2\,\sqrt{M}\,\cos(\theta)\,\alpha - M$ and $\phi_- = 1 - 2\,\sqrt{M}\,\cos(\theta)\,\alpha - M$ where $\alpha = [2 - \cos(\theta)]$, of Equation (5.67) are plotted by $0 \le M \le 1 \times 10^{-5}$ as shown in Figure 4.15. The maximum values of the roots are close to 1 for $M = 1 \times 10^{-5}$. As a result, the temporal spacing is given as

$$k \le \sqrt{\frac{(1 \times 10^{-5})\rho A\,h^4}{(EI)}} \qquad (5.68)$$

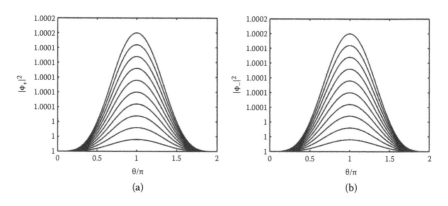

Figure 5.15 Plots of roots $|\phi_+|^2$ versus (θ/π) (a) and $|\phi_-|^2$ versus (θ/π) (b) for the problem of transverse beam deflection by explicit approach. (From H. Li, S. Mulay, and S. See. (2009c). *Computer Modeling in Engineering and Sciences*, 54, 147–199. With permission.)

Equation (5.68) is the stability condition for the deflection of beam with the explicit approach.

5.4.1.2 Consistency analysis for explicit approach of discretisation of transverse beam deflection equation

The terms v from Equation (5.65) are replaced by the corresponding Taylor series expansion (Mulay et al., 2009), and the discretised equation (Appendix C) is given as

$$m\, v_{,tt} + EI\, v_{,xxxx} = -(m\, k)\, v_{,ttt} - (EI\, h^2) v_{,xxxxxx} \tag{5.69}$$

where $m = \rho A$. It is noted from Equation (5.69) that the discretisations of the time and space domains are first and second order accurate, respectively. The terms $v_{,ttt}$ and $v_{,xxxxxx}$ represent the dissipation and dispersion with respect to the time and space terms $v_{,tt}$ and $v_{,xxxx}$, respectively. They have negative coefficients, indicating a dissipation removal from the system with respect to time and a dispersion removal for space. The term h^2 is small and can be neglected. As a result, the final discretised equation is given as

$$m\, v_{,tt} + EI\, v_{,xxxx} = -(m\, k)\, v_{,ttt} \tag{5.70}$$

Note from Equation (5.70) that the original governing equation is consistently discretised as $k \to 0$, $m\, v_{,tt} + EI\, v_{,xxxx} \to 0$, but there will be numerical dispersion due to the −ve term $v_{,ttt}$. It is essential to know the analytical value of dispersion as it can be compared with the numerical results obtained by the RDQ method. Thus, the expressions of the natural frequency (ω_n) and ratio of successive amplitude reduction are derived in the next section.

5.4.1.3 Computation of natural frequency ω_n and ratio of successive amplitude reduction

In this section, the natural frequency of the fixed–fixed beam configuration is derived by the method of separation of variables, and subsequently the ratio of reduction of amplitude is computed (Chandar and Damodaran, 2009). The terms $v(x, t)$ in Equation (5.70) can be substituted as $v(x, t) = \psi(t)\, \xi(x)$ to obtain two ordinary differential equations (ODEs) as

$$\frac{1}{\psi}\left[k\, \frac{\partial^3 \psi}{\partial t^3} + \frac{\partial^2 \psi}{\partial t^2} \right] = -\omega_n^2 \tag{5.71}$$

$$\frac{\partial^4 \phi}{\partial x^4} - \frac{m\phi}{EI}\, \omega_n^2 = 0 \tag{5.72}$$

Equations (5.71) and (5.72) give the variations in the amplitude of the deflection with respect to the time and space domains, respectively. Equation (5.72) is solved for ω_n by assuming the general solution as a linear combination of the trigonometric equations and $k_n^4 = (m\omega_n^2)/(EI)$

$$\phi(x) = C_1[\cos(k_n x) + \cosh(k_n x)] + C_2[\cos(k_n x) - \cosh(k_n x)]$$
$$+ C_3[\sin(k_n x) + \sinh(k_n x)] + C_4[\sin(k_n x) - \sinh(k_n x)]$$

(5.73)

The constants from Equation (5.73) are solved by the boundary conditions $v(0) = 0.0$, $v(1.0) = 0.0$, $\partial v/\partial x|_{(0)} = 0.0$, and $\partial v/\partial x|_{(1)} = 0.0$ for the fixed–fixed beam configuration to obtain

$$C_1 = 0, \ C_3 = 0, \ C_4 = -C_2 \left[\frac{\cos(M) - \cos(bM)}{\sin(M) - \sin(bM)} \right] \quad \text{and} \quad \cos(M)\cos(bM) = 1.0$$

(5.74)

where $M = k_n L$. Equation (5.74) is solved either numerically or by trial and error to get the values of M as 4.73004074486 and 7.85320462 (here solved in MATLAB). The first mode is selected as

$$M = k_n L = 4.73004074486 \Rightarrow \omega_n = 22.3733 \text{ rad/sec. for } L = 1 \quad (5.75)$$

Equation (5.75) gives the natural frequency of the fixed–fixed beam configuration. The value of frequency ω_n from the equation can be verified easily by considering a few uniformly distributed virtual nodes and computing the Eigenvalues of their spatial discretisation matrix by Equation (5.65). Equation (5.71) is used to compute the ratios of reduction in the successive amplitudes of the beam deflection over time. The roots of Equation (5.71) are computed and the general solution (Appendix D) are given as

$$\psi(t) = e^{-v_1 t}[c_2 \sin(\omega_d t)] \quad (5.76)$$

where $\omega_d = [\sqrt{3}/(24 k)] [2\beta - (8/\beta)]$ is the damped natural frequency, $v_1 = (1/12 k) (\beta + (4/\beta) - 4)$, where $\beta = \sqrt[3]{-\alpha - 8 + 12\sqrt{3} (\sqrt{27 \omega_n^2 k^2 + 4})} \, \omega_n k$ and $\alpha = 108 \, \omega_n^2 k^2$. The ratio $\psi[t_1 + (2\pi/\omega_d)]/\psi(t_1)$ for the $t = t_1$ and $[t_1 + (2\pi/\omega_d)]$, namely for the two successive peaks in time, is taken by Equation (5.76) and simplified by the periodicity property of sin function and given as

$$ratio = \exp\left(\frac{-2\pi v_1}{\omega_d} \right) \quad (5.77)$$

Equation (5.77) is used to compute the ratios of successive amplitude reduction in the beam deflection by the explicit approach.

The transverse beam deflection problem is solved here by the explicit approach with the governing equation, boundary, and initial conditions as (Miller, 1971)

$$\frac{\partial^2 v}{\partial t^2} + \frac{\partial^4 v}{\partial x^4} = 0 \quad \text{for } (0 < x < 1) \tag{5.78}$$

$$v(0,t) = 0.0, v(1,t) = 0.0, \frac{\partial v}{\partial x}(0, t) = 0.0, \quad \text{and} \quad \frac{\partial v}{\partial x}(1, t) = 0.0 \tag{5.79}$$

$$v(x,0) = \sin(\pi x) - \pi x(1 - x), \quad \text{and} \quad \frac{\partial v}{\partial x}(x,0) = 0 \tag{5.80}$$

Equation (5.79) indicates two boundary conditions at the nodes on the left and right sides of the domain; therefore one of the boundary conditions is transferred to neighbouring nodes in order to get a single algebraic equation at each virtual node located on the left and right sides of the domain boundaries, when the governing equation is discretised by the RDQ method, as given (Li et al., 2009b)

$$\frac{\partial v}{\partial x}(0, t) = 0 = \frac{v_2 - v_1}{x_2 - x_1} \Rightarrow v_2(x_2, t) = 0.0,$$

$$\frac{\partial v}{\partial x}(1, t) = 0 = \frac{v_{NP-1} - v_{NP}}{x_{NP-1} - x_{NP}} \Rightarrow v_{NP-1}(x_{NP-1}, t) = 0.0 \tag{5.81}$$

where NP is the number of total virtual nodes, and the suffixes 1, 2, $(NP-1)$, and NP indicate the identification of the virtual nodes. The modified boundary conditions are given as

$$v(0, t) = 0.0, \ v(1, t) = 0.0, \ v(x_2, t) = 0.0 \quad \text{and} \quad v(x_{NP-1}, t) = 0.0 \tag{5.82}$$

This problem is solved by the FEM (Reddy, 1993) with two Euler-Bernoulli beam elements; here it is solved by five ($h = 0.25$) uniformly distributed field and virtual nodes for comparison and the temporal spacing $k = 2 \times 10^{-4}$ sec. is computed by Equation (5.68).

The problem is solved initially by the locally applied DQ and RDQ methods with the spatial domain discretised by uniform and random field nodes. The deflection at the central field node $(0.5, t)$ is shown in Figure 5.16a. The

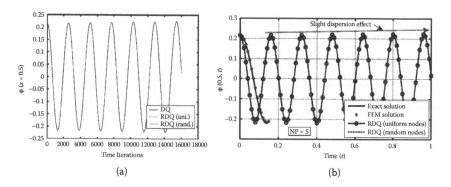

Figure 5.16 Time history plot at field node (0.5, *t*) by locally applied DQ and RDQ methods (a) and comparison of RDQ method with FEM and analytical solutions
(b) for problem of transverse beam deflection. (From H. Li, S. Mulay, and
S. See. (2009c). *Computer Modeling in Engineering and Sciences*, 54, 147–199.
With permission.)

values of the deflection are identical, which indicates that the RDQ method
can effectively handle the distribution of random field nodes.

The problem is further solved by the RDQ method with the spatial
domain discretised by either the uniform or random field nodes coupled
with the uniform virtual nodes, and their results are compared with the
analytical and FEM solutions, as shown in Figure 5.16b. Note in the figure
the slight effect of dispersion over time, as predicted from Equation (5.70).
The successive peak values of the amplitude of beam deflection at the time
$t_1 = 0.1624$ sec. and $t_2 = 0.3248$ sec. are obtained from Figure 5.16b, and
their ratios are found to be $v_2/v_1 = (0.2253)/(0.2199) \approx 1.0245$. The analytical ratio of amplitude reduction computed by Equation (5.77) is found to
be ratio = $\exp\{[-2\pi(-0.050058)]/22.3723\} \approx 1.014$. The numerical ratio of
amplitude reduction closely matches the analytical value.

It is also observed from the ratio that $v_2/v_1 > 1$, which indicates dispersion in the solution. When this problem is solved with the lower temporal
increment $k = 6.25 \times 10^{-5}$ sec., the numerical and analytical ratios of the
amplitude reduction are found to be $v_2/v_1 \approx 1.007$ and 1.005, respectively,
which are also closely matching. As a result, it is concluded that $(v_2/v_1) > 1$
and $(v_2/v_1) \to 1.0$ as $k \to 0$ for the explicit approach.

5.4.2 Implicit approach to solving transverse
beam deflection equation by RDQ method

As seen from Equation (5.68), the explicit approach has a stringent requirement for temporal marching. Therefore the computational cost is very high

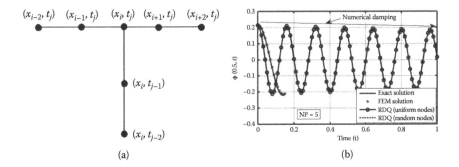

Figure 5.17 Discretisation of time and space domains (a) and comparison of numerical solution of beam deflection at field node (0.5, *t*) with FEM and analytical solutions (b) when problem of transverse beam deflection is solved by implicit approach. (From H. Li, S. Mulay, and S. See. (2009c). *Computer Modeling in Engineering and Sciences*, 54, 147–199. With permission.)

for large numbers of field nodes. Also, Equation (5.70) shows a reduction in the dissipation over time. Therefore the numerical discretisation of the beam deflection equation is consistent only when $k \rightarrow 0$. An implicit approach of time discretisation is studied in this section to overcome these limitations.

The derivative terms from the governing differential equation are discretised at the same time level in the implicit approach, as shown in Figure 5.17a. The discretisation of the beam deflection equation by the RDQ method with the scheme shown in Figure 5.17a is given as

$$
\{b_{j,\,j-2} \quad b_{j,\,j-1} \quad b_{j,\,j}\}
\begin{Bmatrix} v_i^{j-2} \\ v_i^{j-1} \\ v_i^{j} \end{Bmatrix}
$$

$$
+\frac{EI}{\rho A}\{(w_{i,\,i-2})^4 \quad (w_{i,\,i-1})^4 \quad (w_{i,\,i})^4 \quad (w_{i,\,i+1})^4 \quad (w_{i,\,i+2})^4\}
\begin{Bmatrix} v_{i-2}^{j} \\ v_{i-1}^{j} \\ v_i^{j} \\ v_{i+1}^{j} \\ v_{i+2}^{j} \end{Bmatrix} = 0
$$

$$(5.83)$$

The weighting coefficients from Equation (5.83) are computed by Shu's general approach (Shu, 2000) and substituted to get

$$
\{1 \quad -2 \quad 1\}
\begin{Bmatrix} v_i^{j-2} \\ v_i^{j-1} \\ v_i^{j} \end{Bmatrix}
+ \left(\frac{EI\,k^2}{\rho A\,h^4} \right)
\{1 \quad -4 \quad 6 \quad -4 \quad 1\}
\begin{Bmatrix} v_{i-2}^{j} \\ v_{i-1}^{j} \\ v_i^{j} \\ v_{i+1}^{j} \\ v_{i+2}^{j} \end{Bmatrix}
= 0 \quad (5.84)
$$

Rearranging the terms from Equation (5.84) leads to

$$
v_i^j\,(1 + 6\,M) + M\left(v_{i-2}^j - 4\,v_{i-1}^j - 4\,v_{i+1}^j + v_{i+2}^j \right) = 2\,v_i^{j-1} - v_i^{j-2} \qquad (5.85)
$$

where $M = (EI\,k^2)/(\rho A\,h^4)$. The application of Equation (5.85) at all the virtual nodes located in the internal computational domain results in a matrix that can be solved by any of the direct or iterative solvers.

5.4.2.1 Consistency analysis of transverse beam deflection equation discretised by implicit approach

It is essential to know the numerical form of the transverse beam deflection equation that is exactly solved after discretizing it with the implicit approach. Therefore, the consistent equation after expanding the terms v from Equation (5.85) by Taylor series (Mulay et al., 2009; Appendix C) is given as

$$
m\,v_{,tt} + EI\,v_{,xxxx} = m\,k\,v_{,ttt} \qquad (5.86)
$$

where $m = \rho A$. We see from Equation (5.86) that numerical damping with respect to the time domain is added, but still the discretisation is consistent because $m\,v_{,tt} + EI\,v_{,xxxx} \to 0$ as $k \to 0$. As a result, the discretised governing PDE in Equation (5.86) approaches the continuous form in Equation (5.78) with a reduction in the temporal marching increment. In order to analytically compute the numerical dissipation in Equation (5.86), the equation for the ratio of successive amplitude reduction, as also mentioned in Chandar and Damodaran (2009), is developed in the next subsection.

5.4.2.2 Computation amplitude reduction ratio for implicit approach

The terms $v(x, t)$ from Equation (5.86) are substituted by $v(x, t) = \psi(t)\,\xi(x)$, as explained earlier during the method of separation of variables, and it is converted to two ODEs as

$$k\,\frac{\partial^3 \psi}{\partial t^3} - \frac{\partial^2 \psi}{\partial t^2} - \omega_n^2 \psi = 0 \tag{5.87}$$

$$\frac{\partial^4 \phi}{\partial x^4} - \frac{m\phi}{EI}\,\omega_n^2 = 0 \tag{5.88}$$

Equation (5.87) is solved to compute the ratios of the successive amplitude reduction with respect to time. The roots of Equation (5.87) are evaluated and the general solution is given as

$$\psi(t) = e^{\upsilon_1 t}[c_2 \sin(\omega_d\,t)] \tag{5.89}$$

where $\omega_d = [\sqrt{3}/(24k)][2\beta - (8/\beta)]$ is the damped natural frequency, and $\upsilon_1 = (1/12k)[-\beta - (4/\beta) + 4]$, where $\beta = \sqrt[3]{\alpha + 8 + 12\sqrt{3}\left(\sqrt{27\omega_n^2 K^2 + 4}\right)\omega_n k}$ and $\alpha = 108\,\omega_n^2 k^2$. The ratio $\psi[t_1 + (2\pi / \omega_d)] / \psi(t_1)$ for the $t = t_1$ and $[t_1 + (2\pi/\omega_d)]$, namely for two successive peaks in time, is taken by Equation (5.89) and simplified using the periodicity property of sin function as

$$ratio = \exp\left(\frac{2\pi\,\upsilon_1}{\omega_d}\right) \tag{5.90}$$

Equation (5.90) is used to compute the ratios of the successive amplitude reduction by the implicit approach.

The problem of transverse beam deflection is solved here by the implicit approach with the governing differential equation and initial and boundary conditions as given in Equations (5.78), (5.80) and (5.82), respectively. The problem is solved by discretizing the domain with the total of five uniform or random field nodes coupled with a total of five uniform virtual nodes and $k = 2 \times 10^{-4}$ sec to compare the results with those obtained by the explicit approach. Figure 5.17b shows the comparison of results obtained by the RDQ method via the implicit approach with the FEM and analytical solutions. The deflection values at the field node $(0.5, t)$ damp over time, as expected from Equation (5.86), and identical results are obtained by the uniform and random distributions of the field nodes.

The successive peak values are obtained from Figure 5.17b and their ratio is evaluated as $v_2/v_1 = (0.2043)/(0.2094) \approx 0.9756$, and the analytical ratio of amplitude reduction computed by Equation (5.90) is evaluated as $ratio = \exp[2\pi(v_1)/22.3723] \approx 0.9860$. Therefore, the numerical and analytical values of the amplitude reduction ratios match closely and $v_2/v_1 < 1$, which clearly indicates dissipation in the solution. This problem is again solved with the lower temporal increment $k = 6.25 \times 10^{-5}$ sec, similar to the explicit approach. The new numerical and analytical values of the successive amplitude reduction ratio are $(v_2/v_1) \approx 0.992$ and 0.9956, respectively, which also match closely. Therefore, it is concluded that $(v_2/v_1) < 1$ and $(v_2/v_1) \to 1.0$ as $k \to 0$ kfor the implicit approach.

In summary, the stability analysis of the transverse beam deflection equation is performed in this section by both the explicit and implicit approaches. Temporal spacing must be very small with increases in the field nodes to get a stable solution by the explicit approach. This results in high computational cost, and the numerical discretisation of the governing PDE involves the negative dissipation, as shown by Equation (5.70). The implicit approach is used for the temporal discretisation to overcome these problems, and the numerical discretisation of governing PDE is found to be consistent with the continuous form with some numerical damping, as shown by Equation (5.86). There is no direct constraint on temporal spacing from the view of stability. It is also shown that the numerical ratios of successive amplitude reduction match their corresponding analytical values for both the explicit and implicit approaches.

5.5 SUMMARY

First, the stability analysis of the first-order wave equation is performed by the VN and Schur polynomials with five different single and multistep schemes. The stable schemes are identified and further analyzed for additional constraints on r based on the consistency analysis. The observations made from the stability and consistency analyses of the stable schemes are verified by numerically implementing them to solve the first-order wave equation by the methods of locally applied DQ (Sections 5.2.1.1, 5.2.1.4, and 5.2.1.5) and RDQ (Section 5.2.1.5). The results are analyzed by virtual nodes distributed with the uniform and cosine manners in the space domain. Some numerical dissipation is involved when the cosine distribution of virtual nodes is used in the space domain. This observation is analytically verified by deriving the constraint conditions for the scheme in Section 5.2.1.5 with the consistency analysis in Equation (5.47).

Next, the equation of transient heat conduction is solved by the discretisation scheme of forward time and central space with the stable range of temporal spacing k is identified from the stability and consistency analyses. The space domain is discretised with either uniform or random field nodes coupled with uniform or cosine virtual nodes, and their results at the field node $(0.5, t)$ are compared with the FEM and analytical solutions, as shown in Figure 5.11b. The profiles of temperature distribution from the time $t = 0$ to 1 sec are plotted in Figure 5.12 and Figure 5.13. The results are almost identical, indicating that the RDQ method is capable of equally handling the field nodes distributed either uniformly or randomly in the space.

Finally, the equation of the transverse beam deflection is solved by the RDQ method with the explicit and implicit approaches of the discretisation. Even though the explicit approach gives the consistent discretisation of the governing PDE, the numerical dispersion involved makes it unstable. This is verified by comparing the numerically computed ratios of successive amplitude reduction with the corresponding analytical values, and noting that $(v_2/v_1) > 1$ and $(v_2/v_1) \rightarrow 1.0$ as $k \rightarrow 0$. The implicit approach results in the consistent discretisation of the governing PDE with a numerical damping term, but the solution is numerically stable. This is verified by comparing the numerically computed ratios of the successive amplitude reduction with the analytical values, and observing that $(v_2/v_1) < 1$ and $(v_2/v_1) \rightarrow 1.0$ as $k \rightarrow 0$.

In summary, a broad conceptual framework, based on the physical interpretation of mathematical terms, is developed in this chapter to solve transient PDEs by combining stability and consistency analyses. The approach can be successfully applied in solving any transient PDE, as demonstrated by several problems solved in this chapter.

Chapter 6

Adaptive analysis

6.1 INTRODUCTION

In order to capture the local high gradient in a field variable distribution, a higher distribution density of field nodes in a certain computation domain and a lower distribution density of field nodes in the remaining computation domain may be required. A robust meshless method should be able to handle this type of nodal distribution. Thus, an adaptive analysis of a meshless method is essential to ensure that the convergence rates obtained by the adaptive nodal refinement are higher than those obtained by the uniform nodal distribution.

This aspect of a meshless method is illustrated in this chapter by the RDQ method. A numerical method called the adaptive random differential quadrature (ARDQ) method is developed in this chapter. In the ARDQ method, the RDQ method is coupled with an *a posteriori* error estimator based on the relative error norm in the displacement field. An error recovery technique based on least-squares averaging over the local domain of interpolation is proposed, which improves the accuracy of solution as the spacing $h \to 0$.

A novel convex hull approach with the cross product of vectors is proposed in the adaptive refinement to ensure that the newly created nodes are always within the domain of computation. The numerical accuracy of the ARDQ method is successfully evaluated by solving several 1-D and 2-D problems of irregular domain with the local high gradients in the distribution of field variables. It is concluded from the convergence rates that the ARDQ method coupled with the error recovery technique can be effectively used to solve locally high gradient problems of the initial and boundary values.

Adaptive generation of mesh in the FEM has been a hot research area for a long time and many theoretical studies are available. In the classical FEM with the uniform refinement of mesh, the numerical error is monotonically decreased with the element size h, as the elements are said to be quasi-uniform. But, the elements may not be quasi-uniform in the adaptive FEM. As a result, the monotonic decrease of the error may not be guaranteed.

Therefore, it becomes essential to have a good error indicator during the adaptive refinement of mesh that ensures a continuous decrease in errors.

The same requirement is applied to the meshless methods, wherein the field nodes may be randomly scattered in the computational domain. As a result, the nodal spacing h is not constant. A *priori* error estimator gives a rough estimate of error reduction by assuming the asymptotic behaviour of method, and *a posteriori* error estimator evaluates the error based on the residual of discretised governing equation at the field nodes or by comparing the numerical value of any quantity of interest with the corresponding analytical value if the analytical value is known.

In general, the two types of *a posteriori* error estimators are the explicit and implicit (Ainsworth and Oden, 2000). The explicit error estimators evaluate the error based on the currently available data. Therefore the direct solution is available, namely the least-squares error estimator (Ainsworth and Oden, 2000). The implicit *a posteriori* error estimators involve the solution of a linear system of algebraic equations. Some examples of the implicit error estimator are the subdomain residual method and element residual method (Ainsworth and Oden, 2000). The recovery-based error estimators are based on the error evaluation between the numerically computed gradient and the approximation to the gradient; therefore the required order of the approximation of gradient is achieved.

Several adaptive FEM methods based on the error-controlled principle are applied to different engineering issues such as contact, elasto-plasticity, and shell problems under transient loading (Stein, 2003). Babuška (1986) explained the feedback controlled FEM with the h, p and $h - p$ versions, and demonstrated by a 1-D example that the $h - p$ version has the highest rate of convergence. Zienkiewicz and Craig (Babuška, 1986) pointed out that the use of p refinement is better than h refinement for singularities. They also explained the hierarchical FEM, in which there is no need to discard the earlier computed shape functions after each refinement; this leads to the mixed order interpolation.

Several *a posteriori* error estimators based on the different concepts are developed for FEM. Bank and Weiser (1985) presented three *a posteriori* error estimators for FEM in the energy norm, based on solving the local Neumann problem in each element. Babuška and Rheinboldt (1978) derived the error estimators based on an approach similar to the residual method, but it uses the negative norms of Sobolev spaces. Diening and Kreuzer (2008) studied the adaptive FEM (AFEM) for p-Laplacians like PDEs using linear functions and presented the rates of convergence by applying the AFEM to nonlinear Laplace equations.

Zienkiewicz and Zhu (1987) proposed an error estimator based on stress and coupled it with the h-refinement in which the stress field is interpolated by the same shape functions used for the interpolation of displacement. The residual error between the computed and assumed stress fields is reduced over the domain. The error is evaluated by the newly interpolated and the earlier computed values of stress. Babuška and Miller (1984a–c) explained the post-processing of FEM solutions in detail and an adaptive mesh refinement based on *a posteriori* error estimation. Zienkiewicz and Zhu (1992a and b) proposed a stress recovery technique with the values of stress at the Gauss points, as generally they are superconvergent. They considered a patch of elements around a concerned vertex node, and the value of stress at the concerned vertex node is interpolated by the values of stress with the least-squares (LS) procedure at the surrounding Gauss points within the patch of elements. The newly computed value of stress is used to evaluate an error at the concerned node, which in turn is used in the adaptive analysis.

Several *a posteriori* error estimators are developed for meshless methods (Liszka et al., 1996; Duarte and Oden, 1996; Gavete et al., 2001; Han and Meng, 2001; Lee and Shuai, 2007a–c; Rüter and Stenberg, 2008). Duarte and Oden (1996) used a partition of unity based $h - p$ cloud method for adaptive analysis, and derived an *a posteriori* error estimator following the procedure used in the FEM. Gavete et al. (2001) proposed an error indicator for EFGM, in which the FEM shape functions are replaced by the functions of local approximation of LS type. Han and Meng (2001) performed a theoretical analysis of error estimation in the RKPM.

Liszka et al. (1996) tested the h and p adaptivity of the hp meshless method, and discussed in detail the refinement of interpolation domain. Rüter and Stenberg (2008) developed *a posteriori* error estimates for mixed finite elements. As per the formulation, the approximate field of displacement is one order lower than the field of stress. They thus used an averaging technique to improve the displacement field. Lee and Shuai (2007a–c) developed the procedure of adaptive refinement by the RKPM functions and formulated an error estimator based on the superconvergence property of some extraction functions. Several meshless methods are extended to perform adaptive analyses (Liszka and Orkisz, 1980; Kee et al., 2007; Liu et al., 2008b) and solve numerous engineering problems (Demkowicz et al., 1984; Kondratyuk and Stevenson, 2008; Roquet and Saramito, 2008; Champaney et al., 2008; Carstensen, 2009; Blum et al., 2009).

It was observed during computations with the meshless methods that the accuracy of the approximated function at the concerned node is affected by the number of surrounding nodes included in the domain of interpolation. Several studies have been performed to develop a method to obtain the optimal number of support nodes (Duarte and Oden, 1996; Liszka et al., 1996; Nie et al., 2006). In the RDQ method, the number of nodes included in the local interpolation domain affects the accuracy near the locally high gradient computation region but have little effect in the remaining computation region. This phenomenon is called numerical pollution, as a result of which it is extremely difficult to define the number of interpolation nodes to be used near the locally high gradient regions. A new error recovery technique motivated by this problem and based on the LS concept utilizes the ARDQ method to improve the accuracy of solution by recovering the numerical error. The proposed error recovery technique both smooths and improves the values of approximated functions. It can also be incorporated easily into the existing code of the RDQ method due to its simplicity.

The objective of this chapter is to develop an adaptive RDQ method by coupling the RDQ method with an *a posteriori* error estimator based on the l_2 error norm. The local refinement or enrichment is performed at the concerned field node if the computed relative error in the value of the function at the concerned field node exceeds the minimum permitted value. An error recovery technique based on the LS approximation is developed to improve the accuracy of solution. During an adaptive refinement procedure, it is possible to generate a node that falls outside the computational domain. Therefore, to develop a robust adaptive algorithm for the ARDQ method, a convex hull approach with the cross product of vectors is used in a novel manner that ensures all the newly generated field nodes remain within the computational domain.

The proposed approach is demonstrated in this chapter by convex geometry, but it can also be used well for non-convex or irregular geometries, as explained later. The convergence study of the ARDQ method is later performed by solving several 1-D and 2-D locally high gradient problems. Based on this study, the convergence rates obtained after using the error recovery technique are better than those obtained by directly using the numerically computed values of function; also the values of local high gradient are well captured.

A semi-infinite plate with a central hole with an irregular or non-convex geometry along part of the boundary is solved to demonstrate the applicability of the ARDQ method while solving the problems of irregular geometry. For other problems with different geometries of irregular boundaries, the only change in the ARDQ method is the way of creating field and virtual nodes along the irregular boundary. The local parametric equations are developed for the refinement of field nodes during the adaptive iterations. Once this is achieved, the ARDQ method is directly applied as it is independent of the problems and computational domains.

The motivation behind the development of the ARDQ method with an error recovery technique is to solve locally high gradient problems by fine and relatively coarse nodal distributions near the peak and remaining domain, respectively. The potential applications of the method include hydrogel simulation, crack propagation, and other problems involving moving boundaries.

Based on the literature, few other meshless methods were proposed in the past to solve locally high gradient problems. Zhang et al. (2008) proposed a method called the GSM, in which the derivatives of the first and second order are approximated by the gradient smoothing technique. The mesh refinement is performed by the GSM via Delaunay diagram, which is numerically intensive when compared with the proposed ARDQ method, as explained later. The GSM uses conventional mesh, and requires more field nodes as compared with the ARDQ method for an equal reduction in the true error norm. Better convergence rates are observed in the ARDQ method, which can even work well with the initial distribution of random field nodes, while the GSM requires a uniform triangular grid (Zhang et al., 2008).

Duarte and Oden (1996) and Liszka et al. (1996) developed the $h - p$ adaptive and hp meshless methods, respectively, by a partition of unity concept applied over the local cloud of field nodes. They also created four new field nodes in the "refine" stage, as in the ARDQ method. However, these methods (Duarte and Oden, 1996; Liszka et al., 1996) are based on the Galerkin approach. As a result, the integration of domain is performed with a sufficiently dense background structure of cells to ensure the precise integration of function values.

However, no such integration is required in the ARDQ method, as it is based on the strong form approach. Therefore, it is relatively easy to implement and can accurately capture the high local gradients. Kee et al. (2007) developed a regularized LS radial point collocation method (RLS-RPCM) that involves a regularization technique to stabilize the RPCM method. The new nodes are added by Delaunay cells. As a result, the RLS-RPCM is computationally intensive as compared with the ARDQ method.

The ARDQ method offers several advantages as compared with the other existing meshless methods capable of capturing the locally high gradients. It has a simple yet effective algorithm of the adaptive refinement that ensures the newly created field nodes are always within the computational domain. It can work with initial distributions of either uniform or random field nodes, which is not the case with other meshless methods.

The ARDQ method is based on the strong form approach, such that no numerical integration is required. It is easy to implement and offers flexibility while generating new uniformly or randomly distributed field nodes. The ARDQ method coupled with the error recovery technique ensures that the effect of numerical pollution is minimized in the solution, and the function is reproduced up to the required order of approximation. The error recovery technique based on the LS approximation is developed in the next subsection.

6.2 ERROR RECOVERY TECHNIQUE IN ARDQ METHOD

It is observed in the RDQ method, as with other meshless methods, that the numerically computed solution near the region of local high gradient is affected by the number of total field nodes included during interpolation of field variables. Therefore, it may be very difficult to decide the number of nodes required for inclusion in the fixed RKPM interpolation. This issue led to development of an error recovery technique based on LS averaging. After numerical implementation, the newly developed technique both smooths and improves the solutions and it is thus coupled with the ARDQ method.

In the ARDQ method, the nodal parameters u_I at the field node are computed based on the certain order of the monomials included in the fixed RKPM interpolation function. Therefore, if total m nodes are used in the interpolation domain of the kth field node

$$f_k^* = \sum_{i=1}^{m} N_i \, u_i^*$$ (6.1)

where f_k^*, N_i, and u_k^* are the recovered values of the approximate function at the kth field node and the shape function and nodal parameter at the ith field node, respectively. The model expression of the nodal parameter value u_i^* at the ith field node is written as

$$u_i^* = \sum_{j=1}^{n} X_{ij} \beta_j, \quad \text{where} \quad X_{ij} = \{1 \quad x_i \quad y_i \quad \dots \quad y_n\}$$ (6.2)

where $j = 1$ to n are the monomial terms included in the interpolation of the ith field node, $i = 1$ to m are the number of total field nodes used in the local interpolation domain of the kth field node, and X_{ij} and β_j are the vectors of the nth order monomials and unknown coefficients, respectively. Let us take a 2-D problem as an example,

$$X_{ij} = \{1 \quad x_i \quad y_i \quad x_i^2 \quad x_i y_i \quad y_i^2\}, \quad n = 6$$ (6.3)

The residual error at the ith interpolation node is given as

$$r_i = \sum_{j=1}^{n} X_{ij} \beta_j - u_i$$ (6.4)

where u_i is the numerically computed value of the nodal parameter at the ith field node by the RDQ method. The total residual error over the domain of m local interpolation nodes is given as

$$E = \sum_{i=1}^{m} (r_i)^2$$ (6.5)

The minimization of error with respect to β_j results in

$$\frac{\partial E}{\partial \beta_j} = 2 \sum_{i=1}^{m} r_i \frac{\partial r_i}{\partial \beta_j} = 0 \Rightarrow \frac{\partial E}{\partial \beta_j} = 2 \sum_{i=1}^{m} X_{ij} \left[\sum_{j=1}^{n} X_{ij} \beta_j - u_i \right] = 0 \qquad (6.6)$$

Equation (6.6) in the matrix form is written as

$$
\begin{bmatrix}
\sum_{i=1}^{m} X_{i1} X_{i1} & \sum_{i=1}^{m} X_{i1} X_{i2} & \cdots & \sum_{i=1}^{m} X_{i1} X_{in} \\[2mm]
\sum_{i=1}^{m} X_{i2} X_{i1} & \sum_{i=1}^{m} X_{i2} X_{i2} & \cdots & \sum_{i=1}^{m} X_{i2} X_{in} \\[2mm]
\vdots & \cdots & \ddots & \vdots \\[2mm]
\sum_{i=1}^{m} X_{in} X_{i1} & \sum_{i=1}^{m} X_{in} X_{i2} & \cdots & \sum_{i=1}^{m} X_{in} X_{in}
\end{bmatrix}
\begin{Bmatrix}
\beta_1 \\ \beta_2 \\ \vdots \\ \beta_n
\end{Bmatrix}
=
\begin{Bmatrix}
\sum_{i=1}^{m} X_{i1} u_i \\[2mm]
\sum_{i=1}^{m} X_{i2} u_i \\[2mm]
\vdots \\[2mm]
\sum_{i=1}^{m} X_{in} u_i
\end{Bmatrix}
$$

$$(6.7)$$

The values β_j computed by Equation (6.7) are substituted in Equation (6.2) to determine the recovered values of nodal parameter, which are in turn substituted in Equation (6.1) to evaluate the recovered values of the function approximation. It is noted from Equation (6.2) that the recovered values of nodal parameter u_i^* are only valid for the expression of error recovery, as given by Equation (6.1) for the kth field node only. Thus, if a specific i^{th} field node is included in the multiple interpolation domains, the recovered values of the corresponding nodal parameter u_i^* of the ith field node should be evaluated separately for the respective different interpolation domains.

The technique of error recovery developed in this section is little affected by the number of field nodes included in the local interpolation domain of the function at the field nodes, as the computed value u_i^* is multiplied by the shape function N_i in Equation (6.1), thus smoothing u_i^* as per the spatial location of the ith node with respect to the concerned node k. The recovered values of the approximate function are considered for the evaluation of error during the refinement stage of the ARDQ method.

6.3 ADAPTIVE RDQ METHOD

The algorithm of the ARDQ method is broadly divided into two steps, the first for error evaluation and the second for the refinement procedure.

6.3.1 Computation of error in ARDQ method

In order to compute convergence rates, the global error is evaluated by the l_2 error norm averaged over the number of total field nodes and normalized by the maximum analytical value of the function as given (Mukherjee and Mukherjee, 1997)

$$\varepsilon = \frac{1}{\left| f^e \right|_{max}} \sqrt{\frac{1}{NP} \sum_{I=1}^{NP} \left[f_I^{(e)} - f_I^{(n)} \right]^2} \tag{6.8}$$

where ε is a global error in the solution, NP is the number of total field nodes scattered in the domain, and $f_I^{(e)}$ and $f_I^{(n)}$ are the analytical and numerical values of the function at the Ith field node, respectively. The true error norm based on l_2 error norm is given

$$\xi = \sqrt{\frac{\sum_{I=1}^{NP} \left[f_I^{(e)} - f_I^{(n)} \right]^2}{\sum_{I=1}^{NP} \left[f_I^{(e)} \right]^2}} \tag{6.9}$$

In the ARDQ method, as the domain is locally refined where the numerical error is high, the field node spacing h may not remain constant throughout the domain. As a result, an average nodal spacing is evaluated to compute the convergence rates as given (Aluru and Li, 2001)

$$h_x = \frac{L_x}{\sqrt{NP} - 1}, \quad h_y = \frac{L_y}{\sqrt{NP} - 1} \tag{6.10}$$

where h_x and h_y, and L_x and L_y are the average nodal spacings and the domain lengths in the x and y directions, respectively, and NP is the number of total field nodes in the domain. The spacing $h = h_x = h_y$ is used for the purpose of convergence plots.

The relative error in the approximated function at each field node and the global error computed by Equation (6.8) are used as the local refinement and adaptive refinement termination criteria in the ARDQ method, respectively. All the problems solved here to illustrate the ARDQ method have analytical solutions, but when the analytical solution $f_I^{(e)}$ is not available for any specific problem, a resultant function gradient at each field node, $(F_1)_i = \sqrt{f_{,x}^2 + f_{,y}^2 + f_{,xy}^2}$, where $i = 1,2,...,NP$, and $f_{,x}$, $f_{,y}$ and $f_{,xy}$ are the derivatives of function with respect to x and y variables, is used as a criterion of local refinement.

The difference between the values of function at the field nodes obtained during two successive adaptive iterations is computed and used as the refinement termination criterion. Let the previous and current iteration counts of adaptive refinement be n and $(n + 1)$, respectively, having NP and $(4\ NP)$ numbers of total field nodes, respectively, such that $3\ NP$ field nodes are newly created in the $(n+1)^{th}$ iteration. The values of function at the newly created $3\ NP$ field nodes corresponding to the nth iteration are computed by an interpolation using the existing total NP field nodes at the nth level of iteration. As the solution gets convergence, the interpolated values of function at $3\ NP$ field nodes corresponding to the nth adaptive iteration step will closely match their numerically computed values during the $(n+1)^{th}$ adaptive iteration. The second norm of the difference between the values of function at all $4\ NP$ field nodes is computed by their values at the nth and $(n+1)^{th}$ adaptive iterations. This second norm is used as the termination criterion in the adaptive refinement.

6.3.2 Adaptive refinement in ARDQ method

The h-adaptive refinement is implemented in the ARDQ method. It is sometimes possible to generate a node during the adaptive refinement that falls outside the computational domain. To overcome this problem, a novel approach of a convex hull with the cross product of vectors is developed for the ARDQ method. It ensures that the newly created field node is always inside the domain of computation. In order to maintain the adaptive algorithm of the ARDQ method as simple as possible without losing its effectiveness while capturing the values of locally high gradients, two arrangements are adopted to create new field nodes: uniform and random. As explained below, both the arrangements will ensure that the newly created nodes will increase the probability of correctly capturing the locally high gradient values irrespective of their directions. The following algorithm explains the 2-D adaptive refinement for the ARDQ method.

Step A — The governing PDE is solved initially by the RDQ method coupled with the error recovery technique, and the relative error in the approximated function is computed at each field node by $E_i = [(f_i - f_i^b)/f_i] \times 100$, where f_i and f_i^b are the exact and numerical values of function at the ith field node, respectively. The global and true error norms are computed by Equations (6.8) and (6.9), respectively.

Step B — The local refinement around the concerned i^{th} field node is performed if $E_i \geq E_{ip}$, where E_{ip} is the permitted value of the local error. The smallest distance d_m between the ith field node and its surrounding node is computed. If $d_m \geq Min\_Per\_Dist$, where Min_Per_Dist is the minimum

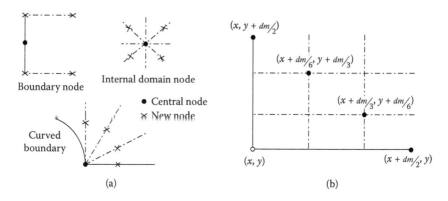

Figure 6.1 Mesh refinement for concerned node when located inside domain or along boundary (a) or at corner of computational domain (b) in ARDQ method. (From S. Mulay, H. Li, and S. See. (2010). *Computational Mechanics*, 45, 467–493. With permission.)

permitted distance between any two nodes in the domain, four new nodes are created around the concerned i^{th} field node, as shown in Figure 6.1, based on the location of the ith field node. If the ith node is within the computational domain, the new nodes are created at

$$\left(x+\frac{d_m}{2}, y+\frac{d_m}{2}\right), \left(x+\frac{d_m}{2}, y-\frac{d_m}{2}\right), \left(x-\frac{d_m}{2}, y+\frac{d_m}{2}\right) \text{ and}$$

$$\left(x-\frac{d_m}{2}, y-\frac{d_m}{2}\right)$$

(6.11)

If the ith field node is along the boundary $x = 0$, as shown in Figure 6.1, the new nodes are created at

$$\left(x, y+\frac{d_m}{2}\right), \left(x+\frac{d_m}{2}, y+\frac{d_m}{2}\right), \left(x+\frac{d_m}{2}, y-\frac{d_m}{2}\right) \text{ and } \left(x, y-\frac{d_m}{2}\right)$$

(6.12)

If there is an existing node at the location of newly created node during the refinement procedure, the next higher d_m is considered. The four nodes are added around the concerned i^{th} field node to ensure that at least one node is created in each of the four quadrants with reference to the concerned i^{th} field node. The motivation behind this approach is to capture the high gradients, $f_{,x}$, $f_{,y}$, and $f_{,xy}$, whether they are in the x or y directions. If the number of total field nodes (newly created plus already existing) exceeds the number of total virtual nodes during the adaptive refinement, depending on the difference between the number of total virtual and field nodes, only one or two new field nodes instead of four are created around only those field nodes having the values of absolute local gradient $(F_l)_i$ more than a certain cut-off value. This precaution is necessary to ensure that the total field nodes are always fewer than or equal to the total virtual nodes to avoid rank deficiency in the final stiffness matrix.

Instead of uniformly distributing the newly generated field nodes around the concerned i^{th} field node, as shown in Figure 6.1, they can also be distributed randomly with the motivation of capturing the high gradients in any direction. This is achieved by initially arranging the entire field nodes in the ascending order by their Euclidean distance from the concerned i^{th} field node. The location of each node with respect to the concerned i^{th} node (x_i, y_i) is checked by this sequence. If the pth node at (x_p, y_p) near the ith field node is falling in its first quadrant such that $x_p \geq x_i$ and $y_p \geq y_i$, then compute $\Delta x = |x_p - x_i|$ and $\Delta y = |y_p - y_i|$. Using the ith node coordinates (x_i, y_i), and Δx and Δy, one can generate a random node with the x and y coordinates as $x = x_i + [\text{rand}()/\text{RND\_MAX}]\Delta x$ and $y = y_i + [\text{rand}()/\text{RND\_MAX}]\Delta y$, where $rand()$ and RND_MAX are the functions of random value generation and the maximum possible generated random value, respectively. If the newly generated node does not overlap any of the existing nodes, it will be included in the refinement, and similarly new nodes are generated in the remaining three quadrants.

Step C — The process of adaptive iteration is terminated, if $\varepsilon \leq E_{gp}$ or $NP > NV$, where E_{gp} is the permitted global error, and NP and NV are the total numbers of field and virtual nodes distributed in the domain, respectively.

The objective of the variable Min_Per_Dist is to avoid any singularities in the moment matrix while computing the shape functions by the fixed RKPM interpolation function. Therefore, its numerical value depends on the total length of the computational domain and the number of total field nodes generated by the adaptive algorithm. For all the test problems in this chapter, it is assigned a numerical value equal to 1×10^{-6}. Therefore, this algorithm ensures that the moment matrix computed in the fixed RKPM interpolation function is not singular by setting the parameter Min_Per_Dist. The adaptive procedure can be easily extended to the problems with 3-D domains.

One of the important tasks in the refinement algorithm of the proposed ARDQ method is checking whether the newly created node falls within the computational domain or outside it. This novel approach involves a convex hull with cross products of vectors; the domain of computation is treated as a convex hull acting as an outermost envelope covering all the given points. As a result, if any two points within the convex hull are joined to form a straight line, all points on the newly created line also fall within the convex hull.

For a given computational domain in the ARDQ method, a convex hull is constructed starting with a node having the smallest x and y coordinates and continuing with the outermost nodes of the domain. For example in the 2-D rectangular domain, the four corner nodes will form a convex hull covering the whole computational domain. Once the convex hull is formed, the newly created node is tested to see whether it falls within the generated convex hull or outside of by taking the cross product of vectors formed by each side of the convex hull with the newly created node. For example, assume the nodes $P_0(x_0, y_0)$, $P_1(x_1, y_1)$, $P_2(x_2, y_2)$ and $P_3(x_3, y_3)$ form a convex hull with the edges given as P_0 to P_1, P_1 to P_2, P_2 to P_3, and P_3 to P_0.

If a new node $P_n(x_n, y_n)$ is generated, the cross product of vectors $\vec{P}_{01} = P_1 - P_0$ and $\vec{P}_{0n} = P_n - P_0$ is computed. If $\vec{P}_{01} \times \vec{P}_{0n} \geq 0$, and similarly if $\vec{P}_{12} \times \vec{P}_{1n} \geq 0$, $\vec{P}_{23} \times \vec{P}_{2n} \geq 0$, and $\vec{P}_{30} \times \vec{P}_{3n} \geq 0$, we conclude that the newly generated node P_n falls within the convex hull formed by the field nodes P_0, P_1, P_2 and P_3. If P_n does not fall in the convex hull, its coordinates must be adjusted with reference to the specific edge of the convex hull for which the cross product was less than zero. This ensures that all the newly created field nodes are within the computational domain. The cross product approach is used in Graham scan algorithms to generate the convex hull, but it is used very differently and innovatively in the ARDQ method. The convex hull approach explained above is shown in Figure 6.2.

Although the algorithm is explained here by the computational domain with a convex or uniform geometry of boundary, it can applied well to an irregular or non-convex geometry, as demonstrated in the test problem of a semi-infinite plate with a central hole. As long as the

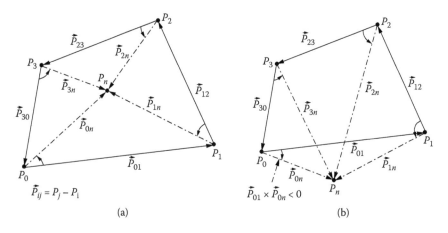

(a)   (b)

Figure 6.2 Working principle of convex hull approach when newly added field node is inside (a) and outside (b) computational domain. (From S. Mulay, H. Li, and S. See. (2010). *Computational Mechanics*, 45, 467–493. With permission.)

computational domain has a well defined boundary, which is normally required for a well-posed problem, the proposed algorithm will work. In case of an irregular or non-convex geometric boundary, a local parametric equation $f[x(t), y(t), z(t)]$ where t is a parameter is developed to model the geometry by the cubic spline or non-uniform rational B-spline (NURBS) curves with the nodes along the irregular boundary as the control points. While checking a newly created node $P_n(x_n, y_n)$ as per the above explained algorithm, the irregular boundary is traced by increasing the parameter t and obtaining the successive nodes on the curve that can be used to compute the cross product of vectors. A similar approach is adopted while solving the problem of semi-infinite plate with a central hole.

In summary, the adaptive refinement procedure for the ARDQ method with the error recovery technique is discussed in this section. It is shown that the refinement algorithm is fairly flexible and can be modified at any level based on the requirement, and readily extended to problems with 3-D domains. Several 1-D and 2-D test problems are solved in the next subsection by the ARDQ method, and it is shown that the ARDQ method yields a reasonable rate of convergence.

6.4 CONVERGENCE ANALYSIS IN ARDQ METHOD

The ARDQ method is applied to solve several locally high gradient problems in this section and their convergence analysis is performed. The values of $E_{ip} = 0.01$ and $E_{gp} = 1 \times 10^{-6}$ are set for all the test problems. The objective of this section is to test how accurately the ARDQ method converges the numerical solution to the corresponding analytical values with each iteration of adaptive refinement.

The analytical solutions for all the present test problems contain at least third or higher order monomial terms. Therefore, it is essential to include maximum up to second order monomial terms in the approximation of function while performing convergence studies. When the convergence rates obtained by the ARDQ method are compared with those by the RDQ method, it is essential for both the methods to maintain the same order of the continuity of function approximation and the same distribution pattern of uniform or random nodes at the beginning of computation.

Therefore, all the test problems are solved here by including up to the complete second order monomial terms in the polynomial basis of the function approximation. It means the approximated function is expected to have second order continuity. The model expression of the nodal parameter value, as given in Equation (6.2), enforces the required order of continuity on the nodal parameter value, and the residual error is computed correspondingly in Equation (6.4). To achieve a consistency between the ARDQ method and error recovery technique while approximating a function, complete monomial terms up to the second order are included in Equation (6.3), i.e., the second order continuity is enforced on the refined values of nodal parameter.

6.4.1 One-dimensional test problems

The first 1-D problem is a mixed boundary condition problem with a fourth order monomial in the analytical solution. The problem is approximated by including up to second order monomial terms in the polynomial basis of function approximation. The governing equation and boundary conditions are given as

$$\frac{d^2f}{dx^2} = \frac{105}{2}x^2 - \frac{15}{2}, \quad (-1 < x < 1) \tag{6.13}$$

$$f(x = -1) = 1, \quad \frac{df}{dx}(x = 1) = 10 \tag{6.14}$$

The analytical solution is given as $f(x) = (35/8)x^4 - (15/4)x^2 + (3/8)$. This problem is solved by the ARDQ method with the domain discretised by 21 uniform and randomly distributed field nodes separately, at the beginning of computation, and combined with 1001 and 641 uniform and cosine distributed virtual nodes. Figure 6.3 shows the successive distributions of the field nodes according to the adaptive algorithm when 21 total random field nodes are used at the beginning of the adaptive refinement.

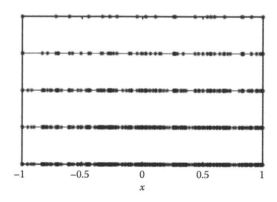

Figure 6.3 Distributions of field nodes during successive adaptive iterations when 21 randomly distributed field nodes at beginning of computation are combined with 1001 uniform virtual nodes for first 1-D problem. (From S. Mulay, H. Li, and S. See. (2010). *Computational Mechanics, 45*, 467–493. With permission.)

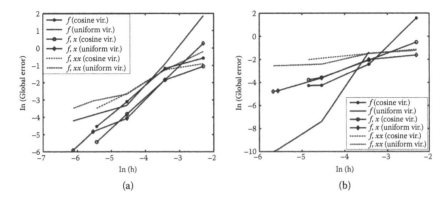

Figure 6.4 Convergence plots when uniformly (a) and randomly (b) distributed field nodes at beginning of computation are combined with uniform and cosine distributed virtual nodes for first 1-D problem of Poisson equation. (From S. Mulay, H. Li, and S. See. (2010). *Computational Mechanics*, 45, 467–493. With permission.)

Figure 6.4a and Figure 6.4b show the convergence plots by the uniform and random distributions of field nodes at the beginning of computation, respectively. The corresponding convergence rates are given in Table 6.1 and Table 6.2. Figure 6.5 shows the reduction in the true error norm and a comparison of the numerical and analytical values of the function. It is seen from Figure 6.4 and Figure 6.5 that the successive values of the global and true error norms are steadily reduced with the best reduction obtained by the random field nodes at the beginning of computation coupled with the uniform virtual nodes, as the true error is reduced from 0.21 to 1.3386×10^{-4}.

The second 1-D problem is a locally high gradient problem solved by the ARDQ method by including up to second order monomial terms in the

Table 6.1 Convergence rates obtained by ARDQ method for first 1-D problem of Poisson equation by combining uniform field notes at beginning of computation with cosine and uniform virtual nodes

Function	Convergence rate (cosine virtual nodes)	Convergence rate (uniform virtual nodes)
f	1.4	1.6
$f_{,x}$	1.43	1.6
$f_{,xx}$	0.9	0.9

Source: S. Mulay, H. Li, and S. See. (2010). *Computational Mechanics*, 45, 467–493. With permission.

Table 6.2 Convergence rates for first 1-D problem of Poisson equation by ARDQ method when random field nodes at beginning of computation are combined with cosine and uniform virtual nodes

Function	Convergence rate (cosine virtual nodes)	Convergence rate (uniform virtual nodes)
f	2.31	3.8
$f_{,x}$	1.32	1.1
$f_{,xx}$	0.4	0.5

Source: S. Mulay, H. Li, and S. See. (2010). *Computational Mechanics*, 45, 467–493. With permission.

polynomial basis of function approximation. The governing equation and boundary conditions are given as

$$\frac{d^2 f}{dx^2} = -6x - \left[\left(\frac{2}{\alpha^2} \right) - 4 \left(\frac{x-\beta}{\alpha^2} \right)^2 \right] \exp \left[-\left(\frac{x-\beta}{\alpha} \right)^2 \right] \quad (0 < x < 1) \quad (6.15)$$

$$f(x=0) = \exp \left[-\left(\frac{\beta^2}{\alpha^2} \right) \right], \quad \frac{df(x=1)}{dx} = -3 - 2 \left(\frac{1-\beta}{\alpha^2} \right) \exp \left[-\left(\frac{1-\beta}{\alpha} \right)^2 \right] \quad (6.16)$$

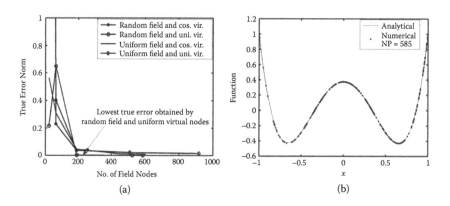

Figure 6.5 Reduction in true error norm (a) where lowest value is obtained by combining random field nodes distributed at beginning with uniform virtual nodes and the comparison of numerical and analytical values of field variable distribution (b) when first 1-D problem is solved by ARDQ method. (From S. Mulay, H. Li, and S. See. (2010). *Computational Mechanics*, 45, 467–493. With permission.)

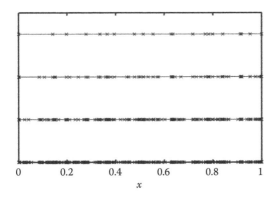

Figure 6.6 Distributions of field nodes during adaptive iterations of ARDQ method when 21 randomly distributed field nodes at beginning of computation are combined with uniform virtual nodes for second 1-D problem. (From S. Mulay, H. Li, and S. See. (2010). *Computational Mechanics*, 45, 467–493. With permission.)

The analytical solution is given as $f(x) = -x^3 + \exp[-[(x-\beta)/\alpha]^2]$. This problem is solved with the domain discretised by 21 uniform and randomly distributed field nodes, at the beginning of computation, separately coupled with 641 uniform and cosine distributed virtual nodes. The successive distributions of the field nodes by the adaptive algorithm are shown in Figure 6.6.

The convergence plots for the uniform and random field nodes at the beginning of computation are shown in Figure 6.7a and Figure 6.7b, respectively, and the corresponding convergence rates are given in Table 6.3 and Table 6.4. The reduction in the true error norm is plotted in Figure 6.8, and the corresponding values are given in Table 6.5 and Table 6.6. Note from Figure 6.8 that the lowest value of error norm is obtained by combining the random field nodes at the beginning of computation with the uniform virtual nodes.

Figure 6.9 shows the comparison between the numerical and analytical values of function and the first and second order derivatives. The figure indicates that the local peak value of the function is correctly captured by the ARDQ method. Comparing the values of global error norms obtained by the ARDQ method coupled with the error recovery technique to results from the RDQ method by uniform increments in the field nodes, we see that the ARDQ method shows a much lower value of the global error norm than the RDQ method for the same number of field nodes in the domain, as also observed from Table 6.7.

In summary, all the solutions of 1-D test problems show that the ARDQ method yields good rates of convergence for locally high gradient

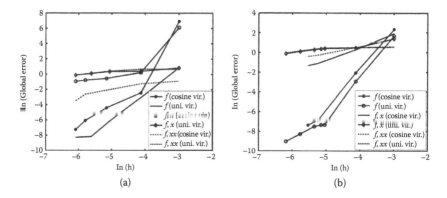

Figure 6.7 Convergence curves when uniformly (a) and randomly (b) distributed field nodes at beginning of computation are combined with uniform and cosine virtual nodes during solution of second 1-D problem by ARDQ method. (From S. Mulay, H. Li, and S. See. (2010). *Computational Mechanics*, 45, 467–493. With permission.)

problems. The local peak values are well captured, possibly because the ARDQ method is a strong form and it discretizes the governing differential equations and boundary conditions at every virtual node within the domain and along the boundaries, respectively. The comparison of the RDQ method with the uniform increment of field nodes and the ARDQ method coupled with the error recovery technique (Figure 6.10) demonstrates that the global error is reduced by implementing the error recovery technique in the ARDQ method, which leads to the reduction in the numerical pollution.

Table 6.3 Convergence rates for second 1-D problem of Poisson equation by combining uniform field nodes at beginning of computation with cosine and uniform virtual nodes when solved by ARDQ method

Function	Convergence rate (cosine virtual nodes)	Convergence rate (uniform virtual nodes)
f	4.22	3.2
$f_{,x}$	2.1	0.5
$f_{,xx}$	0.8	0.5

Source: S. Mulay, H. Li, and S. See. (2010). *Computational Mechanics*, 45, 467–493. With permission.

Table 6.4 Convergence rates for second 1-D problem of Poisson equation by combining random field nodes at beginning of computation with cosine and uniform virtual nodes when solved by ARDQ method

Function	Convergence rate (cosine virtual nodes)	Convergence rate (uniform virtual nodes)
f	4.0	3.5
$f_{,x}$	1.3	0.45
$f_{,xx}$	0.5	0.5

Source: S. Mulay, H. Li, and S. See. (2010). Computational Mechanics, 45, 467–493. With permission.

In order to compare the convergence rates obtained by the ARDQ and RDQ methods, the virtual nodes remain fixed and only variation of the field nodes is allowed. The adaptive algorithm stops the refinement when the number of total field nodes exceeds the total number of virtual nodes. Therefore, if the field nodes are randomly distributed at the beginning, the adaptive nature of the refinement may not be apparent because adaptive refinement stops when the total field nodes exceed the total virtual nodes.

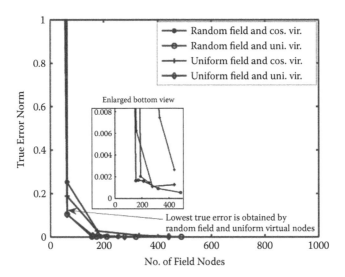

Figure 6.8 Reduction in true error norm with increase in field nodes when second 1-D problem is solved by ARDQ method. (From S. Mulay, H. Li, and S. See. (2010). Computational Mechanics, 45, 467–493. With permission.)

Table 6.5 Reduction in true error norm for second 1-d problem when uniform and random field nodes at beginning of computation are combined with cosine virtual nodes

Adaptive iteration	1	2	3	4	5
Uniform field nodes	21	63	175	331	441
True error norm	2292.1	0.18865	0.02795	0.00743	0.00263
Random field nodes	21	63	189	253	
True error norm	20.5	0.252	0.00203	0.001444	

Source: S. Mulay, H. Li, and S. See. (2010). *Computational Mechanics*, 45, 467–493. With permission.

Comparing the convergence rates obtained by the ARDQ method in this chapter with those by the RDQ method in Chapter 4, we note that the ARDQ method achieves the values of equal or better convergence rates at relatively lower numbers of field nodes. This is a key result as it highlights one objective of the development of the ARDQ method: achieving better convergence rates at relatively lower numbers of field nodes.

Table 6.6 Reduction in true error norm for second 1-d problem when uniform and random field nodes at beginning of computation are separately combined with uniform virtual nodes

Adaptive Iteration	1	2	3	4	5	6	7
Uniform nodes	21	63	159	277	441		
True error norm	5.5	0.1	6.23×10^{-3}	1.1×10^{-3}	1.2×10^{-3}		
Random nodes	21	63	155	173	211	319	487
True error norm	10.6	0.1	1.63×10^{-3}	1.7×10^{-3}	1.6×10^{-3}	9.1×10^{-4}	5.4×10^{-4}

Source: S. Mulay, H. Li, and S. See. (2010). *Computational Mechanics*, 45, 467–493. With permission.

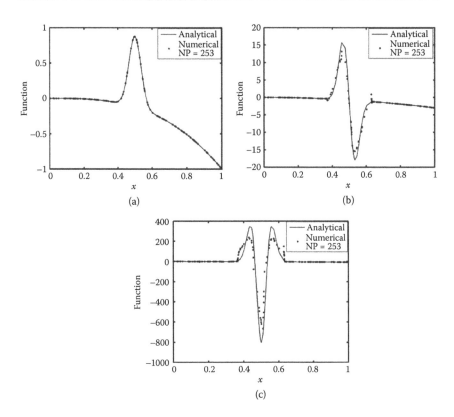

Figure 6.9 Numerical and analytical distributions of values of f (a), (df/dx) (b), and (d^2f/dx^2) (c) when second 1-D problem is solved by ARDQ method. (From S. Mulay, H. Li, and S. See. (2010). *Computational Mechanics*, 45, 467–493. With permission.)

Table 6.7 Reduction in global error norm when second 1-d problem is solved by ARDQ method coupled with error recovery technique and RDQ method with uniform increments in field nodes

Field Nodes	RDQ Method (%)	ARDQ Method (%)
21	1.843×10^2	1.843×10^2
63	13.4	8.526×10^{-2}
175	0.2332	1.21×10^{-2}
331	0.02225	2.363×10^{-3}
441	5.96×10^{-3}	7.238×10^{-4}

Source: S. Mulay, H. Li, and S. See. (2010). *Computational Mechanics*, 45, 467–493. With permission.

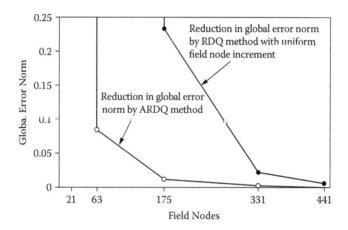

Figure 6.10 Comparison of values of global error norms obtained by solving second 1-D problem using ARDQ method coupled with error recovery technique and RDQ method with uniform refinement of field nodes. (From S. Mulay, H. Li, and S. See. (2010). *Computational Mechanics*, 45, 467–493. With permission.)

6.4.2 Two-dimensional test problems

Several 2-D test problems are solved numerically in this section, and the corresponding convergence analyses are performed. The first 2-D problem concerns steady-state heat conduction without internal heat generation in a slab. There is no temperature gradient in the z-direction, but a temperature gradient exists along the y direction. The governing equation and boundary conditions are given as

$$\frac{d^2T}{dx^2} + \frac{d^2T}{dy^2} = 0 , \quad (0 < x < a) \quad \text{and} \quad (0 < y < b) \tag{6.17}$$

$$T(x=0) = 0, \quad T(x=a) = 0, \quad T(y=0) = 0 \quad \text{and} \quad T(y=b) = 100 \tag{6.18}$$

The analytical solution based on the Fourier series is given as $T(x,y) = T_s \sum_{n=1}^{\infty} [n\pi \sinh(n\pi b/a)]^{-1} 2[1-(-1)^n] \sin(n\pi x/a) \sinh(n\pi y/a)$. This problem is solved by separately combining the uniform and random distributions of the field nodes at the beginning of computation with the cosine and uniform distributions of the virtual nodes. The initial and final distributions for a total of 25 random field nodes at the beginning of adaptive computation are shown in Figure 6.11. The convergence plots of the temperature and curves of reduction in the true error norm are plotted in Figure 6.12a and b, respectively. The

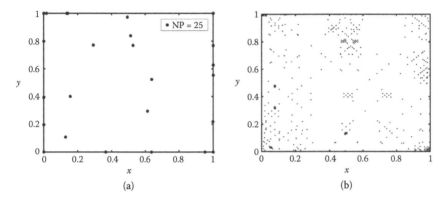

Figure 6.11 Distributions of random field nodes at beginning (a) and end (b) of solution of first 2-D problem of steady-state heat conduction by ARDQ method. (From S. Mulay, H. Li, and S. See. (2010). *Computational Mechanics*, 45, 467–493. With permission.)

corresponding values are given in Tables 6.8 through Table 6.10. Figure 6.13a and b show the distributions of temperature and its contour plots, respectively, corresponding to the final stage shown in Figure 6.11b. From Figure 6.13, we see that the gradient of temperature in the y direction is captured well and the boundary conditions are exactly imposed.

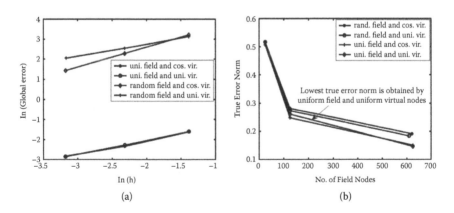

Figure 6.12 Convergence curves when uniform and random field nodes at beginning of computation are combined with cosine and uniform virtual nodes (a) and reduction in true error norm (b) where lowest true error is obtained by uniform field nodes at start combined with uniform virtual nodes for first 2-D problem. (From S. Mulay, H. Li, and S. See. (2010). *Computational Mechanics*, 45, 467–493. With permission.)

Table 6.8 Convergence rates for first 2-D problem when uniform and random field nodes at beginning of computation are combined with cosine and uniform virtual nodes

Node type	Convergence rate (uniform field nodes)	Convergence rate (random field nodes)
Cosine virtual nodes	0.7	1.0
Uniform virtual nodes	0.7	0.7

Source: S. Mulay, H. Li, and S. See. (2010). Computational Mechanics, 45, 467–493. With permission.

Table 6.9 Reduction in true error norm for first 2-D problem when uniform and random field nodes at beginning of computation are combined with cosine virtual nodes

Adaptive Iteration	1	2	3
Uniform field nodes	25	125	625
True error norm	0.506	0.2477	0.1507
Random field nodes	25	125	621
True error norm	0.5136	0.2812	0.191

Source: S. Mulay, H. Li, and S. See. (2010). Computational Mechanics, 45, 467–493. With permission.

Table 6.10 Reduction in true error norm for first 2-D problem when uniform and random field nodes at beginning of computation are combined with uniform virtual nodes

Adaptive Iteration	1	2	3
Uniform field nodes	25	125	625
True error norm	0.516	0.2606	0.1464
Random field nodes	25	125	609
True error norm	0.5172	0.2733	0.1836

Source: S. Mulay, H. Li, and S. See. (2010). Computational Mechanics, 45, 467–493. With permission.

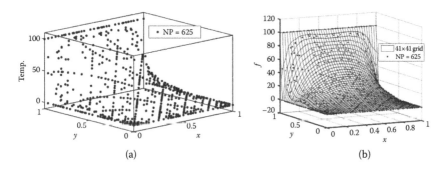

Figure 6.13 Numerical distribution of temperature (a) and temperature contours over uniform grid (b) by 625 field nodes when first 2-D problem is solved by ARDQ method. (From S. Mulay, H. Li, and S. See. (2010). *Computational Mechanics*, 45, 467–493. With permission.)

Figure 6.14a shows the comparison of temperature at the field nodes along (0.5, y) when computed by the RDQ method with uniform increments in the field nodes and the ARDQ method coupled with the error recovery technique. Figure 6.14b shows the global error norms computed by both the RDQ and ARDQ methods. The temperature values computed by the ARDQ method are more accurate than those from the RDQ method.

The second 2-D problem is also with a high local gradient at location (0.5, 0.5). The governing equation and boundary conditions are given as

$$\frac{d^2f}{dx^2} + \frac{d^2f}{dy^2} = -(6x) - (6y) - \left[\left(\frac{4}{\alpha^2}\right) - 4\left(\frac{x-\beta}{\alpha^2}\right) - 4\left(\frac{y-\beta}{\alpha^2}\right)\right]$$

$$\times \exp\left[-\left(\frac{x-\beta}{\alpha}\right)^2 - \left(\frac{y-\beta}{\alpha}\right)^2\right], \quad (0 < x < 1) \text{ and } (0 < y < 1)$$

(6.19)

$$f(x=1,y) = -1 - y^3 + \exp\left[-\left(\frac{(1-\beta)}{\alpha}\right)^2 - \left(\frac{(y-\beta)}{\alpha}\right)^2\right]$$

(6.20)

$$\frac{df}{dy}(x, y=0) = \frac{2\beta}{\alpha^2}\exp\left[-\left(\frac{\beta}{\alpha}\right)^2 - \left(\frac{(x-\beta)}{\alpha}\right)^2\right]$$

(6.21)

$$\frac{df}{dy}(x, y=1) = -3 - 2\left(\frac{(1-\beta)}{\alpha^2}\right)\exp\left[-\left(\frac{(x-\beta)}{\alpha}\right)^2 - \left(\frac{(1-\beta)}{\alpha}\right)^2\right]$$

(6.22)

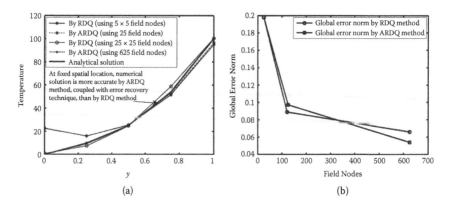

Figure 6.14 Distribution of temperature at field nodes along (0.5, y) (a) and reduction in global error norm (b) where global error obtained by ARDQ method is lower than that obtained by RDQ method for equal number of field nodes when first 2-D problem is solved by ARDQ method. (From S. Mulay, H. Li, and S. See. (2010). *Computational Mechanics*, 45, 467–493. With permission.)

The analytical solution is given as $f(x,y) = -x^3 - y^3 + \exp\{-[(x-\beta)/\alpha]^2 - [(y-\beta)/\alpha]^2\}$. This problem is solved by including up to second order monomials in the polynomial basis of function approximation, and by separately combining the total 25 uniform and random distributions of the field nodes at the start of computation with the total 33 × 33 cosine virtual nodes. The initial and final nodal distributions are shown in Figure 6.15. The

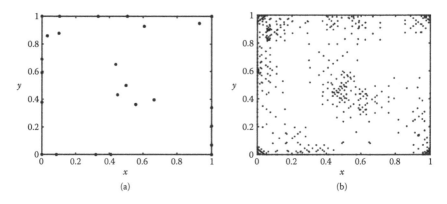

Figure 6.15 Initial 25 random field nodes (a) and final total 625 field nodes (b) during solution of second 2-D problem by ARDQ method. (From S. Mulay, H. Li, and S. See. (2010). *Computational Mechanics*, 45, 467–493. With permission.)

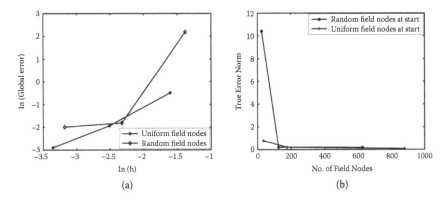

(a)

(b)

Figure 6.16 Convergence curves when uniform and random field nodes at beginning are separately combined with cosine virtual nodes (a) and reduction in true error norm (b) for solution of second 2-D problem by ARDQ method. (From S. Mulay, H. Li, and S. See. (2010). *Computational Mechanics*, 45, 467–493. With permission.)

convergence plots are given in Figure 6.16a, and the values are obtained as 2.3 and 2.4 for the uniform and random field nodes, respectively. The reduction in the true error norm is shown in Figure 6.16b. The comparison of the numerical and analytical values of function and their corresponding derivatives are shown in Figure 6.17 through Figure 6.21.

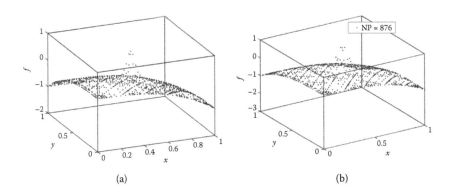

(a)

(b)

Figure 6.17 Comparison of analytical (a) and numerical (b) values of function for second 2-D problem by 876 field nodes obtained at end of adaptive refinement by ARDQ method. (From S. Mulay, H. Li, and S. See. (2010). *Computational Mechanics*, 45, 467–493. With permission.)

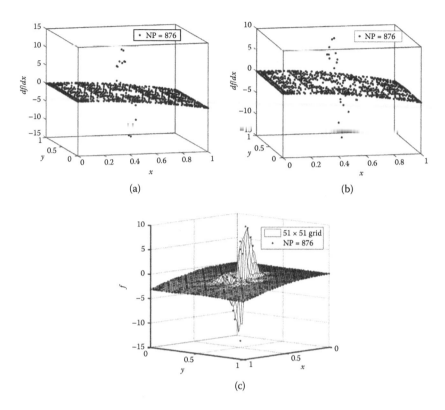

Figure 6.18 Plots of first-order derivative of function with respect to x, as analytical (a) and numerical (b) values and contour plot by uniform grid of 51 × 51 nodes (c) for second 2-D problem of local high gradient when a total of 876 field nodes is obtained at end of adaptive refinement by ARDQ method. (From S. Mulay, H. Li, and S. See. (2010). *Computational Mechanics*, 45, 467–493. With permission.)

The third 2-D problem is of steady-state heat conduction with a heat source. The governing equation and boundary conditions are given as

$$\nabla^2 T = -2\, s^2 \sec h^2[s(y-0.5)]\tanh[s(y-0.5)],\ (0 < x < 0.5) \text{ and } (0 < y < 1)$$

$$(6.23)$$

$$\frac{\partial T(x=0)}{\partial x} = 0, \quad \text{and} \quad \frac{\partial T(x=0.5)}{\partial x} = 0 \tag{6.24}$$

$$T(y=0) = -\tanh\!\left(\frac{s}{2}\right), \quad \text{and} \quad T(y=1) = \tanh\!\left(\frac{s}{2}\right) \tag{6.25}$$

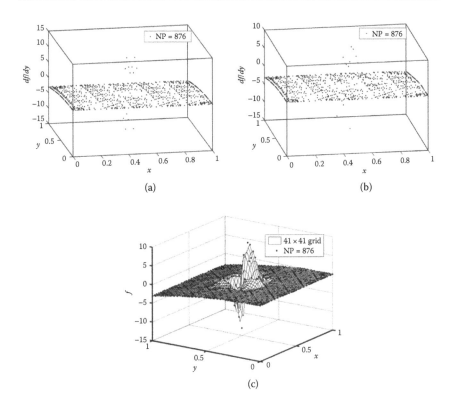

Figure 6.19 Plots of first-order derivative of function with respect to y , as analytical (a) and numerical (b) values and contour plot by uniform grid of 41 × 41 nodes (c) for the second 2-D problem of local high gradient when a total of 876 field nodes is obtained at end of adaptive refinement by ARDQ method. (From S. Mulay, H. Li, and S. See. (2010). *Computational Mechanics, 45,* 467–493. With permission.)

The analytical solution is given as $T = \tanh[s\,(y-0.5)]$. This problem is solved by total 36 uniform and random field nodes at the beginning, separately combined with the cosine and uniform virtual nodes. The initial (36 nodes) and final (896 nodes) distributions of the field nodes are shown in Figure 6.22. The comparison of the convergence rates of the temperature computed with and without the error recovery technique is plotted in Figure 6.23. The reduction in the true error norm is shown in Figure 6.24, and the comparison between the analytical and numerical values of the temperature is shown in Figure 6.25.

The fourth 2-D problem also involves a local high gradient with a peak at (0.5, 0.5). It is solved by including up to second order monomials in the

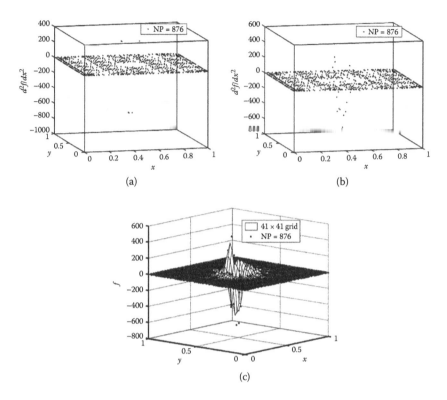

Figure 6.20 Plots of second order derivative of function with respect to *x* , as analytical (a) and numerical (b) values and contour plot by uniform grid of 41 × 41 nodes (c) for second 2-D problem of local high gradient when a total of 876 field nodes is obtained at end of adaptive refinement by ARDQ method. (From S. Mulay, H. Li, and S. See. (2010). *Computational Mechanics*, 45, 467–493. With permission.)

function approximation. The governing equation and boundary conditions are given as

$$\nabla^2 f = \left[-400 + (200\ x - 100)^2 + (200\ y - 100)^2 \right]$$

$$e^{\left[-100\left(x - \frac{1}{2} \right)^2 - 100\left(y - \frac{1}{2} \right)^2 \right]} , x, y \in [0,\ 1] \tag{6.26}$$

$$\frac{\partial f}{\partial n} = 0 \quad \text{along } x = 0 \text{ and } y = 0, \quad \text{and} \quad f = 0 \quad \text{along } x = 1 \quad \text{and} \quad y = 1 \tag{6.27}$$

The analytical solution is given as $f(x,\ y) = e^{[-100\ (x-1/2)^2 - 100\ (y-1/2)^2]}$. This problem is solved by total of 36 field nodes uniformly distributed at the

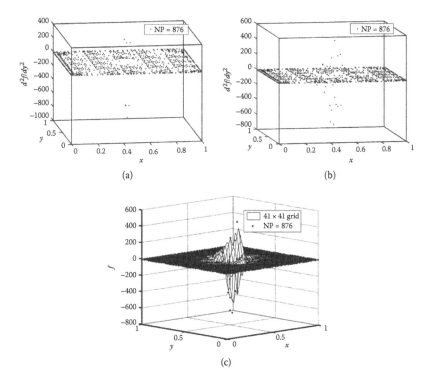

Figure 6.21 Plots of second order derivative of function with respect to y , as analytical (a) and numerical (b) values and contour plot by uniform grid of 41 × 41 nodes (c) for second 2-D problem of local high gradient when total of 876 field nodes is obtained at end of adaptive refinement by ARDQ method. (From S. Mulay, H. Li, and S. See. (2010). *Computational Mechanics*, 45, 467–493. With permission.)

beginning of computation and 33 × 33 cosine virtual nodes. The initial and final distributions of the field nodes are shown in Figure 6.26. The convergence plot of the function is given in Figure 6.27a. It shows that the convergence rate is indeed improved by the error recovery technique. The reduction in the true error norm with an adaptive refinement by the ARDQ method appears in Figure 6.27b. The comparison of the analytical and numerical values of the function is shown in Figure 6.28, and the contour plots of the numerical values of function are given in Figure 6.29.

6.4.3 Semi-infinite plate with central hole

In this section, a semi-infinite plate with a central hole under the action of normal load is solved by the plane stress criterion. Only one quarter of the domain is considered for the actual computation due to symmetric geometry and

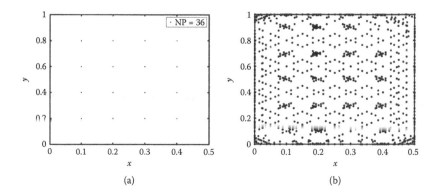

Figure 6.22 Initial total 36 field nodes (a) and final 896 field nodes (b) during solution of
the third 2-D problem by ARDQ method. (From S. Mulay, H. Li, and S. See.
(2010). *Computational Mechanics*, 45, 467–493. With permission.)

loading, as shown in Figure 6.30. The equations of mechanical equilibrium
in the displacement form according to the plane stress condition are given as

$$\frac{E}{(1-v_0^2)}\left[\frac{\partial^2 u}{\partial x^2}+\left(\frac{1-v_0}{2}\right)\frac{\partial^2 u}{\partial y^2}+\left(\frac{1+v_0}{2}\right)\frac{\partial^2 v}{\partial x \partial y}\right]+B_x=0 \qquad (6.28)$$

$$\frac{E}{(1-v_0^2)}\left[\frac{\partial^2 v}{\partial y^2}+\left(\frac{1-v_0}{2}\right)\frac{\partial^2 v}{\partial x^2}+\left(\frac{1+v_0}{2}\right)\frac{\partial^2 u}{\partial x \partial y}\right]+B_y=0 \qquad (6.29)$$

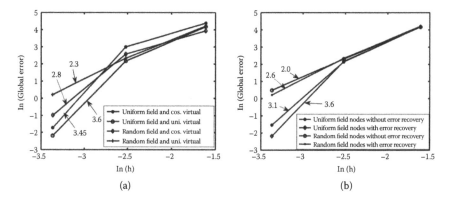

Figure 6.23 Convergence plots when ARDQ method is coupled with error recovery technique (a) and convergence rates obtained with and without using
error recovery technique (b) when third 2-D problem is solved by ARDQ
method. (From S. Mulay, H. Li, and S. See. (2010). *Computational Mechanics*, 45,
467–493. With permission.)

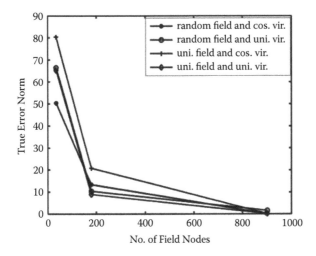

Figure 6.24 Reduction in true error norm with increase in field nodes when third 2-D problem is solved by ARDQ method. (From S. Mulay, H. Li, and S. See. (2010). *Computational Mechanics*, 45, 467–493. With permission.)

where B_x and B_y, and u and v are the body forces and displacements, respectively, in the x and y directions, respectively, and E and v_0 are the modulus of elasticity and Poisson ratio, respectively. The body forces B_x and B_y are neglected in the present results. The analytical solutions of this problem are given in Chapter 3.

This problem is solved by an adaptive refinement with an initial total of 35 uniformly distributed nodes and 961 cosine virtual nodes. The

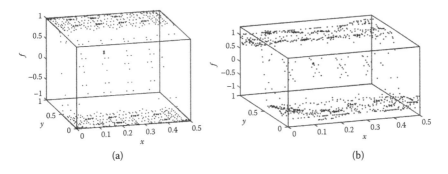

Figure 6.25 Comparison of analytical (a) and numerical (b) distributions of temperature for third 2-D problem. (From S. Mulay, H. Li, and S. See. (2010). *Computational Mechanics*, 45, 467–493. With permission.)

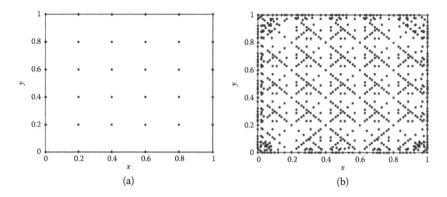

Figure 6.26 Distributions of the initial 36 uniform field nodes (a) and final total 900 field nodes (b) obtained during solution of fourth 2-D problem by ARDQ method. (From S. Mulay, H. Li, and S. See. (2010). *Computational Mechanics*, 45, 467–493. With permission.)

distributions of the initial and final field nodes and the virtual nodes are shown in Figure 6.31. This problem is also solved by uniform refinement with totals of 6×6, 11×11, 21×21, and 31×31 field nodes combined with total 34×34 cosine virtual nodes. The convergence curves corresponding to the uniform and adaptive refinements are plotted in Figure 6.32a, and better convergence rates are obtained by the adaptive refinement.

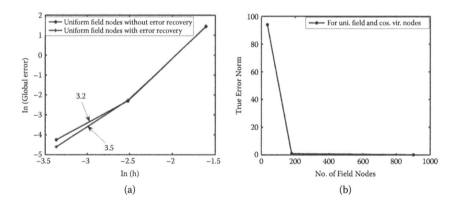

Figure 6.27 Convergence curves with and without using error recovery technique (a) and reduction in true error norm (b) when fourth 2-D problem is solved by ARDQ method. (From S. Mulay, H. Li, and S. See. (2010). *Computational Mechanics*, 45, 467–493. With permission.)

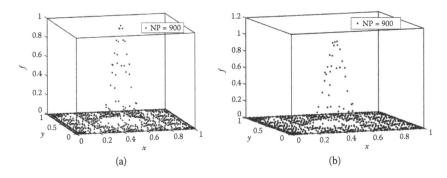

Figure 6.28 Comparison of analytical (a) and numerical (b) values of function for fourth 2-D problem of local high gradient solved by ARDQ method. (From S. Mulay, H. Li, and S. See. (2010). *Computational Mechanics*, 45, 467–493. With permission.)

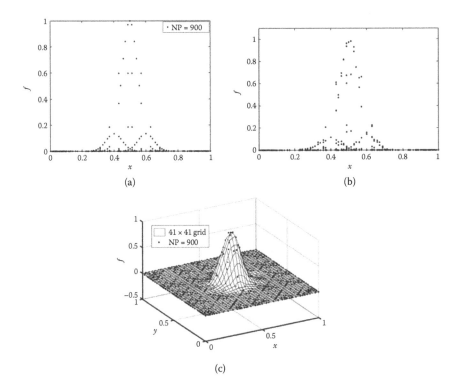

Figure 6.29 Analytical (a) and numerical (b) plots of field variable distribution and contour plot by 41 × 41 uniform grid (c) for fourth 2-D problem. (From S. Mulay, H. Li, and S. See. (2010). *Computational Mechanics*, 45, 467–493. With permission.)

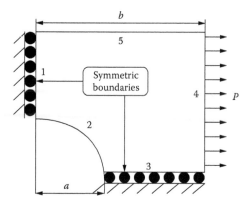

Figure 6.30 Computational domain for problem of semi-infinite plate with central hole. (From S. Mulay, H. Li, and S. See. (2010). *Computational Mechanics*, 45, 467–493. With permission.)

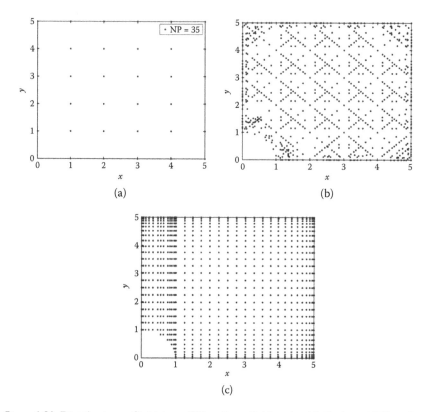

Figure 6.31 Distributions of initial total 35 uniform field nodes (a), final total 875 nodes (b), and total 961 cosine virtual nodes (c) for solution of problem of semi-infinite plate with central hole. (From S. Mulay, H. Li, and S. See. (2010). *Computational Mechanics*, 45, 467–493. With permission.)

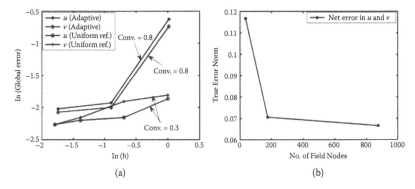

Figure 6.32 Convergence plots of displacements (a) and decrease in resultant true error norms (b) by ARDQ method where convergence rates obtained by ARDQ method based on adaptive refinement are higher than for RDQ method with uniform refinement of nodes. (From S. Mulay, H. Li, and S. See. (2010). *Computational Mechanics*, 45, 467–493. With permission.)

The resultant error in the displacements u and v is computed by $\sqrt{\varepsilon_u^2 + \varepsilon_v^2}$, where ε_u and ε_v are the global errors in the displacements u and v, respectively, as plotted in Figure 6.32b. The normal stress σ_{xx} along the boundary $x = 0$ is plotted in Figure 6.33a, and the contour plots of the analytical and numerical values, at the field nodes with relative errors within ±2%, are given in Figure 6.33b and Figure 6.33c, respectively. Based on Figure 6.33a and c, the local peak value of the normal stress σ_{xx} is well captured. The numerical and analytical values of the stress σ_{yy} along the boundary $y = 0$ are plotted in Figure 6.34.

In summary, based on numerical results of 2-D test problems, the ARDQ method well captures the local peak values and satisfies one of the objectives behind the development. It is evident from Figure 6.23b and Figure 6.27a that the convergence rates achieved by the ARDQ method are improved by the error recovery technique. This indicates that the numerical pollution effect is minimized by the error recovery technique. Based on Figure 6.10, Figure 6.14, and Figure 6.32, the global error is reduced with an improvement in the values of the field variable distribution when a solution obtained by the ARDQ method with an adaptive refinement coupled with the error recovery technique is compared a solution by the RDQ method with uniform increments of field nodes. The various contour plots shown in Figure 6.13, Figure 6.18 to Figure 6.21, and Figure 6.29 indicate that the ARDQ method well captures the values of function and the first and second order derivatives.

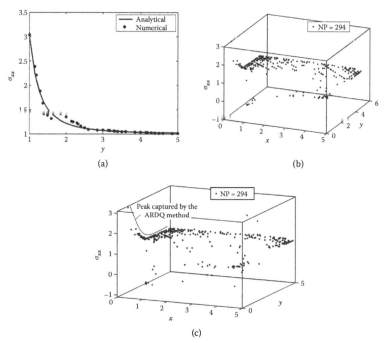

Figure 6.33 Comparison of numerical and analytical values of stress σ_{xx} along the boundary $x = 0$ (a) and analytical (b) and numerical (c) values of the stress σ_{xx}, where the relative error is within ±2% during the final distribution of field nodes. (From S. Mulay, H. Li, and S. See. (2010). *Computational Mechanics,* 45, 467–493. With permission.)

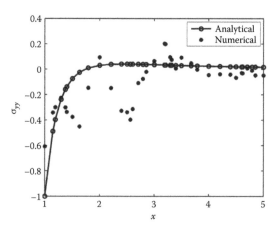

Figure 6.34 Comparison of numerical and analytical values of stress σ_{yy} along boundary $y = 0$ for problem of semi-infinite plate with central hole by ARDQ method. (From S. Mulay, H. Li, and S. See. (2010). *Computational Mechanics,* 45, 467–493. With permission.)

6.5 SUMMARY

First, a novel strong form adaptive meshless method known as the ARDQ method is presented. In this method, the governing PDE is solved by the RDQ method after each step of adaptive refinement. An error recovery technique based on LS averaging is proposed to overcome a difficulty in deciding the number of interpolation nodes to include in the domain of fixed RKPM interpolation for the approximation of function near the local peak region. The refinement procedure in the ARDQ method is easy to implement and can be extended directly to 3-D problems with the difference of the shape of an interpolation domain from circular (2-D) to spherical (3-D).

Further, a novel approach with the cross product of vectors along with the convex hull property of the computational domain is used in the ARDQ method to ensure that all the newly created field nodes are always within the computational domain. The proposed approach only requires a closed boundary (regular or irregular). In the case of an irregular or non-convex boundary, a local parametric equation of the curve is developed and can be used in conjunction with the proposed approach of the cross product. This has been demonstrated by solving the problem of a semi-infinite plate with a central hole with an irregular boundary.

Finally, the detailed convergence analysis of the ARDQ method is performed by solving several 1-D and 2-D test problems. The convergence rates were improved by the proposed error recovery technique. The various contour plots demonstrate that the ARDQ method well captures the values of local peak and first and second order derivatives.

While performing numerical computations, the absolute values of the global error norms are as important as convergence rates. For some problems solved in this chapter, the convergence rates obtained by the ARDQ method are not significantly high, compared those with the RDQ method. A closer look at the global error norm reveals that, with the same number of field nodes, the global error by the ARDQ method is lower than that by the RDQ method, as given in Table 6.7. Also, the adaptive refinement gives accurate result with a smaller number of field nodes than required by the RDQ method. For example, for the first 1-D problem, 641 uniform field nodes give the global error equal to 1.0×10^{-3} by the RDQ method, while the 585 adaptive field nodes result in the global error equal to 4.6×10^{-5}. Therefore, it is concluded that the adaptive refinement technique with the ARDQ method can work well.

Overall, we conclude that the ARDQ method coupled with the error recovery technique and the convex hull approach of the computational domain is a robust and good convergent adaptive meshless method.

Chapter 7

Engineering applications

7.1 INTRODUCTION

Several practical engineering problems relating to microelectromechanical systems (MEMS) and marine devices are simulated in this chapter. First, the fixed–fixed and cantilever microswitches are studied for their pull-in instability voltage under the influence of electrostatic force by meshless νPIM. Second, microoptoelectromechanical systems (MOEMS) devices are also simulated by meshless νPIM. Third, microtweezers are analyzed by the LoKriging method and the meshless Hermite-cloud method is applied to simulate nonlinear fluid-structure analysis of near-bed submarine pipelines under current. Finally the 2-D simulation of pH-sensitive hydrogels is performed by the RDQ method. The numerical solutions are compared with published results and found to match closely.

7.2 APPLICATION OF MESHLESS METHODS TO MICROELECTROMECHANICAL SYSTEM PROBLEMS

Because of the rapid development of MEMS, the numerical simulation is already an important tool for the analysis and design of practical devices. Simulation of MEMS is usually a very complex problem. Types of MEMS device properties include multiple coupled energy domains and media, dimensional scaling, and nonlinear problems. Therefore, the traditional analysis techniques (e.g., FEM) become difficult or ineffective. Meshless methods can avoid the disadvantages of conventional numerical methods such as FEM. For that reason, meshless methods exhibit very good potential for numerical simulation of MEMS devices.

Several typical meshless methods are introduced and then used to simulate some MEMS devices. The numerical solutions demonstrate that the meshless method is accurate, efficient, and convenient for simulating

MEMS devices. Thus, meshless methods have already shown a very good potential for this application. Of course, further research is still required, especially to handle nonlinearity.

Attention to MEMS devices in recent years led to their phenomenal growth in engineering applications such as in automotive systems (transducers and accelerometers), avionics and aerospace (microscale actuators and sensors), manufacturing and fabrication (micro smart robots), and medicine and bioengineering (DNA and genetic code analysis and synthesis, drug delivery, diagnostics, and imaging) according to Lyshevski (2002). Numerical simulation has already become an important tool for the analysis and design of a practical MEMS device.

The development of increasingly complicated MEMS demands sophisticated simulation techniques (Senturia, 1998; Hung and Senturia, 1999). The conventional FEM and the computer-aided design (CAD) systems are two widely used numerical simulation tools. In mathematical modeling and numerical simulation, the problem for MEMS devices has the following properties.

1. MEMS devices typically involve multiple coupled energy domains and media.
2. Dimensional analysis and scaling are commonly required.
3. Nonlinear problems are often considered in the analysis of MEMS devices.

Traditional analysis techniques (e.g. FEM, FDM) are becoming increasingly difficult or ineffective for the analysis of MEMS. The reason is that mesh generation is computationally expensive and mesh refinement is very difficult, especially for devices with complicated geometries and multiphysics tasks such as coupling in solid mechanics, electromagnetics, heat transfer, and fluid diffusion field equations. Furthermore, highly nonlinear phenomena cause additional simulation difficulties for MEMS devices and it thus became necessary to find alternatives to traditional methods.

Meshless methods can avoid the disadvantages of conventional numerical methods such as FEM. They thus have very good potential for numerical simulation of MEMS devices. Several MEMS devices are simulated in this chapter using meshless methods developed by the authors. The devices include two microswitches, one microoptoelectromechanical systems (MOEMS) device, and one microtweezer. The primary focus in simulation is the electrostatic pull-in characteristic—a well known sharp instability exhibited by an elastically supported structure subjected to parallel plate electrostatic actuation (Osterberg and Senturia, 1997). Therefore, the static or/and dynamic behaviours of MEMS devices subjected to electrostatic voltage changes are studied.

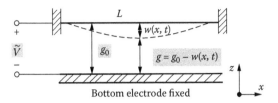

Figure 7.1 Microswitch simplified as fixed–fixed beam. (Modified from Q.X. Wang, H. Li, and K.Y. Lam. (2007). *Computational Mechanics*, 40, 1–11.)

7.2.1 Fixed–fixed microswitches

The static and dynamic behaviours of fixed–fixed microswitches are studied by the vPIM. In this simulation, the microswitch is simplified as a fixed–fixed beam as shown in Figure 7.1. The parameters of the microswitch are taken as (Ananthasuresh et al., 1996): 80 μm long, 10 μm wide, and 0.5 μm thick. The initial gap between the beam and the bottom electrode is 0.7 μm. The Young's modulus E is 169 GPa, the Poisson ratio is 0.3, the shear correction coefficient is 0.833, and the mass density ρ is 2231 kg/m³.

The nonlinear electrostatic force f acting on the microswitch can be obtained from Osterberg and Senturia (1997), and takes the following form

$$f = -\frac{\varepsilon_0 \tilde{V}^2 \tilde{w}}{2g^2}\left(1+0.65\frac{g}{\tilde{w}}\right) \tag{7.1}$$

where ε_0 is the permittivity of vacuum, $\tilde{V}$ applied voltage, $\tilde{w}$ beam width, and g the gap between the beam and bottom electrode, $g = g_0 - w(x,t)$, where g_0 is the distance between the beam and electrode before the beam deflects as shown in Figure 7.1. The static governing equation of the thin beam based on the thin beam theory (Timoshenko and Goodier, 1970; Popov, 1990) is given as

$$E I \frac{\partial^4 w(x)}{\partial x^4} = -\rho \frac{\partial^2 w(x)}{\partial t^2} + q(x) \tag{7.2}$$

where $q(x)$ is an applied load, $E I$ is flexural rigidity, and w is beam deflection. In analyzing this fixed–fixed beam, 41 distributed nodes are used. When the time derivative is omitted from Equation (7.2), the static governing equation of the beam can be obtained. Figure 7.2 plots the result of the static deflection along the beam under different applied voltages. As the applied voltage increases, the deflection of the beam increases, and the gap between the beam and the bottom electrode decreases. When the applied

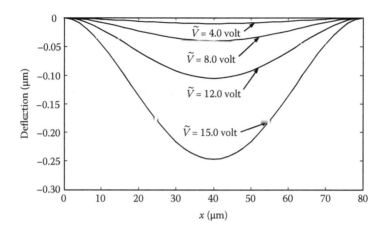

Figure 7.2 Static deflection of fixed–fixed beam under different applied voltages. (From H. Li, Q.X. Wang, and K.Y. Lam. (2004d). *Engineering Analysis with Boundary Elements*, 28, 1261–1270. With permission.)

voltage increases to one certain value, the beam becomes unstable and the centre of the beam just touches the bottom electrode.

This process is defined as the pull-in behavior and the certain value is defined as the quasi-static critical pull-in voltage. In the implementation of the program, if the deflection of the centre of the beam (peak deflection) equals the initial gap, the corresponding critical pull-in voltage is obtained. In this example, the critical pull-in voltage is 15.1 volts. In comparison a result of 15.17 volts was obtained through the experiment and simulation in Ananthasuresh et al. (1996); the difference between these two values is only ~0.5%.

In the dynamic simulation for the fixed–fixed beam, the time step is taken as 1×10^{-3} µs. Figure 7.3 shows the dynamic response of the centre of the beam, i.e., peak deflection, under different applied voltages. From this figure, two main remarks can be concluded. First, when the applied voltage increases, the peak deflection of the beam increases nonlinearly. Second, when the applied voltage increases, the fundamental frequency of the beam decreases. Similarly, in the dynamic simulation, one dynamic critical pull-in voltage can be defined. Through calculation, the dynamic critical pull-in voltage of the fixed–fixed beam is 13.8 volts. This value matches the result in Ananthasuresh et al. (1996) for the same beam.

Furthermore, from the static and dynamic results, we find that the quasi-static critical pull-in voltage is larger than the dynamic critical pull-in voltage by about 9%. Figure 7.4 plots the dynamic pull-in process at different applied voltages. As the applied voltage exceeds the dynamic critical pull-in voltage, the beam collapses onto the bottom electrode very quickly.

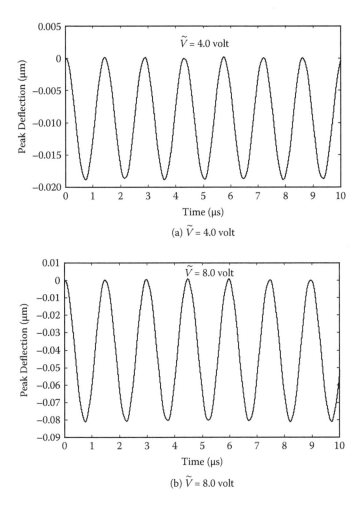

Figure 7.3 Dynamic response of fixed–fixed beam. (From H. Li, Q.X. Wang, and K.Y. Lam. (2004d). *Engineering Analysis with Boundary Elements*, 28, 1261–1270. With permission.)

Figure 7.5 demonstrates the relationship between the critical pull-in voltage and the initial gap for both the static and dynamic analyses of the beam. It is shown that with the increase of the initial gap, both quasi-static critical pull-in voltage and the dynamic critical pull-in voltage increase. When the initial gap is less than 0.5 μm, the difference between the quasi-static and the dynamic critical pull-in voltages is very small. However, as the initial gap exceeds 0.5 μm, this difference becomes larger.

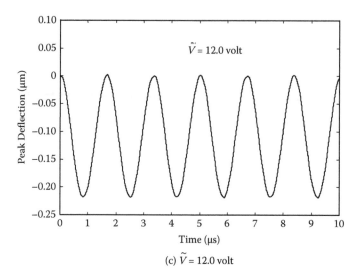

(c) $\tilde{V}$ = 12.0 volt

Figure 7.3 (continued) Dynamic response of fixed–fixed beam. (From H. Li, Q.X. Wang, and K.Y. Lam. (2004d). *Engineering Analysis with Boundary Elements,* 28, 1261–1270. With permission.)

7.2.2 Cantilever microswitches

The considered cantilever beam is shown in Figure 7.6. The left end of the beam is fixed and the right end is free. The dimensional and the material parameters for the cantilever beam are identical to the fixed–fixed microswitch with 41 nodes employed to discretize it.

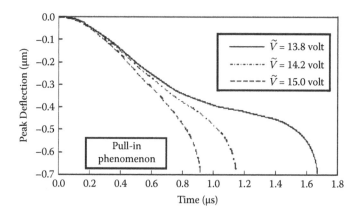

Figure 7.4 Pull-in process of fixed–fixed beam. (From H. Li, Q.X. Wang, and K.Y. Lam. (2004d). *Engineering Analysis with Boundary Elements,* 28, 1261–1270. With permission.)

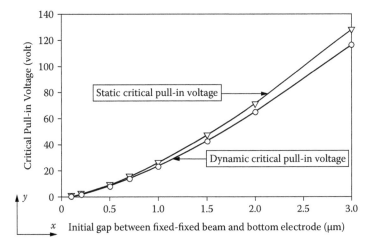

Figure 7.5 Relationship of critical pull-in voltage. (From H. Li, Q.X. Wang, and K.Y. Lam. (2004d). *Engineering Analysis with Boundary Elements*, 28, 1261–1270. With permission; modified from Q.X. Wang, H. Li, and K.Y. Lam. (2007). *Computational Mechanics*, 40, 1–11.)

Figure 7.7 shows the static deflection along the beam for different applied voltages. As the applied voltage increases, the deflection of the beam increases. Furthermore, the deflection of the free end increases greatly. When the applied voltage reaches a certain value, the free end of the beam touches the bottom electrode. Similarly, this voltage is defined as the quasi-static critical pull-in voltage. This value for the considered cantilever beam is 2.33 volts.

The dynamic simulation is also performed for the cantilever beam. The time step in this computation is taken as 5×10^{-3} μs. The peak deflection at the right end of the cantilever beam is studied under different applied voltages, as shown in Figure 7.8. The peak deflection of the beam increases nonlinearly as the applied voltage increases. The fundamental frequency of the cantilever beam decreases with an increase in the applied voltage. The dynamic critical pull-in voltage for the cantilever beam is 2.12 volts. The

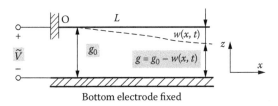

Figure 7.6 Microswitch simplified as cantilever beam. (Modified from Q.X. Wang, H. Li, and K.Y. Lam. (2007). *Computational Mechanics*, 40, 1–11.)

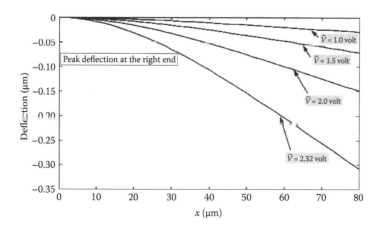

Figure 7.7 Static deflection of cantilever beam. (From H. Li, Q.X. Wang, and K.Y. Lam. (2004d). *Engineering Analysis with Boundary Elements*, 28, 1261–1270. With permission.)

error between this value and the quasi-static critical pull-in value is about 9%. The dynamic pull-in process at different applied voltages is shown in Figure 7.9. The same conclusion, as the fixed–fixed beam, is made: as the applied voltage exceeds the dynamic critical pull-in voltage, the beam collapses onto the bottom electrode very quickly.

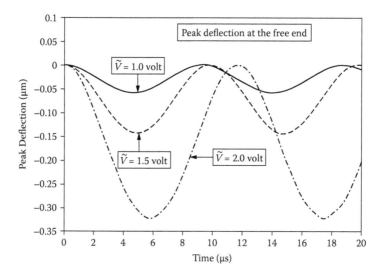

Figure 7.8 Dynamic response of cantilever beam. (From H. Li, Q.X. Wang, and K.Y. Lam. (2004d). *Engineering Analysis with Boundary Elements*, 28, 1261–1270. With permission; modified from Q.X. Wang, H. Li, and K.Y. Lam. (2007). *Computational Mechanics*, 40, 1–11.)

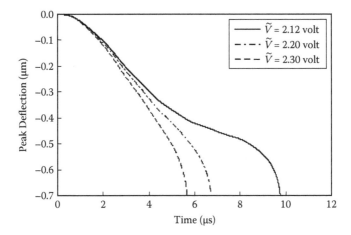

Figure 7.9 Pull-in process of cantilever beam. (From H. Li, Q.X. Wang, and K.Y. Lam. (2004d). *Engineering Analysis with Boundary Elements*, 28, 1261–1270. With permission.)

The relationship of the critical pull-in voltage and the initial gap for both the static and dynamic analyses of the cantilever beam is shown in Figure 7.10. The static and dynamic critical pull-in voltages increase as the initial gap increases. When the initial gap is less than 0.5 μm, the difference between the quasi-static and dynamic critical pull-in voltages is very small. However, as the initial gap increases, this difference becomes larger. This conclusion is also similar to that for the fixed–fixed beam.

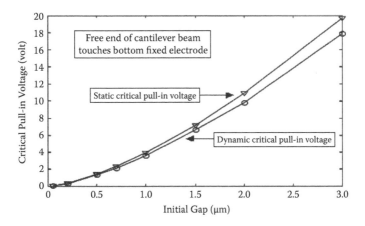

Figure 7.10 Relationship of critical pull-in voltage. (From H. Li, Q.X. Wang, and K.Y. Lam. (2004d). *Engineering Analysis with Boundary Elements*, 28, 1261–1270. With permission; modified from Q.X. Wang, H. Li, and K.Y. Lam. (2007). *Computational Mechanics*, 40, 1–11.)

7.2.3 Microoptoelectromechanical systems devices

Figure 7.11 is a schematic of the MOEMS device. Two suspensions shown as cantilever beams are connected by a mirror. The mirror acts as a bar and is assumed to be linear elastic. When voltage is applied, the two beams will deflect in opposite directions and the free ends of the beams are subjected to external forces induced by the bar extension. Following the deflection of two suspensions, the tilt angle of the mirror will change. This principle can be used to change the direction of reflected light. Each suspension is 10 μm long, 1 μm wide, and 0.5 μm thick. The Young's modulus is 169 GPa, the Poisson ratio is 0.3, and the initial gap between the suspension and the bottom or up electrode is 0.5 μm. The mirror is 1 μm long, and its width, thickness and the Young's modulus are the same as the suspension.

Due to the symmetry of the device, only one suspension and half of the mirror connecting with this suspension are analyzed. In the simulation by the meshless νLPIM, the beam is discretised by 41 nodes. Because the beam is analyzed in a 1-D model, it only deforms in the vertical direction; no deformation exists in the transversal direction (along the axial line). However, the mirror will extend and produce strains along its axial line.

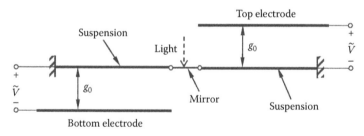

(a) Before Deflection.

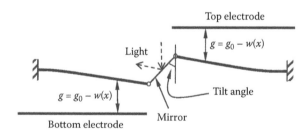

(b) After Deflection.

Figure 7.11 MOEMS device. (From H. Li, Q.X. Wang, and K.Y. Lam. (2004d). *Engineering Analysis with Boundary Elements*, 28, 1261–1270. With permission.)

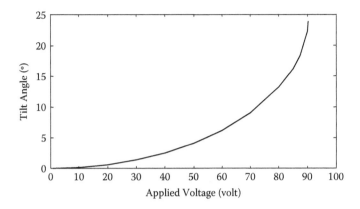

Figure 7.12 Relationship of tilt angle and applied voltage. (From H. Li, Q.X. Wang, and K.Y. Lam. (2004d). *Engineering Analysis with Boundary Elements, 28*, 1261–1270. With permission.)

This strain will induce a nonlinear force that imposes on the free end of the beam.

Figure 7.12 presents the curve of the tilt angle and the applied voltage. Note that the tilt angle of the mirror increases with the increase of the applied voltage, and the increase is nonlinear. Table 7.1 lists the peak deflection, tilt angle, and mirror strain at different voltages. We can conclude from the table that the demand for the different tilt angle of the mirror can be satisfied by changing the applied voltage.

Table 7.1 Results of peak deflection, tilt angle, and mirror strain under different applied voltages for MOEMS device

Applied voltage (V)	Deflection (μm)	Tilt Angle (°)	Mirror strain
1.00	0.000013	0.00149	0.3383E-09
10.00	0.001304	0.14954	0.3406E-05
20.00	0.005274	0.60434	0.5563E-04
30.00	0.012080	1.38402	0.2918E-03
40.00	0.022055	2.52569	0.9724E-03
50.00	0.035772	4.09221	0.2556E-02
60.00	0.054264	6.19402	0.5872E-02
70.00	0.079578	9.04319	0.1259E-01
80.00	0.116931	13.16273	0.2698E-01
90.00	0.204788	22.27279	0.8063E-01

Source: H. Li, Q.X. Wang, and K.Y. Lam. (2004d). *Engineering Analysis with Boundary Elements, 28*, 1261–1270. With permission.

7.2.4 Microtweezers

As shown in Figure 7.13a, another MEMS device known as a microtweezer is analyzed by the meshless LoKriging method. It is made of tungsten and has a simplified geometry. The tweezer arms are simplified as cantilever beams in Figure 7.13b. In the simulation, the effect of the coating layers is neglected.

The microtweezer arms are 200 μm long, 2.7 μm wide, and 2.5 μm thick. The Young's modulus is 410 GPa, and the mass density ρ is 19300 kg/m³. The initial opening of the two arms is 3 μm. It was designed and simulated by MacDonald et al. (1989) and Shi et al. (1995). In practical applications, there may be an initial angle between the arms and the central line of the microtweezer, as shown in Figure 7.13c.

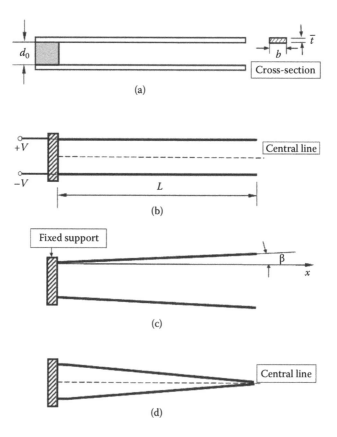

Figure 7.13 Microtweezer simplified geometry (a). Arms simplified as cantilever beams (b). Arms with angles (c). Arms closed (d). (Modified from Q.X. Wang, H. Li, and K.Y. Lam. (2007). *Computational Mechanics*, 40, 1–11.)

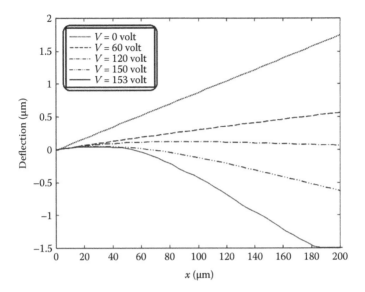

Figure 7.14 Deformed shape of microtweezer arm at different applied voltages. (Modified from Q.X. Wang, H. Li, and K.Y. Lam. (2007). *Computational Mechanics*, 40, 1–11.)

For generalization, a microtweezer with an initial angle $\beta = 0.5°$ is simulated using the present LoKriging method. When the applied voltage is imposed on the arms, they deflect and move to the central line. As the applied voltage increases, the deflection of the arms becomes larger. When the voltage reaches a certain value, the tips of two arms contact each other as shown in Figure 7.13d. This critical voltage is defined as pull-in or closing voltage. Only one arm is considered in the simulation due to the symmetry of the microtweezer. The closing voltages for the static and dynamic analyses are summarized in Table 7.2. The static results agree very well with the experimental and other numerical results.

Figure 7.14 plots the deflection of the microtweezer arm at different applied voltages. When the applied voltage is smaller than 150 volt, the two arms do

Table 7.2 Comparison of effects of methods on critical closing voltage of microtweezer

	Experimental method (MacDonald, et al., 1989) (volts)	FEM (Shi et al., 1995) (volts)	LoKriging method (volts)
Static simulation	150	156 to 157	153
Dynamic simulation	–	–	125.2 to 125.3

Source: Q.X. Wang, H. Li, K.Y. Lam et al. (2004). *Journal of the Chinese Institute of Engineers*, 27, 573–583. With permission.

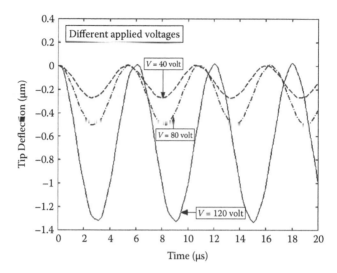

Figure 7.15 Dynamic simulation of microtweezer arm at different applied voltages. (Modified from Q.X. Wang, H. Li, and K.Y. Lam. (2007). *Computational Mechanics*, 40, 1–11.)

not contact each other. When the voltage is increased slightly beyond 150 volts to 153 volts, a large deflection of the arm tip is generated, leading to the contact of the two arms. The dynamic response of the arm tip is obtained and plotted in Figure 7.15. In dynamic analysis, the time step is taken as $\Delta t = 1 \times 10^{-3}$ μs. Similar to the microswitch discussed above, the fundamental frequency decreases with an increase in the applied voltage. In addition, the dynamic closing voltage is also obtained and listed in Table 7.2. Compared with the static closing voltage, the dynamic closing voltage is smaller.

Table 7.3 Numerical comparisons of Hermite-cloud method and hM-DOR method (Ng et al., 2003) for cantilever beam subjected to shear end load

Point distribution	Global error (ξ) for u		Global error (ξ) for $u_{,x}$	
$(N_x \times N_y)$	hM-DOR	Hermite-cloud	hM-DOR	Hermite-cloud
5×11	35.9%	35.9%	30.9%	30.9%
11×11	17.9%	17.7%	15.0%	14.9%
21×11	6.16%	5.92%	5.38%	5.19%
65×11	4.28%	1.95%	3.92%	2.13%
81×11	4.01%	0.384%	3.90%	1.92%

Note: $E = 3.0 \times 10^7, D = 1, L = 8,$ and $\mu = 0.25$.

Source: H. Li, J.Q. Cheng, T.Y. Ng et al. (2004b). *Engineering Structures*, 26, 531–542. With permission.

7.3 APPLICATION OF MESHLESS METHOD IN SUBMARINE ENGINEERING

The Hermite-cloud method developed and described earlier is applied to submarine engineering problem described as a nonlinear fluid–structure interaction of near-bed submarine pipelines under a current. The behaviour of near-seabed submarine pipelines is greatly affected by environment factors such as ocean currents, earthquakes, and underwater explosions. This section investigates the influence of ocean currents on submarine pipelines.

When the distance between a supported submarine pipeline and the seabed is very large, the current around the pipeline is hardly affected by the seabed and the flow is symmetrical to the pipeline. In the vertical direction, no net force is applied on the pipeline. However, when the distance becomes small, the velocity of current between the pipeline and seabed becomes higher than that above the pipeline. Based on Bernoulli's equation, the current pressure above the pipeline is higher than that between the pipeline and seabed.

A downward net force caused by the pressure gradient is thus produced and the pipeline is expected to deform under the external load. The deflection of pipeline increases with an increase in the current velocity. When the velocity reaches a certain critical value, the pipeline will fail to work. Two types of failure patterns are considered in the present work. One is caused by instability, in which the pipeline will rest on the seabed completely. The other results from the limits of the materials when the stress or deflection caused by the external load reaches the allowable stress or deformation of the constitutive materials. The variation of critical velocities with the distance from pipelines to seabed in various failure patterns is obtained. The distribution of pipeline stress and the relation between the pipeline deflection and current velocity are simulated. The computed results are discussed in detail and indicate a good agreement with published work (Liew et al., 2002).

7.3.1 Numerical implementation of Hermite- cloud method

In general, engineering partial differential boundary value (PDBV) problems can be written as,

$$Lf(x,y) = P(x,y) \qquad \text{PDEs in interior domain } \Omega \qquad (7.3)$$

$$f(x,y) = Q(x,y) \qquad \text{Dirichlet boundary condition on } \Gamma_D \qquad (7.4)$$

$$\frac{\partial f(x,y)}{\partial n} = R(x,y) \qquad \text{Neumann boundary condition on } \Gamma_N \qquad (7.5)$$

where L is a differential operator and $f(x,y)$ an unknown real function. By the point collocation technique and taking $\tilde{f}(x,y)$ as the approximation of $f(x,y)$, the problem is discretised and expressed approximately by

$$L\tilde{f}(x_i,y_i) = P(x_i,y_i) \qquad\qquad i = 1,2,\ldots, N\Omega \tag{7.6}$$

$$\tilde{f}(x_i,y_i) = Q(x_i,y_i) \qquad\qquad i = 1,2,\ldots, N_D \tag{7.7}$$

$$\frac{\partial \tilde{f}(x_i,y_i)}{\partial n} = R(x,y) \qquad\qquad i = 1,2,\ldots, N_N \tag{7.8}$$

where N_Ω, N_D, and N_N are the numbers of scattered points in the interior domain and along the Dirichlet and Neumann boundary edges, respectively. The total number of scattered points is thus $N_T = (N_\Omega + N_D + N_N)$. Approximating Equations (7.6) to (7.8) by the Hermite-cloud method, a set of discrete algebraic governing equations with respect to the unknown point values f_i, g_{xi}, and g_{yi} is obtained and can be written in matrix form as

$$[H_{ij}]_{(N_T+2N_S)\times(N_T+2N_S)}\{F_i\}_{(N_T+2N_S)\times1} = \{d_i\}_{(N_T+2N_S)\times1} \tag{7.9}$$

where $\{d_i\}$ and $\{F_i\}$ are (N_T+2N_S)-order column vectors, with

$$\{F_i\}_{(N_T+2N_S)\times1} = \{\{f_i\}_{1\times N_T}, \{g_{xi}\}_{1\times N_S}, \{g_{yi}\}_{1\times N_S}\}^T \tag{7.10}$$

$$\{d_i\}_{(N_T+2N_S)\times1} = \{\{P(x_i,y_i)\}_{1\times N_\Omega}, \{Q(x_i,y_i)\}_{1\times N_D}, \{R(x_i,y_i)\}_{1\times N_N}, \{0\}_{1\times 2N_S}\}^T \tag{7.11}$$

and $[H_{ij}]$ is a $(N_T+2N_S)\times(N_T+2N_S)$ coefficient square matrix

$$[H_{ij}] = \begin{bmatrix} [LN_j(x_i,y_i)]_{N_\Omega\times N_T} & \left[L((x_i-\sum_{n=1}^{N_T}N_n(x_i,y_i)x_n)M_j(x_i,y_i))\right]_{N_\Omega\times N_S} & \left[L((y_i-\sum_{n=1}^{N_T}N_n(x_i,y_i)y_n)M_j(x_i,y_i))\right]_{N_\Omega\times N_S} \\ [N_j(x_i,y_i)]_{N_D\times N_T} & [0]_{N_D\times N_S} & [0]_{N_D\times N_S} \\ [0]_{N_N\times N_T} & [M_j(x_i,y_i)]_{N_N\times N_S} & [M_j(x_i,y_i)]_{N_N\times N_S} \\ [N_{j,x}(x_i,y_i)]_{N_S\times N_T} & \left[-(\sum_{n=1}^{N_T}N_{n,x}(x_i,y_i)x_n)M_j(x_i,y_i)\right]_{N_S\times N_S} & \left[-(\sum_{n=1}^{N_T}N_{n,x}(x_i,y_i)y_n)M_j(x_i,y_i)\right]_{N_S\times N_S} \\ [N_{j,y}(x_i,y_i)]_{N_S\times N_T} & \left[-(\sum_{n=1}^{N_T}N_{n,y}(x_i,y_i)x_n)M_j(x_i,y_i)\right]_{N_S\times N_S} & \left[-(\sum_{n=1}^{N_T}N_{n,y}(x_i,y_i)y_n)M_j(x_i,y_i)\right]_{N_S\times N_S} \end{bmatrix} \tag{7.12}$$

Solving numerically the complete set of linear algebraic equations in (7.9), $(N_T + 2N_S)$ point values $\{F_i\}$, consisting of the N_T point values $\{f_i\}$ and $2N_S$ point values $\{g_{xi}\}$ and $\{g_{yi}\}$, are obtained and the approximate solution $\tilde{f}(x, y)$ and its first-order derivatives $\tilde{g}_x(x, y)$ and $\tilde{g}_y(x, y)$ of the PDBV problem can be computed through Hermite-based interpolation equations given in Chapter 3.

7.3.2 Numerical study of near-bed submarine pipeline under current

When a submarine pipeline is placed near a seabed, the horizontal current is not symmetrical to the pipeline because of the seabed. As a result, a downward external force is produced and applied on the pipeline, resulting in its deformation. This problem can be simplified as a beam with different boundary support conditions. In a published work, Lam et al. (2002) studied the static behaviour of a pipeline under a static nonlinear external force. The pipeline was simplified as a Bernoulli-Euler beam by the assumption that there was no transverse shear strain. However, the method is not applicable in the case where the ratio of length to diameter of the pipeline is not very large, where the effect of shear deformation should be included. To break through the limit, the Timoshenko beam theory (Reddy, 1993) is introduced here to include the transverse shear deformation. Based on that theory, the general governing equations of the pipeline are written as

$$\frac{\partial}{\partial x}\left[GAk_s\left(\frac{\partial w}{\partial x} + \theta\right)\right] + f = 0 \tag{7.13}$$

$$\frac{\partial}{\partial x}\left(EI\frac{\partial \theta}{\partial x}\right) - GAK_s\left(\frac{\partial w}{\partial x} + \theta\right) = 0 \tag{7.14}$$

where w is the deflection of the pipeline, θ the rotation, G the shear modulus, A the cross section area, k_s the shear correction coefficient, E the elasticity modulus, and I the moment of inertia. The general boundary conditions for the above problem are given as

$$w(x_0) = \bar{w}, \quad \text{on } \Gamma_w \tag{7.15}$$

$$\theta(x_0) = \bar{\theta}, \quad \text{on } \Gamma_\theta \tag{7.16}$$

$$M(x_0) = EI\frac{\partial \theta}{\partial x}\bigg|_{x=x_0} = \bar{M}, \quad \text{on } \Gamma_M \tag{7.17}$$

$$V(x_0) = GAk_s\left(\theta + \frac{\partial w}{\partial x}\right)\bigg|_{x=x_0} = \bar{V}, \quad \text{on } \Gamma_V \tag{7.18}$$

in which Γ_w, Γ_θ, Γ_M, and Γ_V are the boundaries where w, θ, M, and V satisfy, respectively. In Equation (7.13), f is the fluid force caused by the current pressure difference. Although the problem has an analytical solution, it converges slowly. To accelerate the computation, Lam et al. (2002) used the boundary element method (BEM) to construct the approximation of the exact solution. The results are in very good agreement. The expression of fluid force $f(x)$ is given as

$$f(x) = \frac{1}{2}\rho A U_0^2 c(d) \tag{7.19}$$

where ρ is the mass density, U_0 the current velocity, c and d are dimensionless coefficients and has a relation as

$$c(d) = \frac{2.23d^2 + 12.54d + 0.02}{0.77d^3 + 0.44d^2 + 0.02d} \tag{7.20}$$

in which d is defined as $d = (D_0 - w(x) - R_s)/(2R_s)$ and D_0 is the distance between the central line of pipeline at initial status and the seabed and R_s the pipe outer radius.

Note from Equations (7.19) and (7.20) that the fluid force $f(x)$ is a non-linear term related to the deflection $w(x)$. The governing equations (7.13) and (7.14) for the pipeline deformation thus are the coupled nonlinear functions. It is also observed that when the materials and dimensions of pipelines are given, the fluid force $f(x)$ is a function of both the current velocity U_0 and the gap D_0. These two parameters have significant influence on the deformation behaviours of near-bed submarine pipelines.

In order to demonstrate the influences in detail, a practical example is designed here. A circular steel pipeline is placed near the seabed under a static current, as shown in Figure 7.16. Several physical and material parameters are given: pipeline length L = 10 m, R_s = 0.5 m, thickness of steel pipeline t_s = 0.02 m, D_0 = 0.7 m, E_s = 2.11 × 10^{11} N/m², Poisson ratio v = 0.3, ρ_s = 7800 kg/m³, steel yield stress Y_s = 2.5 × 10^8 N/m², and allowable deflection/span ratio Y_w = 0.004. The L/R_s ratio is not very large. The pipeline can be simplified as a thick beam fixed at two ends. By the Timoshenko beam theory and Hermite-cloud meshless method, numerical analysis is carried out and discussions are also presented.

The variation of mid-point deflection $w_{L/2}$ with the current velocity U_0 is illustrated in Figure 7.17. With an increase in the current velocity, the deflection increases until a critical value U_{cb} is reached. When the velocity is smaller than the critical value, the deflection converges with the iterations and the pipeline reaches stable equilibrium. However, if the velocity is over the critical level, the deflection does not converge and this is defined as

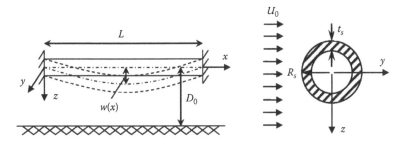

Figure 7.16 Submarine pipeline deformation under current. *(From H. Li, J.Q. Cheng, T.Y. Ng et al. (2004b). Engineering Structures, 26, 531–542. With permission.)*

instability, i.e., the pipeline has fallen into the seabed. Instability is the first failure pattern of pipelines.

The influence of the distance D_0 between the pipeline and seabed under the critical current velocities caused by instability U_{cb} is revealed in Figure 7.18. It is shown that U_{cb} is a monotonically increasing function with D_0. As indicated in Equations (7.19) and (7.20), if other variables are fixed, the fluid force $f(x)$ decreases with the enlargement of D_0 and increases with the increasing U_0. Therefore, to obtain the same critical fluid force causing the pipeline to lose stability when the gap D_0 becomes large, it is reasonable to require a higher critical velocity U_{cb}. It is also expected that the critical velocity is infinite when the gap becomes infinite, since the effect of seabed on the current is then zero and the flow symmetry is not changed at all.

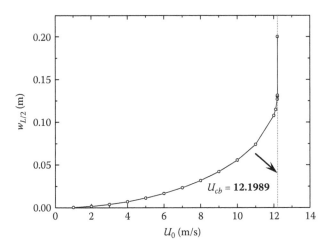

Figure 7.17 Variation of deflection of mid-point of pipeline with respect to current velocity U_0 (when $D_0 = 0.7$ m). (From H. Li, J.Q. Cheng, T.Y. Ng et al. (2004b). *Engineering Structures*, 26, 531–542. With permission.)

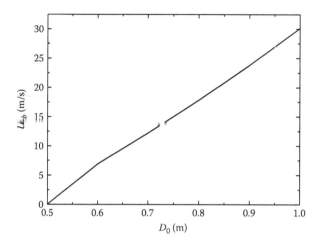

Figure 7.18 Effect of gap D_0 on critical velocity U_{cb} of instability failure. (From H. Li, J.Q. Cheng, T.Y. Ng et al. (2004b). *Engineering Structures*, **26**, 531–542. With permission.)

As well known, sometimes the failure due to material properties occurs before instability failure. It is thus necessary to discuss material failure. The distribution of stress along a pipeline is plotted in Figure 7.19. The curve is found to be similar to that of a two-end fixed beam under uniform load, but actually the curves are different. For the beam under uniform load, the maximal absolute value of stresses (moments) along the beam occurs at the

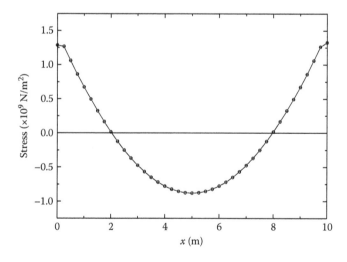

Figure 7.19 Distribution of stress along pipeline (when $U_0 = 10$ m/s and $D_0 = 0.7$ m). (From H. Li, J.Q. Cheng, T.Y. Ng et al. (2004b). *Engineering Structures*, **26**, 531–542. With permission.)

mid-point, but the present maximal absolute value of stresses (moments) appears at the end points, as shown in Figure 7.19. This phenomenon results from the different external load distributions. In this problem, the fluid force $f(x)$ is a decreasing function of pipeline deflection $w(x)$, and it is clear that the deflection of two ends is smaller than that of the mid-point. This implies that the fluid force at two ends is larger than that at mid-point. Obviously the effect is different for uniform load distribution.

Figure 7.20 presents the critical velocities caused by the two types of material failures: strength failure and deflection failure. As noted above, the end-point stress and mid-point deflection first reach the respective

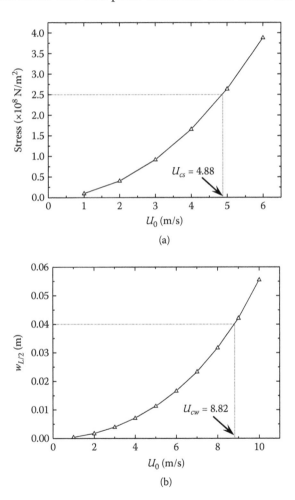

Figure 7.20 Critical velocities of (a) strength failure and (b) deflection failure (when $D_0 =$ 0.7 m). (From H. Li, J.Q. Cheng, T.Y. Ng et al. (2004b). *Engineering Structures*, 26, 531–542. With permission.)

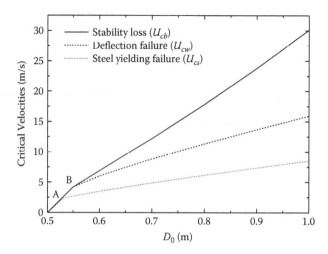

Figure 7.21 Comparison of distributions of respective critical velocity with respect to gap D_0 in various failure patterns. (From H. Li, J.Q. Cheng, T.Y. Ng et al. (2004b). *Engineering Structures*, 26, 531–542. With permission.)

critical values along the whole pipeline. Thus their corresponding critical current velocities are defined as the critical values of the whole system. In Figure 7.20a and b, both the end-point stress and mid-point deflection enlarge with the increase of current velocity U_0, and when they reach the yield stress and allowable deflection, the corresponding critical velocities U_{cs} and U_{cw} are obtained.

Similar to instability failure, the distance between the pipeline and sea-bed D_0 has an influence on the critical velocities of material failure. To compare these two failure patterns, the distributions of respective critical velocities (U_{cb} due to instability failure, U_{cs} due to strength failure, and U_{cw} due to deflection failure) with respect to the gap D_0 are synthetically plotted in Figure 7.21. There are bifurcations at points A and B. In other words, beyond point A, the pipeline will not break down due to stability loss because the strength failure always appears before the instability failure. The results agree with Lam et al. (2002). In their work, the pipeline is simplified as a Bernoulli-Euler beam.

7.4 APPLICATION OF- RDQ METHOD FOR 2-D SIMULATION OF PH-SENSITIVE HYDROGEL

The objective of the present work is to simulate the response of a 2-D hydrogel when subjected to the varying pH of a buffer solution. This is one of the earliest attempts of 2-D simulation of pH-responsive hydrogels by the novel strong form meshless RDQ method. The ionic diffusion between the hydrogel and solution is simulated by the system of Poisson-Nernst-Plank (PNP) equations, and the hydrogel swelling is captured by the mechanical equilibrium equations. To date, this problem has been solved by simplifying it to the 1-D hydrogel deformation only. This simplification holds well if the hydrogel domain is regular and uniform. However, it is incorrect to assume the 1-D deformation for irregular domains, and the problem becomes a truly 2-D deformation.

At first, the PNP equations are studied for type as the elliptic, parabolic, or hyperbolic PDE. A novel approach is proposed to correctly impose the Neumann boundary condition for the nonuniform boundary. The effects of the solution pH and initial fixed-charge concentration are also investigated on the swelling of the hydrogel. The simulation results are in good qualitative agreement with the physics of the problem and the experimental results.

Extensive research by theory and experiment has been performed by different research groups to determine the behaviours of polyelectrolyte gels and hydrogels responsive to different environmental stimuli such as solution temperature, pH, ionic strength etc. The key motivation is that the hydrogels are excellent candidates because they can convert different forms of chemical and electrical potentials into mechanical work. Because of their responsive behaviours, hydrogels can be effectively used for controlled drug release, micro-fluidic flow controls, sensors, and other devices.

The hydrogels generally are triphasic mixtures containing polymer crosslinked networks, interstitial fluid and fixed charge ions, as shown in Figure 7.22. Depending on the nature of fixed-charge (anions or cations), the hydrogels are called acidic or basic. In the present work, the acidic

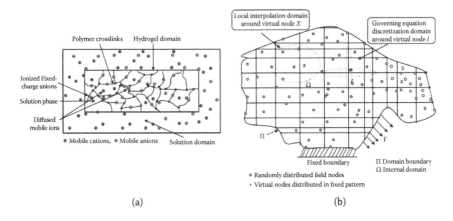

(a)   (b)

Figure 7.22 Charged hydrogel and solution (a) and working principle of RDQ method by irregular computational domain (b). (From S. Mulay and L. Li. *Modelling and Simulation in Materials Science and Engineering*, DOI:10.1088/0965-0393/19/6/065009. With permission; modified from H. Li and S. Mulay. (2007). *Computational Mechanics*, DOI: 10.1007/s004466-011-0622-5.)

hydrogel responsive to the solution pH is studied. In the pH-responsive acidic hydrogels, the anionic fixed charge is bound to the polymer cross-links, and the surrounding solution contains various salt ions (Na^+ and Cl^-) along with the H^+ ions. Due to the different values of ionic concentrations inside and outside the hydrogel at the beginning, the electrical potential develops and initiates the diffusion of mobile ions from the solution into the hydrogel through the porous surface.

Some diffused H^+ ions combine with the fixed charge anions. The ionic diffusion between the hydrogel and solution causes the difference in the ionic concentration, which leads to the generation of osmotic pressure. This osmotic pressure serves as a surface traction, and then causes the deformation of hydrogel. This deformation is stored as the strain energy in the crosslinked polymers that then try to restrict the hydrogel deformation by their stiffness. The hydrogel is said to be in equilibrium with the surrounding solution when it stops expanding, as the external traction is balanced by the hydrogel internal stress.

In 1943, Flory and Rehner (1943a and b) developed a statistical model to represent the polymer crosslinked network and its interaction with the solvent without considering diffusion. Hon et al. (1999) developed a set of equations for the mixture theory based on the triphasic system from the generalized first law of thermodynamics in which ionic concentration is expressed by chemical energy.

De and Aluru (2004) used the mechanical equilibrium equation to compute hydrogel deformation. After computing the concentration of H^+ ions, the other ionic concentrations are computed by Donnan membrane theory, and subsequently the osmotic pressure is computed. Li et al. (2005a and b) developed a chemoelectromechanical model with the PNP equations by incorporating the relationship between the concentrations of ionized fixed charge groups and the diffusive hydrogen ion via the Langmuir isotherm. They also validated the model by 1-D simulation via the meshless Hermite-cloud method.

Hong et al. (2008) formulated a theory of mass transport and large deformation with the free energy of a hydrogel and solution system. The free energy of the gel comes from two molecular processes: stretching the polymer network and mixing the polymer and small molecules. The small molecules and long polymers are considered incompressible by enforcing a constraint via the Lagrange multipliers.

Zhang et al. (2009) developed the FEM model for transient analysis of large deformation and mass transport in hydrogels. They used the free energy of system by following the model of Hong et al. (2008) and developed the governing equations in the weak form with the equations of mass conservation and mechanical equilibrium. Using the developed FEM model, they solved several numerical examples such as the free swelling of gel cube, free swelling of a thin sheet, and swelling of a partially constrained gel. Although several models for ionic transportation and gel deformation have been developed, none fully captures the whole highly complex phenomenon.

Several researchers performed the numerical simulations of hydrogels by considering different working conditions. Wallmersperger et al. (2001, 2008) performed the FEM simulation of 1-D and 2-D hydrogel deformations by a system of PNP and mechanical equilibrium equations. They also studied the effects of chemical and mechanical stimulations on gel deformation. In the work of Wallmersperger et al. (2008, 2009), the chemoelectromechanical model was used to analyze time-dependent effect due to electrical stimulation, where the larger difference in the electric potential across the hydrogel results in a larger change in the values of ionic concentrations over time.

Samson et al. (1999) numerically solved the transient PNP equations by two different iterative schemes, i.e., the Picard iteration and the Newton-Raphson (NR) iteration and concluded that the Picard technique cannot be used to solve the PNP equations for all the cases, while all the PNP equations can be easily coupled and solved for any number of ionic species with different valence numbers by the NR method. This is a very important result because the PNP equations are nonlinear PDEs that must be solved iteratively by a stable iterative technique.

Li (2009), Li and Luo (2009), Li and Yew (2009) and Li et al. (2004a, e, and f, 2007a and b, 2009a) simulated the chemoelectromechanical model of the 1-D hydrogel subject to different solution stimuli, such as temperature, pH, ionic strength, and electrical potential. They extensively investigated the behaviours of hydrogels by changing the various simulation parameters such as the solution pH, initial fixed charge concentration inside the gel, Young's modulus, ionic strength of solution, and externally applied electrical potential. They also conducted 1-D transient analysis of hydrogels subjected to external electric fields.

Luo and Li (2009) and Luo et al. (2007, 2008) studied the responsive behaviour of a 1-D hydrogel due to the effect of the pH, external electric stimulus, and ionic strength of solution. They also developed the PNP and mechanical equilibrium equations in Lagrangian coordinates and compared the results with the same set of equations in Eulerian coordinates. The deformations of the hydrogel computed by the governing equations developed in the Eulerian and Lagrangian coordinate systems are almost equal for the lower values of pH, but the differences appear for the higher values of pH. The additional results about the 1-D hydrogel can be found (Lai et al., 1991; Brock and Li, 1994; Guelch et al., 2000; Iordanskii et al., 2000; De et al., 2001; Šnita et al., 2001; Zhou et al., 2002; De et al., 2002; Chatterjee et al., 2003; Traitel et al., 2003; Ostroha et al., 2004; Wu et al., 2004; Chen et al., 2005; Lam et al., 2006; Wang et al., 2006; Lebedev et al., 2006; Mann et al., 2006; Yew et al., 2007; Yuan et al., 2007; Yu et al., 2007-2008; Ballhause and Wallmersperger, 2008; Zhou et al., 2008; Birgersson et al., 2008; Kang et al., 2008; Suthar et al., 2008; Kvarnström et al., 2009; Joseph and Aluru, 2009; Zhao and Suo, 2009).

The simulation of pH-sensitive hydrogel to this point in the literature is performed by simplifying it as a 1-D problem. This simplification is based on the fact that the hydrogel is constrained at the top and bottom surfaces in the microfluidic flow channel. The cross-section of the hydrogel is essentially circular but it may deform unequally in different directions when the cross-sectional geometry of a hydrogel is nonuniform. As a result, it becomes truly a 2-D problem.

The objectives of the present work are to perform the multiphysics 2-D simulation of pH-sensitive hydrogel with the nonuniform boundary of a domain and study the effects of changes in the values of initial fixed charge concentrations, Young's modulus of hydrogels, and solution pH on the deformation of hydrogels. It is also interesting to see how the geometrical shape of a hydrogel at the dry or initial state affects the final deformation under the various values of solution pH.

Three different geometries of the hydrogel disc are studied for the various values of solution pH and Young's moduli. In the present work, the RDQ method is implemented to solve the PNP equations over the hydrogel and solution domains for the simulation of ionic diffusion and solve the mechanical equilibrium equations over the hydrogel domain for the simulation of the hydrogel deformation. Since the PNP equations are nonlinear PDEs, they are solved iteratively by the NR method until convergence is reached.

The motivation behind the present work is to demonstrate the capability of the RDQ method for the simulation of complex multiphysics problems such as the 2-D simulation of pH-sensitive hydrogels with nonuniform moving boundaries. There are several challenges in this problem, such as the nonlinear and complex nature of the 2-D PNP equations, the nonuniform and moving boundaries of hydrogels, and the sudden jump in the values of the ionic concentrations and electrical potential across the interface between hydrogel and solution.

The numerical method must be consistent, convergent, stable, and capable of effectively handling local high gradients over the irregular boundaries of the domain to overcome these challenges, as demonstrated by the RDQ method (Mulay and Li, 2009; Mulay et al., 2009; Li et al., 2009b-c; Mulay et al., 2010). This problem thus demands a much more careful approach in performing numerical simulation, as described in subsequent sections.

7.4.1 Model development of two-dimensional pH-sensitive hydrogel

The RDQ method is implemented in this section to simulate the problem of 2-D pH-sensitive acidic hydrogels where the ionic diffusion is simulated by solving 2-D PNP equations over the hydrogel and solution domains.

The diffusion of each ionic species is captured by the Nernst-Plank (NP) equation corresponding to individual ionic species. The electric field due to uneven ionic distribution over the hydrogel and solution domains is simulated by developing a Poisson equation for the electrical potential. After the concentrations of all the mobile ions are obtained, the osmotic pressure is computed and imposed as the surface traction on the boundary of the hydrogel. Swelling of the hydrogel due to the osmotic pressure is computed by solving the mechanical equilibrium equations in the plane stress state. The 2-D computational domain of the hydrogel is obtained by cutting a disc from the full hydrogel so that the thickness of the disc is much smaller than the length and height. As a result, it becomes a 2-D plane stress problem. The general model to solve the 2-D hydrogel can be found (Li et al., 2005a; Li et al., 2007a).

Figure 7.23a shows the hydrogel and solution domains where L, W, and R are the hydrogel length, width, and boundary radius, respectively. The hydrogel domain is nonuniform due to the curved boundary. Since the hydrogel and solution domains are symmetric about the central x and y direction axes, as shown in Figure 7.23a, only a quarter of the whole domain

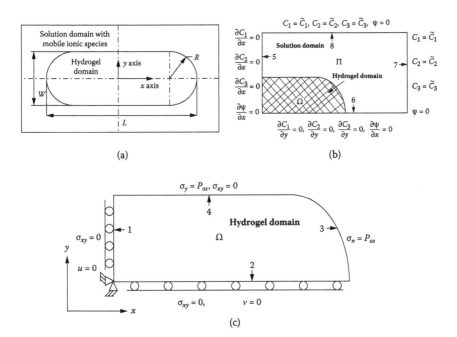

(a)

(b)

(c)

Figure 7.23 Hydrogel and solution domain (a). Computational domain of hydrogel and solution with solution boundary conditions (b). Hydrogel computational domain along with boundary conditions (c). (Modified from H. Li and S. Mulay. (2007). *Computational Mechanics*, DOI: 10.1007/s004466-011-0622-5.)

is considered for the actual computation. Figure 7.23b shows the computational domains of the hydrogel and solution along with the symmetric boundary conditions for the solution domain. These multi-domains are utilized to solve the 2-D PNP equations for the concentrations of all the ionic species and the electrical potential. Figure 7.23c shows the hydrogel domain along with the symmetric boundary conditions used to solve the mechanical equilibrium equations for the hydrogel deformation in the xy plane.

7.4.1.1 Model of hydrogel

In the present work, the acidic hydrogel is modeled with the anions as the ionized fixed-charge groups bound to the polymeric chains of the hydrogel. The bath solution contains Na$^+$, H$^+$, and Cl$^-$ mobile ions as $C_1 = C_{Na^+}$, $C_2 = C_{H^+}$, and $C_3 = C_{Cl^-}$, and ψ is the electrical potential, as shown in Figure 7.23b. In order to use the NR method, the governing equations and boundary conditions are written as the residual equations in the nondimensional form to avoid any singularity in the Jacobean matrix during the NR iterations.

The equation of mass conservation for the ionic species C_1 by the ionic diffusion is developed via the chemical potential and electrical flux due to the gradient in the electrical potential. By Reynolds' transport theorem (Li et al., 2007b), we have

$$
R_1 = 0 \Rightarrow \frac{\partial^2 \bar{C}_1}{\partial \bar{x}^2} + \frac{\partial^2 \bar{C}_1}{\partial \bar{y}^2} + (\eta\, Z_1)\left[\left(\frac{\partial \bar{C}_1}{\partial \bar{x}} \frac{\partial^2 \bar{\psi}}{\partial \bar{x}}\right) + \left(\frac{\partial \bar{C}_1}{\partial \bar{y}} \frac{\partial^2 \bar{\psi}}{\partial \bar{y}}\right)\right]
$$

$$
+ (\eta\, Z_1\, \bar{C}_1)\left(\frac{\partial^2 \bar{\psi}}{\partial \bar{x}^2} + \frac{\partial^2 \bar{\psi}}{\partial \bar{y}^2}\right) = 0 \tag{7.21}
$$

where $\bar{C}_1 = C_1/C_{ref}$, $\bar{x} = x/L_{ref}$, $\bar{y} = y/L_{ref}$, and $\bar{\psi} = (\psi\, F)/(\eta\, R\, T)$ are the nondimensionalized values of the concentration of ion C_1, x and y coordinates, and electrical potential, respectively, and C_{ref}, L_{ref}, F, η, R, and T are the reference concentration, reference length, Faraday constant, weighted coefficient, gas constant, and temperature, respectively. Equation (7.21) is the 2-D NP equation for the ionic species C_1. Similarly, the residual equations R_2 and R_3 can be written for the ionic species C_2 and C_3, respectively.

All the governing equations are solved with the nondimensionalized field variables to avoid any singularity in the Jacobean matrix of the NR iteration due to different scalings of the parameters. At the end, the nondimensional solutions are converted back to their original forms with the conversion equations given earlier. The flux due to convection is neglected while developing Equation (7.21), as the surrounding solution is assumed to be stationary. There are several possible ways to describe the spatial charge.

One of them, the electrical potential governed by 2-D Poisson equation (Li et al., 2007b), is adopted in the present work as

$$R_4 = 0 \Rightarrow \frac{\partial^2 \overline{\psi}}{\partial \overline{x}^2} + \frac{\partial^2 \overline{\psi}}{\partial \overline{y}^2} = \frac{-F^2 \ L_{ref} \ C_{ref}^2}{\varepsilon_r \ \varepsilon_0} \ (Z_1 \ \overline{C}_1 + Z_2 \ \overline{C}_2 + Z_3 \ \overline{C}_3 + Z_f \overline{C}_f)$$

$$(7.22)$$

where ε_r and ε_0 are the relative and vacuum permittivities of material, respectively, Z_1, Z_2, Z_3, and Z_f are the valances of ions C_1, C_2, C_3, and C_f respectively, and $\overline{C}_f$ is the nondimensionalized ionized concentration of the hydrogel fixed charge. The residual equations R_1 to R_4 are applied over the internal computational domain $(\pi + \Omega)$, as shown in Figure 7.23b. Equation (7.21) is applied at the virtual node (x_i, y_j) located in the internal domain of the solution as

$$R_1(x_i,y_j) = \{b_{ik}\}\left\{C_1^b(x_k,y_j)\right\} + \{b_{jl}\}\left\{C_1^b(x_i,y_l)\right\} + (Z_1\eta)\Big[\{a_{ik}\}\left\{C_1^b(x_k,y_j)\right\}$$

$$\{a_{im}\}\left\{\psi^b(x_m,y_j)\right\} + \{a_{jl}\}\left\{C_1^b(x_i, \ y_l)\right\} \ \{a_{jn}\} \ \left\{\psi^b(x_i, \ y_n)\right\}\Big] \qquad (7.23)$$

$$+\Big(Z_1 \ \eta \ C_1^b(x_i, \ y_j)\Big)\Big[\{b_{im}\} \ \left\{\psi^b(x_m, \ y_j)\right\} + \{b_{jn}\} \ \left\{\psi^b(x_i, \ y_n)\right\}\Big]$$

where $m = 1, \ 2, \ \cdots, \ N_x$ and $n = 1, \ 2, \ \cdots, \ N_y$ indices are used for the discretisation of ψ in the x and y directions, respectively. C_1^b and ψ^b are the values of approximate function. The discretised equations $R_2(x_i, \ y_j)$ and $R_3(x_i, \ y_j)$ are also written similarly to Equation (7.23). The discretised equation $R_4(x_i, \ y_j)$ from (7.22) is given as

$$R_4(x_i, \ y_j) = [\{b_{im}\} \ \{\psi^b(x_m, \ y_j)\} + \{b_{jn}\} \ \{\psi^b(x_i, \ y_n)\}]$$

$$= \lambda[Z_1 C_1^b(x_i, \ y_j) + Z_2 C_2^b(x_i, \ y_j) + \qquad (7.24)$$

$$Z_3 C_3^b(x_i, \ y_j) + Z_f C_f(x_i, \ y_j)]$$

The residual equations R_5 to R_{20} for the domain boundaries 5 to 8, as shown in Figure 7.23b, are discretised at the virtual node $(x_i, \ y_j)$ as

$$\left.\begin{aligned} R_5 = 0 \Rightarrow \left(\frac{\partial C_1}{\partial x}\right)_{(x_i,y_j)} &= \sum_{k=1}^{N_x} a_{ik} \ C_1^b(x_k,y_j) = 0 \\ R_6 = 0 \Rightarrow \left(\frac{\partial C_2}{\partial x}\right)_{(x_i,y_j)} &= \sum_{k=1}^{N_x} a_{ik} \ C_2^b(x_k,y_j) = 0 \end{aligned}\right\} \qquad (7.25)$$

$$R_7 = 0 \Rightarrow \left(\frac{\partial C_3}{\partial x} \right)_{(x_i, y_j)} = \sum_{k=1}^{N_x} a_{ik} \, C_3^b(x_k, y_j) = 0$$

$$R_8 = 0 \Rightarrow \left(\frac{\partial \psi}{\partial x} \right)_{(x_i, y_j)} = \sum_{k=1}^{N_x} a_{ik} \, \psi^b(x_k, y_j) = 0$$

(7.26)

$$R_9 = 0 \Rightarrow \left(\frac{\partial C_1}{\partial y} \right)_{(x_i, y_j)} = \sum_{l=1}^{N_y} a_{jl} \, C_1^b(x_i, y_l) = 0$$

$$R_{10} = 0 \Rightarrow \left(\frac{\partial C_2}{\partial y} \right)_{(x_i, y_j)} = \sum_{l=1}^{N_y} a_{jl} \, C_2^b(x_i, y_l) = 0$$

(7.27)

$$R_{11} = 0 \Rightarrow \left(\frac{\partial C_3}{\partial y} \right)_{(x_i, y_j)} = \sum_{y=1}^{N_y} a_{jl} \, C_3^b(x_i, y_l) = 0$$

$$R_{12} = 0 \Rightarrow \left(\frac{\partial \psi}{\partial y} \right)_{(x_i, y_j)} = \sum_{l=1}^{N_l} a_{jl} \, \psi^b(x_i, y_l) = 0$$

(7.28)

$$R_{15} \text{ and } R_{19} = 0 \Rightarrow C_1^b(x_i, y_j) - \tilde{C}_3(x_i, y_j) = 0$$

$$R_{16} \text{ and } R_{20} = 0 \Rightarrow C_2^b(x_i, y_j) - \tilde{C}_2(x_i, y_j) = 0$$

(7.29)

$$R_{15} \text{ and } R_{19} = 0 \Rightarrow C_3^b(x_i, y_j) - \tilde{C}_1(x_i, y_j) = 0$$

$$R_{16} \text{ and } R_{20} = 0 \Rightarrow \psi^b(x_i, y_j) = 0$$

(7.30)

where $k = 1, 2, \cdots, N_x$ and $l = 1, 2, \cdots, N_y$. Equations (7.25) through (7.28) are applied over boundaries 5 and 6, respectively, and Equations (7.29) and (7.30) are applied over boundaries 7 and 8.

The 2-D plane-stress mechanical equilibrium equation for the hydrogel domain is as

$$\nabla \bullet [(-P_{os}) \, I + \sigma] = 0$$

(7.31)

where I and $\sigma = \sigma_{ij}$ are the identity matrix and Cauchy stress tensor, respectively, and P_{os} is the osmotic pressure computed by the concentration values of the mobile ionic species across the interface between the hydrogel and solution as

$$\bar{P}_{os} = R\,T\,C_{ref}\sum_{k=1}^{3}\left(\bar{C}_k - \bar{C}_{k0}\right) \tag{7.32}$$

where $\bar{P}_{os}$ is nondimensionalized osmotic pressure, and $\bar{C}_k$ and $\bar{C}_{k0}$ are the nondimensionalized concentrations of the kth mobile ion in the hydrogel and solution domains, respectively. Equation (7.31) can be expanded as

$$\left(\frac{E}{1-v_0^2}\right)\left[\frac{\partial^2 u}{\partial x^2} + \left(\frac{1-v_0}{2}\right)\frac{\partial^2 u}{\partial y^2} + \left(\frac{1+v_0}{2}\right)\frac{\partial^2 v}{\partial x\partial y}\right] = \frac{\partial P_{os}}{\partial x} \tag{7.33}$$

$$\left(\frac{E}{1-v_0^2}\right)\left[\frac{\partial^2 v}{\partial y^2} + \left(\frac{1-v_0}{2}\right)\frac{\partial^2 v}{\partial x^2} + \left(\frac{1+v_0}{2}\right)\frac{\partial^2 u}{\partial x\partial y}\right] = \frac{\partial P_{os}}{\partial y} \tag{7.34}$$

where u and v are the displacements of hydrogel in the x and y directions, respectively. Equations (7.33) and (7.34) are solved over the hydrogel internal domain Ω along with the symmetric boundary conditions, as shown in Figure 7.23c, and given as

$u = 0$ and $\sigma_{xy} = 0$ along Boundary 1; $v = 0$ and $\sigma_{xy} = 0$ along Boundary 2

$$\tag{7.35}$$

$$\sigma_n = P_{os} \text{ along Boundary 3} \tag{7.36}$$

$$\therefore T_x = P_{os}\,n_x = \sigma_n\,n_x = \sigma_{xx}\,n_x + \sigma_{yx}\,n_y,$$
$$T_y = P_{os}\,n_y = \sigma_n\,n_y = \sigma_{xy}\,n_x + \sigma_{yy}\,n_y \tag{7.37}$$

where T_x and T_y, and n_x and n_y are the traction components and surface normals in the x and y directions, respectively, and σ_{xx}, σ_{yy}, and σ_{xy} are the normal and shear stress components, respectively. Along boundary 3, the surface normals for the curved boundary are given as $n_x = \cos(\theta)$ and $n_y = \sin(\theta)$.

$$\sigma_{xy} = 0 \text{ and } \sigma_{yy} = P_{os} \text{ along boundary 4} \tag{7.38}$$

The plane stress equations for the stresses (Timoshenko and Goodier, 1970) are given as:

$$\sigma_{xx} = \left(\frac{E}{1-v_0^2}\right)\left[\frac{\partial u}{\partial x} + v_0 \frac{\partial v}{\partial y}\right],$$

$$\sigma_{yy} = \left(\frac{E}{1-v_0^2}\right)\left[\frac{\partial v}{\partial y} + v_0 \frac{\partial u}{\partial x}\right], \qquad (7.39)$$

$$\sigma_{xy} = \left(\frac{E}{2(1+v_0)}\right)\left[\frac{\partial v}{\partial x} + \frac{\partial u}{\partial y}\right],$$

Equations (7.35) to (7.39) are applied over the hydrogel boundaries, and the final system of equations is solved for the displacements u and v at the field nodes of the hydrogel.

It is essential to know the classifications of the PNP equations, namely elliptic, parabolic, or hyperbolic, to decide the discretisations for the derivative terms from the governing equations. Therefore, the PNP system is initially studied for the classification, and then the spatial discretisation stencils are decided.

7.4.1.2 Classification of PNP equations

In this section, the PNP equations are studied for their nature, namely elliptic, parabolic, or hyperbolic PDEs. Consider the residual equations R_1 and R_4, as given in Equations (7.21) and (7.22), respectively. To ensure the continuity of $C_{1,x}$, $C_{1,y}$, $\psi_{,x}$, and $\psi_{,y}$ along the characteristic line where $,x$ and $,y$ indicate the derivatives with respect to the x and y directions, respectively, the total derivatives are given as

$$dC_{1,x} = \frac{\partial C_{1,x}}{\partial x}dx + \frac{\partial C_{1,x}}{\partial y}dy \Rightarrow C_{1,xx}\,dx + C_{1,xy}dy \qquad (7.40)$$

$$dC_{1,y} = \frac{\partial C_{1,y}}{\partial x}dx + \frac{\partial C_{1,y}}{\partial y}dy \Rightarrow C_{1,yx}\,dx + C_{1,yy}\,dy \qquad (7.41)$$

$$d\psi_{,x} = \frac{\partial \psi_{,x}}{\partial x}dx + \frac{\partial \psi_{,x}}{\partial y}dy \Rightarrow \psi_{,xx}\,dx + \psi_{,xy}dy \qquad (7.42)$$

$$d\psi_{,y} = \frac{\partial \psi_{,y}}{\partial x}dx + \frac{\partial \psi_{,y}}{\partial y}dy \Rightarrow \psi_{,yx}\,dx + \psi_{,yy}dy \qquad (7.43)$$

It is written from Equation (7.21) to (7.22), and (7.40) to (7.43) that

$$
\begin{bmatrix}
1 & 0 & 1 & \lambda_1 C_1 & 0 & \lambda_1 C_1 \\
0 & 0 & 0 & 1 & 0 & 1 \\
dx & dy & 0 & 0 & 0 & 0 \\
0 & dx & dy & 0 & 0 & 0 \\
0 & 0 & 0 & dx & dy & 0 \\
0 & 0 & 0 & 0 & dx & dy
\end{bmatrix}
\begin{Bmatrix}
C_{1,xx} \\
C_{1,xy} \\
C_{1,yy} \\
\psi_{,xx} \\
\psi_{,xy} \\
\psi_{,yy}
\end{Bmatrix}
=
\begin{Bmatrix}
R_1 \\
R_2 \\
R_3 \\
R_4 \\
R_5 \\
R_6
\end{Bmatrix}
\Rightarrow A f = R
$$

$$(7.44)$$

where $\lambda_1 = Z_1(F/RT)Z_1$. As the second order derivative may be discontinuous, $|A| = 0$

$$\therefore |A| = 0 \Rightarrow dx^4 + 2dx^2 \, dy^2 + dy^4 = 0 \tag{7.45}$$

$$\therefore \left(\frac{dy}{dx}\right)^4 + 2\left(\frac{dy}{dx}\right)^2 + 1 = 0 \Rightarrow M^2 + 2M + 1 = 0 \tag{7.46}$$

where $(dy/dx)^2 = M$. The solution of Equation (7.46) is given as

$$M = -1 \Rightarrow \frac{dy}{dx} = i \tag{7.47}$$

As the roots are imaginary, the residual equations R_1 and R_4 have elliptic nature. Therefore, as the PNP equations are coupled together, they give the elliptic PDE.

It is concluded from this section that, the PNP equations are elliptic in nature and there is no need to choose a specific technique of spatial pattern (namely upwind or downwind). As a result, two virtual nodes from the sides of the concerned virtual node and the concerned virtual node itself are chosen in the present work for discretizing the governing equations and boundary conditions at the concerned virtual node.

7.4.1.3 Derivation of analytical displacements along moving boundaries 3 and 4 of hydrogel

In this section, the analytical expressions of the displacements u and v are derived by the semi-inverse method (Timoshenko and Goodier, 1970). Let ϕ be the second order Airy stress function as

$$\phi = a\,x^2 + b\,y^2 \tag{7.48}$$

The stresses along the hydrogel boundaries are given as

$$\sigma_{xx} = \frac{\partial^2 \phi}{\partial y^2} = 2\,b = P_x \quad \therefore \Rightarrow b = \frac{P_x}{2}, \quad \sigma_{yy} = \frac{\partial^2 \phi}{\partial x^2} = 2\,a = P_y \quad \therefore \Rightarrow a = \frac{P_y}{2}$$

(7.49)

where P_x and P_y are the components of osmotic pressure in the x and y directions, respectively. The substitution $\sigma_x = P_x$ and $\sigma_y = P_y$ is valid only for boundaries 3 and 4, respectively, when they are uniform.

$$\therefore \phi = \frac{P_y}{2} x^2 + \frac{P_x}{2} y^2$$

(7.50)

It is verified by Equation (7.50) that the compatibility equation $\nabla^4 \phi = 0$ is satisfied. The strains are then computed as

$$\varepsilon_{xx} = \frac{\partial u}{\partial x} = \frac{1}{E}[\sigma_x - \nu_0\,\sigma_y] \Rightarrow \varepsilon_{xx} = \frac{1}{E}[P_x - \nu_0\,P_y]$$

(7.51)

Integrating Equation (7.51) results in

$$u = \frac{x}{E}\Big[P_x - \nu_0 P_y\Big] + c_1$$

(7.52)

$$u = 0 \text{ at } x = 0 \Rightarrow c_1 = 0, \Rightarrow u = \frac{x}{E}\Big[P_x - \nu_0 P_y\Big]$$

(7.53)

The displacement in the y direction is similarly computed as

$$\varepsilon_{yy} = \frac{\partial v}{\partial y} = \frac{1}{E}[\sigma_y - \nu_0\,\sigma_x] \Rightarrow \varepsilon_{yy} = \frac{1}{E}[P_y - \nu_0\,P_x]$$

(7.54)

Integrating Equation (7.54) results in

$$v = \frac{y}{E}[P_y - \nu_0\,P_x] + c_2$$

(7.55)

$$v = 0 \text{ at } y = 0 \Rightarrow c_2 = 0, \Rightarrow v = \frac{y}{E}[P_y - \nu_0\,P_x]$$

(7.56)

Equations (7.53) and (7.56) give the displacements u and v, respectively, over the field nodes along hydrogel boundaries 3 and 4, respectively. We see that they are first-order continuous when P_x and P_y are constant, namely not the functions of x or y nodal coordinates. This observation will be verified in discussion of the simulation results for the square geometry of the hydrogel. In summary, the analytical solutions of the displacements u and v over the nodes along hydrogel boundaries 3 and 4 are derived for the uniform boundary and noted the displacements are first-order continuous.

7.4.1.4 Discretisation of PNP equations and formulation of Jacobian matrix

In this section, the discretisation of 2-D PNP equations and the formulation of Jacobian matrix in the NR iteration are discussed. To obtain a valid solution at each iteration step, it is ensured that the Jacobian matrix is not singular such that its inverse exists. Therefore, the direct inverse of Jacobian matrix is avoided in the present work, and instead it is solved by the Gauss elimination method.

The residual equations R_1 to R_{20} given by Equations (7.23) to (7.30) are used to formulate the Jacobian matrix in the NR method. Let N be the number of total virtual nodes in the computational domain of the PNP equations as shown in Figure 7.23b, NV be the number of total virtual nodes in the internal computational domain (i.e., excluding boundaries) of the PNP equations, and NV_5, NV_6, NV_7 and NV_8 be the numbers of total virtual nodes along the boundaries of solution 5 to 8, respectively, as shown in Figure 7.23b. The Jacobian matrix is formulated by taking the derivative of each residual equation discretised at N virtual nodes. Let the residual equation R_1 be discretised at all NV virtual nodes and represented as R_1^j, where $j \in [1, NV]$. The portion of the Jacobian matrix corresponding to the discretised equations R_1^j is given as

$$
\begin{bmatrix}
\dfrac{\partial R_1^1}{\partial C_1^1} & \cdots & \dfrac{\partial R_1^1}{\partial C_1^N} & \dfrac{\partial R_1^1}{\partial C_2^1} & \cdots & \dfrac{\partial R_1^1}{\partial C_2^N} & \dfrac{\partial R_1^1}{\partial C_3^1} & \cdots & \dfrac{\partial R_1^1}{\partial C_3^N} & \dfrac{\partial R_1^1}{\partial \psi^1} & \cdots & \dfrac{\partial R_1^1}{\partial \psi^N} \\[2ex]
\dfrac{\partial R_1^2}{\partial C_1^1} & \cdots & \dfrac{\partial R_1^2}{\partial C_1^N} & \dfrac{\partial R_1^2}{\partial C_2^1} & \cdots & \dfrac{\partial R_1^2}{\partial C_2^N} & \dfrac{\partial R_1^2}{\partial C_3^1} & \cdots & \dfrac{\partial R_1^2}{\partial C_3^N} & \dfrac{\partial R_1^2}{\partial \psi^1} & \cdots & \dfrac{\partial R_1^2}{\partial \psi^N} \\[2ex]
\vdots & \cdots & & & & & & & & & \cdots & \vdots \\[2ex]
\dfrac{\partial R_1^{NV}}{\partial C_1^1} & \cdots & \dfrac{\partial R_1^{NV}}{\partial C_1^N} & \dfrac{\partial R_1^{NV}}{\partial C_2^1} & \cdots & \dfrac{\partial R_1^{NV}}{\partial C_2^N} & \dfrac{\partial R_1^{NV}}{\partial C_3^1} & \cdots & \dfrac{\partial R_1^{NV}}{\partial C_3^N} & \dfrac{\partial R_1^{NV}}{\partial \psi^1} & \cdots & \dfrac{\partial R_1^{NV}}{\partial \psi^N}
\end{bmatrix}_{NV \times 4N}
$$

$$(7.57)$$

An individual term from Equation (7.57) is given as

$$
\frac{\partial R_i^j}{\partial f^k}
$$

$$(7.58)$$

where i, j and k represent the residual equation number such that $i \in [1, 20]$, index of the virtual node from the internal computational domain of the PNP equations such that $j \in [1, NV]$, and the overall index of virtual nodes such that $k \in [1, N]$, respectively. Therefore, Equation (7.58) is the $(R_i)^{th}$ residual equation discretised at the jth virtual node from the internal computational domain, and its derivative is taken with respect to the function value at the kth virtual node from the full computational domain. The Jacobean matrix for the remaining residual equations is constructed as Equation (7.57), to get the complete Jacobean matrix $[J]_{(4 \ N \times 4 \ N)}$. Let the residual equation R_1, as given in Equation (7.23), be discretised at the ith virtual node, then Equation (7.23) is modified as

$$R_1(x_i, \ y_i) = R_1^i = \{b_{ik}\} \ \left\{C_1^b(x_k, \ y_k)\right\} + \{b_{il}\} \ \left\{C_1^b(x_l, \ y_l)\right\} + (Z_1 \ \eta) \ \left[\{a_{ik}\} \ \left\{C_1^b(x_k, \ y_k)\right\}\right.$$

$$\times \ \{a_{im}\} \ \left\{\psi^b(x_m, \ y_m)\right\} + \{a_{il}\} \ \left\{C_1^b(x_l, \ y_l)\right\} \ \{a_{in}\} \ \left\{\psi^b(x_n, \ y_n)\right\}\right] + \left(Z_1 \ \eta \ C_1^b(x_i, \ y_i)\right)$$

$$\times\left[\{b_{im}\} \ \left\{\psi^b(x_m, \ y_m)\right\} + \{b_{in}\} \ \{\psi^b(x_n, \ y_n)\}\right]$$

$$(7.59)$$

Three cases are possible to compute $\left(\partial R_1^i \ / \ \partial C_1^j\right)$. First, if $j \in k$ and l, where $k = 1, 2, \ ..., \ N_x$ and $l = 1, 2, \ ..., \ N_y$

$$\frac{\partial R_1^i}{\partial C_1^j} = b_{ij} + b_{ij} + (Z_1 \ \eta) \ \left[a_{ij} \ \{a_{im}\} \ \left\{\psi^b(x_m, \ y_m)\right\} + a_{ij} \ \{a_{in}\} \ \left\{\psi^b(x_n, y_n)\right\}\right] +$$

$$(Z_1 \ \eta \ \delta_{ij}) \ \left[\{b_{im}\} \ \left\{\psi^b(x_m, \ y_m)\right\} + \{b_{in}\} \ \left\{\psi^b(x_n, \ y_n)\right\}\right]$$

$$(7.60)$$

The second case is that, if $j \in k$

$$\frac{\partial R_1^i}{\partial C_1^j} = b_{ij} + (Z_1 \ \eta) \ \left[a_{ij} \ \{a_{im}\}\left\{\psi^b(x_m, \ y_m)\right\}\right] +$$

$$(Z_1 \ \eta \ \delta_{ij}) \ \left[\{b_{im}\} \ \left\{\psi^b(x_m, \ y_m)\right\} + \{b_{in}\} \ \left\{\psi^b(x_n, \ y_n)\right\}\right]$$

$$(7.61)$$

The third case is that if $j \in l$

$$\frac{\partial R_1^i}{\partial C_1^j} = b_{ij} + (Z_1 \ \eta) \ \left[a_{ij}\{a_{in}\}\left\{\psi^b(x_n, y_n)\right\}\right] +$$

$$(Z_1 \ \eta \ \delta_{ij}) \ \left[\{b_{im}\} \ \left\{\psi^b(x_m, \ y_m)\right\} + \{b_{in}\} \ \left\{\psi^b(x_n, \ y_n)\right\}\right]$$

$$(7.62)$$

Three cases are possible to compute $(\partial R_1^i / \partial \psi^j)$ as well. The first case is that if $j \in m$ and n, where $m = 1, 2, \ldots, N_x$ and $n = 1, 2, \ldots, N_y$

$$\frac{\partial R_1^i}{\partial \psi^j} = (Z_1 \; \eta) \left[a_{ij} \; \{a_{ik}\} \{C_1^b(x_k, y_k)\} + a_{ij} \{a_{il}\} \{C_1^b(x_l, y_l)\} \right] + \tag{7.63}$$

$$(Z_1 \; \eta \; C_1^b(x_i, y_i)) \; [b_{ij} + b_{ij}]$$

The second case is that if $j \in m$

$$\frac{\partial R_1^i}{\partial \psi^j} = (Z_1 \; \eta) \left[a_{ij} \; \{a_{ik}\} \{C_1^b(x_k, y_k)\} \right] + \left[Z_1 \; \eta \; C_1^b(x_i, y_i) \; b_{ij} \right] \tag{7.64}$$

The third case is that if $j \in n$

$$\frac{\partial R_1^i}{\partial \psi^j} = (Z_1 \; \eta) \left[a_{ij} \; \{a_{il}\} \{C_1^b(x_l, y_l)\} \right] + \left[Z_1 \; \eta \; C_1^b(x_i, y_i) \; b_{ij} \right] \tag{7.65}$$

Similar to the derivatives of equations R_1^i, as given from Equations (7.60) to (7.65), the derivatives of the residual equations R_2^i and R_3^i are computed. Let the residual equation R_4 as in Equation (7.24) be discretised at the ith virtual node. Three cases are possible to compute $\left(\partial R_4^i / \partial \psi^j \right)$. The first is if $j \in m$ and n

$$\frac{\partial R_4^i}{\partial \psi^j} = b_{ij} + b_{ij} \tag{7.66}$$

The second and third cases are that if $j \in m$ or $j \in n$, respectively, as

$$\frac{\partial R_4^i}{\partial \psi^j} = b_{ij} \tag{7.67}$$

Other derivatives of R_4^i are computed as

$$\frac{\partial R_4^i}{\partial C_1^j} = -k \; Z_1 \; \delta_{ij}, \; \frac{\partial R_4^i}{\partial C_2^j} = -k \; Z_2 \; \delta_{ij}, \; and \; \frac{\partial R_4^i}{\partial C_3^j} = -k \; Z_3 \; \delta_{ij} \tag{7.68}$$

Let the residual equation R_s be discretised at the ith virtual node such that if $j \in k$

$$\frac{\partial R_5^i}{\partial C_1^j} = a_{ij} \tag{7.69}$$

The derivatives of the residual equations R_6^i to R_{12}^i are similarly computed. The derivatives of the residual equations R_{13} to R_{16} applied at the ith virtual node are

$$\frac{\partial R_{13}^i}{\partial C_1^j} = \delta_{ij}, \frac{\partial R_{14}^i}{\partial C_2^j} = \delta_{ij}, \frac{\partial R_{15}^i}{\partial C_3^j} = \delta_{ij}, \text{ and } \frac{\partial R_{16}^i}{\partial \psi^j} = \delta_{ij} \tag{7.70}$$

The derivatives of the residual equations R_{17} to R_{20} are similarly computed and the complete Jacobian matrix is assembled. New values of the field variables at the virtual nodes are obtained after solving the Jacobian matrix as

$$\left(C_1^j\right)^{n+1} = (1 - SOR)\left(C_1^j\right)^n + SOR\ \Delta C_1^j$$

$$\left(C_2^j\right)^{n+1} = (1 - SOR)\left(C_2^j\right)^n + SOR\ \Delta C_2^j$$

$$\left(C_3^j\right)^{n+1} = (1 - SOR)\left(C_3^j\right)^n + SOR\ \Delta C_3^j \tag{7.71}$$

$$\left(\psi^j\right)^{n+1} = (1 - SOR)\left(\psi^j\right)^n + SOR\ \Delta\psi^j$$

where n and SOR are the previous iteration count and successive relaxation factor, respectively, and ΔC_1^j, ΔC_2^j, ΔC_3^j, and $\Delta\psi^j$ are the incremental values of the field variables obtained by solving the Jacobian matrix with the Gauss elimination method. Since the PNP equations are nonlinear with the locally high gradient distributions of the field variables, they are prone to become unstable during the NR iterations. Therefore the relaxation factor is used for the stable computation. The formulation of the Jacobian matrix for the NR iteration is discussed in this section. The results obtained by this approach are presented in the subsequent sections.

7.4.1.5 Exact imposition of Neumann boundary conditions

In order to correctly solve the PNP equations, the Neumann boundary conditions along boundaries 5 and 6 are imposed exactly as

$$\frac{\partial C_1^i}{\partial x} = \frac{C_1^{next} - C_1^i}{\Delta x} = 0 \implies C_1^i = C_1^{next}, \text{ at boundary } x = 0 \tag{7.72}$$

$$\frac{\partial C_1^i}{\partial y} = \frac{C_1^{next} - C_1^i}{\Delta y} = 0 \quad \Rightarrow \quad C_1^i = C_1^{next}, \text{ at boundary } y = 0 \qquad (7.73)$$

where i and *next* are the indices of the virtual node on the Neumann boundary and the immediate next virtual node in the specific x or y direction, respectively, and Δx and Δy are the spacings between the ith and *next* nodes in the x and y directions, respectively. The Neumann boundary condition is first converted to the Dirichlet boundary condition in this way, and then applied.

For the hydrogel domain shown in Figure 7.23c, the last virtual node along the boundary $y = 0$ does not have any virtual node in the y direction; therefore following strategy is adopted while computing the gradient $\partial/\partial y$ at this virtual node. Consider the last virtual node along the boundary $y = 0$ identified as 6 and the last virtual node along the curved boundary as 11. The end portion of the hydrogel is enlarged, as shown in Figure 7.24. Virtual node 6 is connected with virtual node 11 and the Euclidean length between them is computed as $S = \sqrt{(\Delta x)^2 + (\Delta y)^2}$ as shown in Figure 7.24. The gradient between virtual nodes 6 and 11 along the direction S is given as

$$\left(\frac{\partial f}{\partial S}\right) = \frac{f^{11} - f^6}{S} \qquad (7.74)$$

The gradient at the virtual node 6 in the y direction is given by Equation (7.74) as

$$\left(\frac{\partial f}{\partial y}\right)_6 = \left(\frac{\partial f}{\partial S}\right)_{6,11} \cos(\theta) \quad \Rightarrow \quad \left\{ \begin{array}{cc} -\dfrac{\cos(\theta)}{S} & \dfrac{\cos(\theta)}{S} \end{array} \right\} \left\{ \begin{array}{c} f^6 \\ f^{11} \end{array} \right\} \qquad (7.75)$$

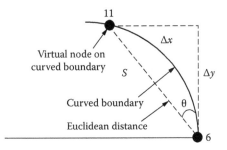

Figure 7.24 Computation of derivative in y direction at last virtual node along boundary y = 0 of hydrogel. (Modified from H. Li and S. Mulay. (2007). *Computational Mechanics*, DOI: 10.1007/s004466-011-0622-5.)

As a result, Equation (7.75) is used to compute the gradient $\partial/\partial y$ at virtual node 6.

7.4.1.6 Implementation of RDQ method for 2-D simulation of pH-sensitive hydrogels

In this section, implementation of the RDQ method is discussed in detail to solve the problem of 2-D pH-sensitive hydrogels. Since the PNP equations are solved iteratively by the NR method, the convergence is detected by the true error norm in the percentage as

$$\xi = 100 \sqrt{\sum_{i=1}^{N} \frac{[(\Delta C_1^i)^2 + (\Delta C_2^i)^2 + (\Delta C_3^i)^2 + (\Delta \psi^i)^2]}{[(C_1^i)^2 + (C_2^i)^2 + (C_3^i)^2 + (\psi^i)^2]^n}} \% \quad (7.76)$$

where N and n indicate the number of total virtual nodes in the solution domain and the previous iteration count, respectively. The following procedure is adopted while iteratively solving the coupled PNP and mechanical equilibrium equations. The hydration H of the hydrogel is computed as

$$H = \frac{\text{current hydrogel volume}}{\text{initial dry solid hydrogel volume}} \quad (7.77)$$

The value $H = 1$ is taken at the beginning of NR iterations, since it is assumed that the volumes of the hydrogel in the current and initial dry states are identical. The concentration values of the ionized fixed-charge at the field and virtual nodes of the hydrogel are computed by Langmuir monolayer adsorption theory (Li et al., 2005a) as

$$Z_f \ C_f = -\frac{1}{H} \left[\frac{C_{f,s}^0 \ K_d}{(K_d + C_{H^+})} \right] \text{(for acidic hydrogel)} \quad (7.78)$$

where K_d is the dissociation constant of hydrogel, and $C_{f,s}^0$ and C_{H^+} are the concentration values of the initial fixed charge and mobile hydrogen ions within the hydrogel, respectively. The residual equations R_1 to R_{20} are applied over the virtual nodes in the solution domain to solve the PNP equations by the NR iterative method. Convergence of the NR method is ensured when $\xi \leq k^{PNP}$ where k^{PNP} is the predefined true error norm in the percentage. The values of the nodal parameters at the field nodes are computed after the convergence of PNP equations as

$$\{U^F\}_{n\times1} = \bar{A}^{-1}\{f^V\}_{m\times1} \tag{7.79}$$

$$\text{where } A = \begin{cases} [N]_{m\times n} & \text{if } m \equiv n \\ [N^T]_{n\times m}[N]_{m\times n} & \text{if } m \neq n \end{cases} \quad \text{and} \quad \bar{A}^{-1} = \begin{cases} A^{-1} & \text{if } m \equiv n \\ A^{-1}[N^T]_{n\times m} & \text{if } m \neq n \end{cases}$$

$$\tag{7.80}$$

where $[N]$ is the linear transformation matrix interpolating the values of the function at the virtual nodes with the values of nodal parameters at the surrounding field nodes. $\{U^F\}$ and $\{f^V\}$ are the vectors of the nodal parameter values at the field nodes and the field variable values at the virtual nodes, respectively, and m and n are the numbers of total virtual and field nodes in the solution domain, respectively. Due to the lack of Kronecker delta property $f^i \neq u^i$ where $i\in[1, n]$, the values of function at the field nodes are computed by interpolating each field node by the surrounding field nodes with the fixed RKPM interpolation as $f^h(x_i, y_i) = \sum_{k=1}^{NP} N_k u_k$ where the total NP field nodes are used in the local interpolation domain of the ith field node such that $NP \subseteq N$, where N is the total field nodes in the whole domain, and N_k and u_k are the values of the shape function and nodal parameter at the kth field node, respectively. All the values of functions, C_1, C_2, C_3 and ψ, at the field nodes are thus computed.

The values of the osmotic pressures at the field nodes located along hydrogel boundaries 3 and 4 are computed by Equation (7.32). The mechanical equilibrium equations are solved for the hydrogel swelling by imposing the osmotic pressure as the Neumann boundary condition.

The expanded hydrogel volume is evaluated by numerically computing the hydrogel area and then multiplying it by the hydrogel thickness. The field nodes along hydrogel boundaries 3 and 4 are utilized for tracking the moving boundaries by projecting them on boundary 2. The deformed area of the hydrogel is subdivided where each portion is formed by the nodes along the boundaries 3 and 4 and their projections on boundary 2. Figure 7.25 shows the portion of the deformed hydrogel domain with the field nodes 1 to 4 along the corners. The area covered by nodes 1 to 4 is subdivided into two areas, i.e., A_1 and A_2 that are computed by assuming them to be a rectangle and triangle, respectively, and given as

$$A_1 = (diff\_y)(x_4 - x_3) \quad \text{and} \quad A_2 = \frac{1}{2}|(y_2 - y_1)|(x_4 - x_3) \tag{7.81}$$

The total area between nodes 1 to 4 is given as $\Delta A = A_1 + A_2$. Similarly, the remaining area of the deformed hydrogel is computed by the subsequent

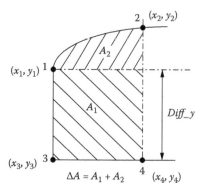

Figure 7.25 Approach for computing expanded hydrogel area by field nodes along hydrogel boundaries 3 and 4 and their projection on boundary 2. (From S. Mulay and L. Li. *Modelling and Simulation in Materials Science and Engineering*, DOI:10.1088/0965-0393/19/6/065009. With permission.)

strip portions, and the total area is computed by adding all the individual areas ΔA. The new value of the hydration H is computed that is used during the next iteration of PNP equations. In this way, the PNP equations coupled with mechanical equilibrium equations are solved iteratively until they converge. The true error norm ζ for the values of the hydrogel displacements u and v is computed in the percentage as

$$\zeta = 100 \sqrt{\sum_{i=1}^{N} \frac{\left(u_i^{new} - u_i^{old}\right)^2 + \left(v_i^{new} - v_i^{old}\right)^2}{\left(u_i^{old}\right)^2 + \left(v_i^{old}\right)^2}} \ \% \tag{7.82}$$

where *new* and *old* indicate the values of deformation corresponding to the current and previous iterations, respectively, at N field nodes in the hydrogel. The convergence is detected when $\zeta \leq k^{PNPME}$, where k^{PNPME} is the predefined true error norm in the percentage.

In summary, the implementation of the RDQ method to solve the 2-D hydrogel for the pH response is discussed in this section. The stability of the simulation depends on how the values k^{PNP} and k^{PNPME} are chosen suitably. For the results presented here, the values $k^{PNP} = 1 \times 10^{-4}$ and $k^{PNPME} = 1 \times 10^{-2}$ are used. During the iterations between the PNP and mechanical equilibrium equations, the displacement values of the hydrogel field nodes are utilized for computing the expanded hydrogel volume and the hydration H, which consequently gives the new values of the concentration C_{ef} of the fixed charge at the field nodes of the hydrogel. However, the PNP and mechanical equilibrium equations are always solved by the virtual and field nodes in the original reference configuration X, such that no new discretisation is required during

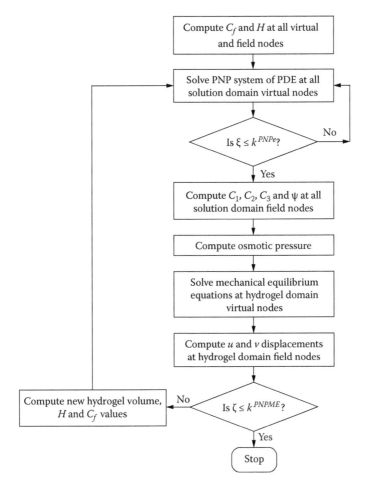

Figure 7.26 Complete flowchart of 2-D simulation of pH-responsive hydrogel by RDQ method. (From S. Mulay and L. Li. *Modelling and Simulation in Materials Science and Engineering*, DOI:10.1088/0965-0393/19/6/065009. With permission.)

the iterations. The simulation procedure of 2-D pH-sensitive hydrogel by the RDQ method is presented in Figure 7.26 as a flowchart.

7.4.2 Two-dimensional simulation of pH-sensitive hydrogels by RDQ method

The steady-state simulation results of 2-D pH-responsive hydrogels are presented in this section. To demonstrate the capability of the RDQ method, a nonuniform hydrogel domain is defined as shown in Figure 7.23a. The hydrogel and solution domains are discretised by the uniform or random field nodes that are separately coupled with the uniform virtual nodes. The dimensions of the hydrogel domain are considered as $W = 2$ mm, $L = 4$ mm and $R = 1$ mm, while the parameters of the solution domain are used as $W = 10$ mm and $L = 10$ mm, according to Figure 7.23a. The simulation results are initially presented for the solution pH = 3 with the remaining parameters, as given in Table 7.4. The coupled PNP and mechanical equilibrium equations are solved by first discretizing the solution and hydrogel domains by the uniform field nodes coupled with the uniform virtual nodes. Figure 7.27 shows the distributions of the field variables, C_f, C_1, C_2, C_3, and ψ, for the solution pH = 3 when the hydrogel and solution are in equilibrium with each other.

As the simulation is performed for the acidic hydrogel, the positive mobile ions diffuse into the hydrogel and combine with the negative fixed charge groups leading to the higher values of the concentration of mobile ions $C_{Na^+} = C_1$ and $C_{H^+} = C_2$ in the hydrogel domain as compared with the solution domain, as seen in Figures 7.27b and c, respectively. Simultaneously,

Table 7.4 Parameters used in 2-D simulation of pH-responsive hydrogels by RDQ method

Parameter	Value	Unit
Na^+ mobile ion valance (Z_1)	+1	–
H^+ mobile ion valance (Z_2)	+1	–
Cl^- mobile ion valance (Z_3)	–1	–
C_f fixed-charge ion valance (Z_f)	–1	–
Na^+ mobile ion concentration $(\tilde{C}_1)$	10.0	mM
H^+ mobile ion concentration $(\tilde{C}_2)$	1.0	mM
Cl^+ mobile ion concentration $(\tilde{C}_3)$	11.0	mM
$C^0_{f,s}$ initial fixed-charge ion concentration	1.050	mM
Temperature (k)	298	K
Gas constant (R)	8.314	J/(mol K)
Vacuum permittivity (ε_0)	8.8542×10^{-12}	C2/(N m^2)
Relative permittivity (ε_r)	80	–
Dissociation constant (k_d)	0.02	mM
Faraday constant (F)	96485.3399	C/mol
Young's modulus (E)	29	MPa
Poisson's ratio (v_0)	0.45	–

Sources: S. Mulay and L. Li. Modelling and Simulation in Materials Science and Engineering, DOI:10.1088/0965-0393/19/6/065009. With permission; modified from H. Li and S. Mulay. (2007). Computational Mechanics, DOI: 10.1007/s004466-011-0622-5.

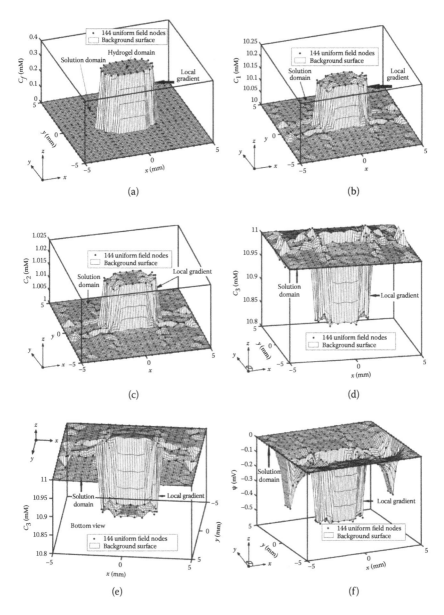

Figure 7.27 Schematic of the profiles of various field variables after solving the 2-D pH-responsive hydrogel for pH – 3, $E = 29$ and uniform field nodes, where the different profiles are given as C_f (a), C_1(b), C_2(c), top view of C_3 (d), bottom view of C_3 (e) and ψ(f). (Modified from H. Li and S. Mulay. (2007). *Computational Mechanics*, DOI: 10.1007/s004466-011-0622-5.)

the negative mobile ion $C_{Cl^-} = C_3$ has a lower value of concentration inside the hydrogel when compared with the solution, as shown in Figure 7.27d and e, due to the repulsive force between the anionic fixed charge and the mobile ion C_3.

Also, as the anionic fixed charge groups are bound to the hydrogel cross-linked network of polymers, the hydrogel has a higher negative chemical potential than the surrounding solutions, which further reduces the diffusion of mobile ions C_3 into the hydrogel. Figure 7.28 shows the distribution of cumulative concentrations of all the mobile ions $(C_1 + C_2 + C_3)$ in the computational domain. While the solution almost satisfies the electroneutrality condition, the hydrogel has an average concentration value of mobile ions as 0.35 mM, which closely matches with the average negative fixed-charge concentration value of 0.36 mM inside the hydrogel, as shown in Figure 7.27a.

The net difference between all the ionic concentrations in the hydrogel leads to the constant electrical potential inside the hydrogel as shown in Figure 7.27f. However, the electrical potential is almost equal to zero in the

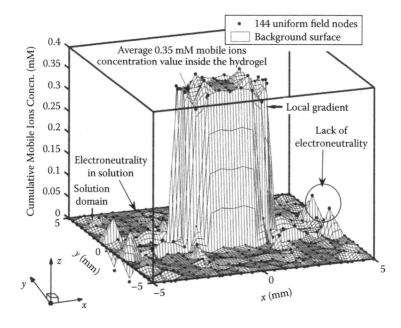

Figure 7.28 Cumulative concentration values of mobile ions distributed in hydrogel and solution domains, where electroneutrality condition is satisfied in solution and average value of 0.35 mM is obtained in hydrogel domain, closely matching average concentration value of hydrogel fixed charge of 0.36 mM. (Modified from H. Li and S. Mulay. (2007). *Computational Mechanics*, DOI: 10.1007/s004466-011-0622-5.)

surrounding solution, since there is an electroneutrality for the total concentrations of all the ionic species in the solution. It is seen from Figure 7.28 that the electroneutrality condition is not exactly satisfied at a few field nodes in the solution domain. As a result, a negligible value of the electrical potential is present at these locations as seen in Figure 7.27f.

Figure 7.29 shows the swelling of hydrogel until equilibrium is reached between the solution and hydrogel. The displacements u and v at the field nodes of the hydrogel and the comparison of the initial and deformed configurations of the hydrogel are shown in Figure 7.30 and the Dirichlet boundary conditions are imposed exactly. The profiles of the displacements u and v almost look like first-order continuous, except at the field nodes near the curved boundary.

The coupled PNP and mechanical equilibrium equations are again solved by discretizing the solution and hydrogel domains with the random field nodes coupled with the uniform virtual nodes. The random field nodes are generated with the uniform probability by the linear congruential generator algorithm (Christoph, 1997; Mulay et al., 2009). Figure 7.31 illustrates various profiles of the ionic concentrations in the hydrogel and solution. Figure 7.31e shows that the average cumulative concentration value, 0.29 mM of mobile ions in the hydrogel matches well with the concentration

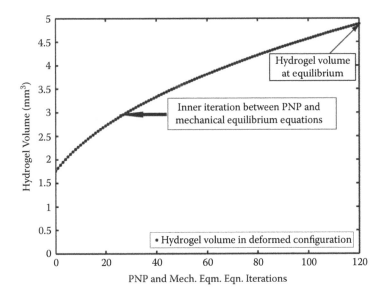

Figure 7.29 Increase in hydrogel volume during iterations of coupled PNP and mechanical equilibrium equations, where initial configuration volume was 1.75 mm². (Modified from H. Li and S. Mulay. (2007). *Computational Mechanics*, DOI: 10.1007/s004466-011-0622-5.)

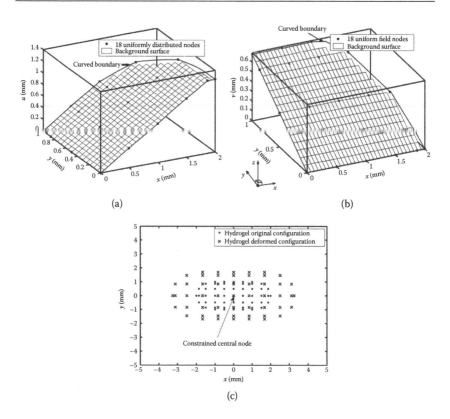

Figure 7.30 Displacements u (a) and v (b) and comparison of original and deformed hydrogel (c) for curved boundary hydrogel with uniform distribution of field nodes where central node is fully constrained and Dirichlet boundary conditions are exactly imposed. (Modified from H. Li and S. Mulay. (2007). *Computational Mechanics*, DOI: 10.1007/s004466-011-0622-5.)

value 0.29 mM of fixed charge inside the hydrogel, as shown in Figure 7.31a. The displacements u and v at the field nodes of the hydrogel and the configurations of the hydrogel in the initial and swelled states are shown in Figure 7.32.

In summary, the PNP and mechanical equilibrium equations are solved iteratively for the solution pH = 3 with the computational domains discretised by either uniform or random field nodes. We see from the various distributions of the field variables that the RDQ method is capable of capturing the jump in the field variables across the interface between multi-domain boundaries. It is observed by comparing Figure 7.27 and Figure 7.31 that the peak values of the concentrations of the mobile ions are almost identical when the domains are discretised by either uniform or random field nodes. This demonstrates that the RDQ method evenly handles uniform and random field nodes.

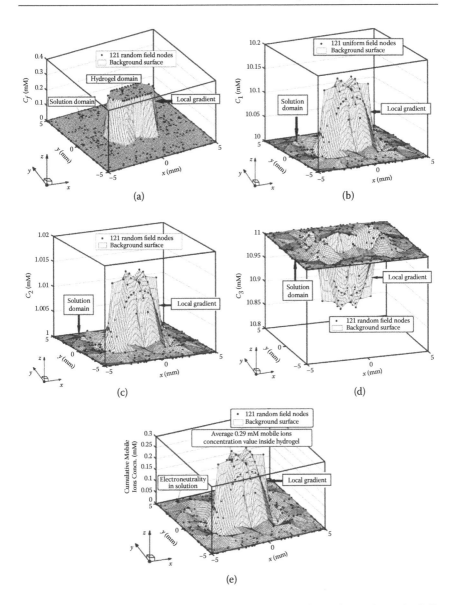

Figure 7.31 Schematic of the profiles of various field variables after solving the 2-D PH-sensitive hydrogel for pH = 3 and random field nodes, where the different profiles are given as C_f (a), C_1 (b), C_2 (c), C_3 (d), and the distribution of cumulative concentrations of mobile ions (e). (Modified from H. Li and S. Mulay. (2007). *Computational Mechanics*, DOI: 10.1007/s004466-011-0622-5.)

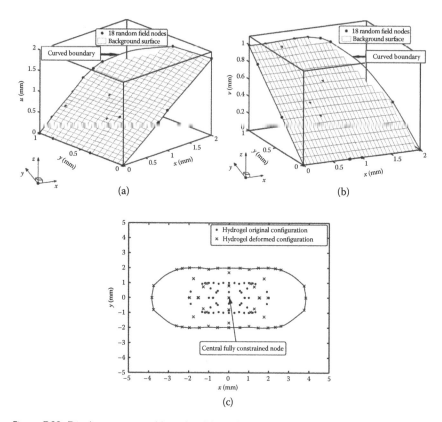

Figure 7.32 Displacements u (a) and v (b), and comparison of original and deformed hydrogel (c) for curved boundary hydrogel with randomly distributed field nodes. (Modified from H. Li and S. Mulay. (2007). *Computational Mechanics*, DOI: 10.1007/s004466-011-0622-5.)

7.4.3 Effects of solution pH and initial fixed-charge concentration on swelling of two-dimensional hydrogel

To appropriately understand the swelling behaviour of 2-D pH-responsive hydrogels, it is essential to study the effects of solution pH and the initial fixed charge concentration. Therefore, the PNP and mechanical equilibrium equations are solved together in this section by changing the values of solution pH to 3, 7 and 11 and the initial fixed charge as 0.5, 1.05, and 1.5 mM inside the hydrogel. Figure 7.33 shows the distributions of all the mobile ions and electrical potential with the different values of initial fixed charge concentration inside the hydrogel at solution pH = 3. It is easily verified from Figure 7.33f that, while the solution satisfies the electroneutrality condition, the average cumulative concentration value of mobile ions inside the hydrogel matches closely with the

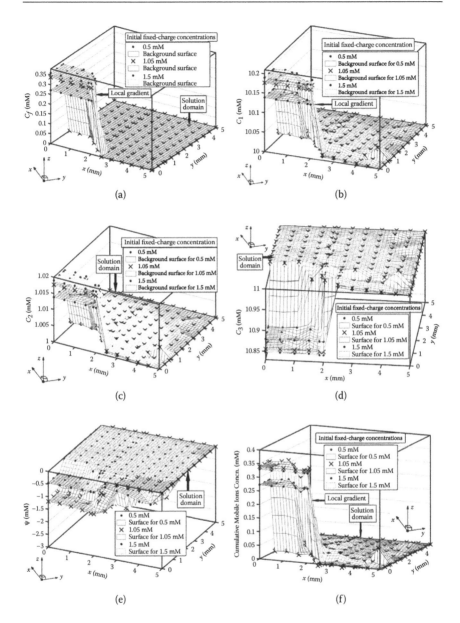

Figure 7.33 Schematic of the profiles of various field variables after solving the hydrogel for different values of initial fixed-charge and pH = 3, where the different profiles are as C_f (a), C_1 (b), C_2 (c), C_3 (d), ψ (e) and cumulative mobile ions distribution (f). (Modified from H. Li and S. Mulay. (2007). *Computational Mechanics*, DOI: 10.1007/s004466-011-0622-5.)

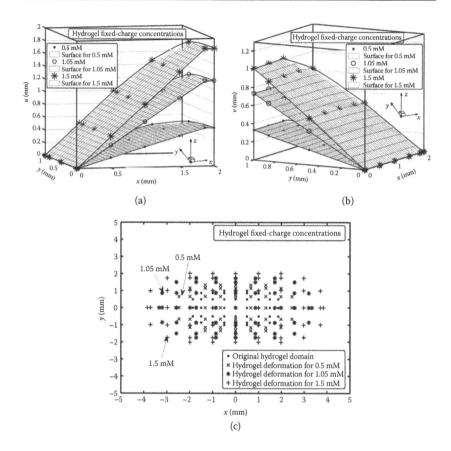

Figure 7.34 Displacements *u* (a) and *v* (b), and comparison of original and deformed hydrogel (c) for different concentration values of initial fixed charge with solution pH = 3. (Modified from H. Li and S. Mulay. (2007). *Computational Mechanics*, DOI: 10.1007/s004466-011-0622-5.)

average concentration of fixed-charge inside the hydrogel, as shown in Figure 7.33a.

Figure 7.34 shows the displacements *u* and *v*, and the swelling of the hydrogel by different values of initial fixed charge concentration at solution pH = 3. We see from the figure that the displacements of the hydrogel are almost linear, except at the field nodes near the curved boundary, even at the higher concentrations of the initial fixed charge inside the hydrogel. Figures 7.35a through d show the displacements *u* and *v*, respectively, at the hydrogel nodes corresponding to the solution pH = 7 and 11, respectively. It is observed by comparing Figure 7.34 and Figure 7.35 that increasing the pH of the solution and the concentration of the initial fixed charge results

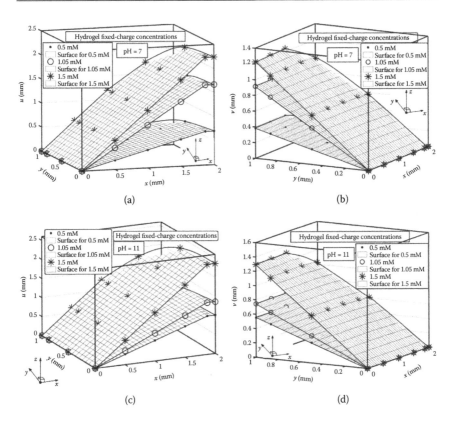

Figure 7.35 Displacements *u* (a) and *v* (b) for solution pH = 7 and displacements *u* (c) and *v* (d) for solution pH = 11 at different initial fixed charge concentrations. (Modified from H. Li and S. Mulay. (2007). *Computational Mechanics*, DOI: 10.1007/s004466-011-0622-5.)

in larger swelling of the hydrogel. This is an expected behaviour according to the nature of the problem.

The concentration of H⁺ mobile ions reduces with an increase in the pH of solution. More Na⁺ ions have to diffuse into the hydrogel to maintain the electroneutrality by combining with the increasing values of the initial fixed charge concentration, resulting in higher gradients of the different concentrations across the interface between the solution and hydrogel. This results in higher osmotic pressure and then larger swelling of the hydrogel. Figure 7.36 demonstrates an increase in the swelling of the hydrogel with an increase in the values of the pH and initial fixed charge concentration, as explained earlier.

A square hydrogel domain is considered with the dimensions $W = 2$ mm and $L = 2$ mm to compare the analytical and numerical values of the displacement

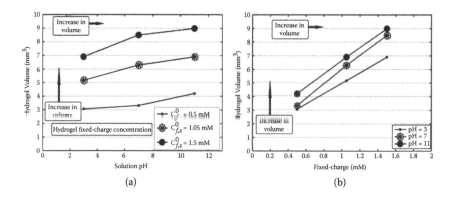

Figure 7.36 Change in volume of hydrogel at pH = 3, 7, and 11 as function of initial fixed charge concentration (a), and change in volume of hydrogel with respect to initial fixed charge concentrations of 0.5, 1.05, and 1.5 mM as function of solution pH (b). (Modified from H. Li and S. Mulay. (2007). *Computational Mechanics*, DOI: 10.1007/s004466-011-0622-5.)

at the field nodes along boundaries 3 and 4. The PNP and mechanical equilibrium equations are solved over the square domain of hydrogel with pH = 3 and $C_{f,s}^0 = 1.05$ mM. Figure 7.37 shows the x and y coordinates of the deformed hydrogel and the analytical values corresponding to the field nodes along boundaries 3 and 4 by Equations (7.53) and (7.56), respectively. It is observed in Figure 7.37a that the final numerical values of x coordinate at the field nodes along boundary 3 closely match the corresponding analytical values, similarly the final numerical values of y coordinate at the field

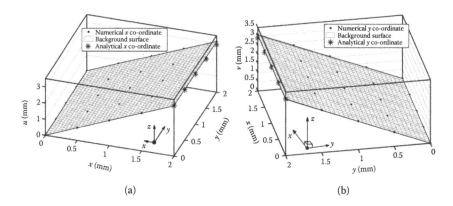

Figure 7.37 Comparison of numerical and analytical values of x (a) and y (b) coordinates along boundaries 3 and 4, respectively, of deformed square hydrogel, where both deformations closely match. (Modified from H. Li and S. Mulay. (2007). *Computational Mechanics*, DOI: 10.1007/s004466-011-0622-5.)

nodes along boundary 4 closely match the corresponding analytical values in Figure 7.37b.

In summary, the effects of the solution pH and concentration of initial fixed charge inside hydrogels are studied in this section. Hydrogel swelling increases with the increase of the solution pH and the initial concentration of fixed charge inside the hydrogel.

7.4.4 Effects of Young's modulus and geometrical shape of hydrogel at dry state on swelling

In this subsection, the 2-D simulation of a pH-responsive hydrogel is performed by Young's moduli $E = 25$, 29, and 36 and three different geometrical shapes of hydrogels as shown in Figure 7.38, with the values of ratio $r = W/L$ as 1:1, 1:2, and 1:3.

The hydrogel with the geometrical shape in Figure 7.38a, is already solved for the solution pH = 3 and Young's modulus $E = 29$, and the results are given from Figure 7.27 to Figure 7.30. It is solved further in this subsection for the strains and stresses inside the hydrogel. The expansion of the hydrogel causes strains and stresses due to the stiffness of the crosslinked polymer network. As a result, all the strains and stresses inside the hydrogel are plotted in Figure 7.39 and Figure 7.40, respectively.

Note from Figure 7.39 that the absolute nodal values of the strains are almost constant with the difference that ε_{xx} increases angularly from the y to x axis, and ε_{yy} increases angularly from the x to y axis. This is expected due to the Dirichlet boundary conditions along the x and y axes. Based on Figure 7.40, the maximum values of the normal stresses occur at the central fully constrained node.

We can see in Figure 7.40c that the boundary condition of the shear stress $\sigma_{xy} = 0$ is exactly imposed. The computed state of stress inside the hydrogel is different at each field node due to the unequal values of the osmotic pressure along the curved boundary. As a result, by considering the state of stress at each field node individually, the corresponding orientation of the principal plane is determined by Mohr's circle (Timoshenko and Goodier, 1970) and the surface normals to the principle plane are plotted in Figure 7.40d. The figure verifies that the field nodes along the boundaries $y = 0$ and x = 0 have the principal planes normal to the x and y axes, respectively, which indicates $\sigma_{xy} = 0$ along these boundaries. This is expected due to the symmetric boundary conditions.

This problem is solved with the hydrogel cross-section shown in Figure 7.38a by the solution pH = 3, 7 and 11, and Young's moduli $E = 25$, 29 and 36, with the remaining parameters given in Table 7.4. All the profiles of the various field variables for pH = 3 and $E = 25$, 29, and 36 are shown in Figure 7.41. The profiles of the mobile ions are as per the nature of the problem and the solution satisfies the electroneutrality condition.

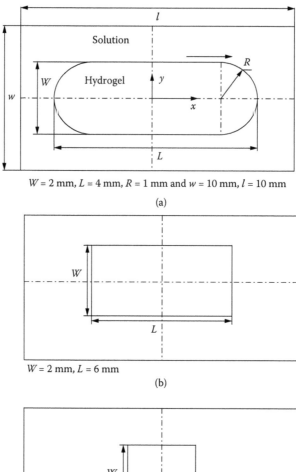

W = 2 mm, L = 4 mm, R = 1 mm and w = 10 mm, l = 10 mm

(a)

W = 2 mm, L = 6 mm

(b)

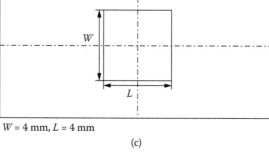

W = 4 mm, L = 4 mm

(c)

Figure 7.38 Solution and various hydrogel domains In which three geometrical shapes of hydrogel are considered with different values of ratio r = W/L as r = 1:2 (a), r = 1:3 (b) and r = 1:1 (c). (From S. Mulay and H. Li. *Modelling and Simulation in Materials Science and Engineering*, DOI:10.1088/0965-0393/19/6/065009. With permission.)

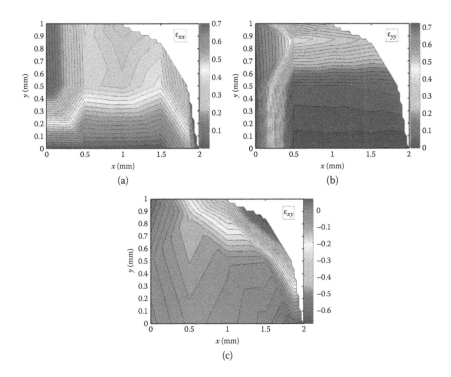

Figure 7.39 Profiles of strains ε_{xx}(a), ε_{yy}(b), and ε_{xy}(c) when 2-D pH-sensitive hydrogel is solved for pH = 3 and $E = 29$.

The cumulative average value of the concentration of the mobile ions inside the hydrogel closely matches with the concentration value of the fixed charge group, as seen from Figure 7.41a and f. The displacements and expansion of the hydrogel are shown in Figure 7.42, and it is noted that the hydrogel swelling decreases with an increase in the value of the normalized Young's modulus $\varepsilon = E/E_0$, where $E_0 = 25$. This is expected since the hydrogel crosslinked polymers offer more resistance to deformation for the higher values of E. Figure 7.43 shows various profiles of the field variables for pH = 7, and $E = 25$, 29 and 36.

It is observed from Figure 7.43 that the concentration values of the mobile ions inside the hydrogel increased slightly as compared with Figure 7.41 due to the increase in the pH of the solution. Figure 7.43f shows that the solution satisfies the electroneutrality condition, while the cumulative average concentration value of mobile ions inside the hydrogel closely matches the concentration value of negative fixed charge ions, as seen in Figure 7.43a.

Figure 7.44 shows the displacements at the field nodes of the hydrogel, and it is easily verified that the hydrogel swelling reduces with an increase in the value of E. Figure 7.45 shows all the profiles of the field variables for pH = 11 and E = 25, 29 and 36. We can see from Figure 7.45c the jump in

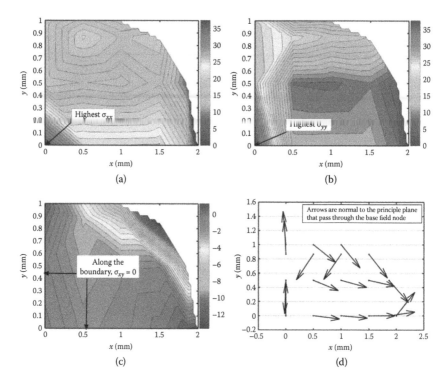

Figure 7.40 Schematic of the profiles of various field variables after solving the 2-D pH-sensitive hydrogel for pH = 3, and E = 25, 29 and 36, where the different profiles are given as C_f (a), C_1 (b), C_2 (c), C_3 (d), ψ (e) and the distribution of cumulative mobile ions (f). (From S. Mulay and L. Li. *Modelling and Simulation in Materials Science and Engineering*, DOI:10.1088/0965-0393/19/6/065009. With permission.)

the concentration value of ion C_2 at the hydrogel boundary for E = 36, while the electroneutrality condition is satisfied in the solution.

Figure 7.46 shows the displacements at the field nodes of the hydrogel and the expansion for the different values of the ratio ε. It is noted again that the expansion of the hydrogel reduces with an increase in the value of E. Figure 7.47a shows the influence of the solution pH on the swelling of the hydrogel for the different ratios ε. For a given constant value of ε, swelling of the hydrogel increases with an increase in the pH of solution. The swelling of the hydrogel decreases with an increase in the *values* ε. Figure 7.47b shows the influence of ε on the swelling of the hydrogel, and again it is noted that swelling reduces with an increase in the value of ε at a constant pH. The influence of ε on the hydrogel becomes more prominent at the higher values of pH, as observed from Figure 7.47a. The principal (major and minor) stresses are now computed at each field node of the hydrogel by considering the state of stress at each field node (Timoshenko and Goodier,

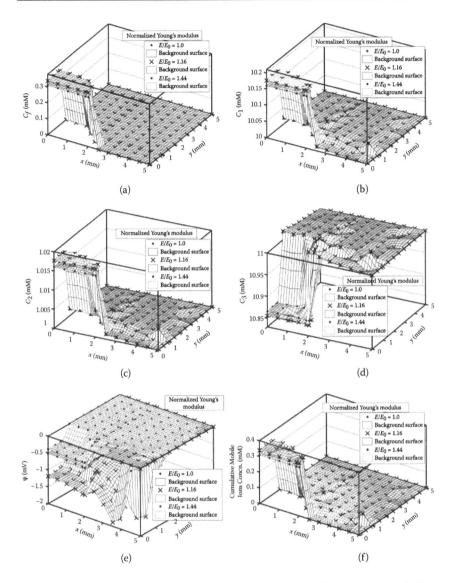

Figure 7.41 Profiles of various field variables after solving 2-D pH-sensitive hydro-gel for pH = 3, and E = 25, 29, and 36, where profiles are given as (a), (b), (c), (d), ψ (e), and distribution of cumulative mobile ions (f). (From S. Mulay and L. Li. *Modelling and Simulation in Materials Science and Engineering*, DOI:10.1088/0965-0393/19/6/065009. With permission.)

1970), and it is found that the maximum stress exists at the central fully constrained field node.

The maximum values of the principal stresses in the hydrogel with the different values of pH are plotted in Figure 7.48a. The total strain energy,

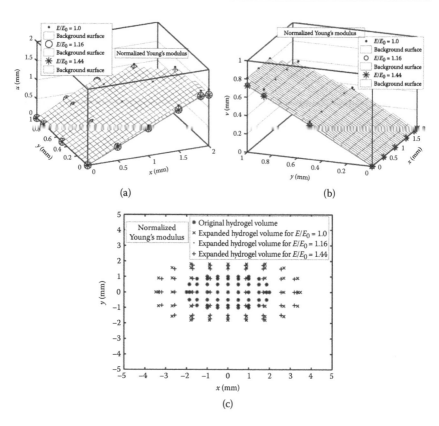

Figure 7.42 Displacements *u* (a) and *v* (b), and original and deformed shapes (c) for curved boundary hydrogel with pH = 3, and *E* = 25, 29, and 36, where hydrogel expansion decreases with increase in ratio of normalized Young's modulus ε. (From S. Mulay and L. Li. *Modelling and Simulation in Materials Science and Engineering*, DOI:10.1088/0965-0393/19/6/065009. With permission.)

$U = 0.5\,\sigma_{ij}\,\varepsilon_{ij}$, over the hydrogel domain is computed and plotted in Figure 7.48b. We conclude from the figure that the values of strain and stress inside the hydrogel generally increase with an increase in the values of ε. Correspondingly, the strain energy also increases.

Finally, to study the influence of geometrical shape of a hydrogel on the distributions of different ions and electrical potential, the 2-D simulation of a pH-sensitive hydrogel is performed with different geometrical shapes as shown in Figure 7.38b and c. At first, the hydrogel with *r* = 1:3 as shown in Figure 7.38b is solved and the various profiles of the field variable distributions are plotted in Figure 7.49. It is seen from Figure 7.49f that the solution satisfies the electroneutrality condition, and all jumps in the values of the field variables across the interface between the hydrogel and solution are smoothly captured by the RDQ method.

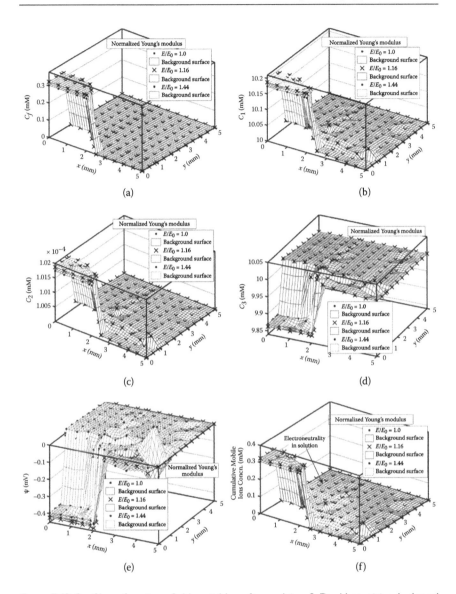

Figure 7.43 Profiles of various field variables after solving 2-D pH-sensitive hydrogel for pH = 7, and E = 25, 29, and 36, where different profiles are given as (a), (b), (c), (d), ψ (e), and distribution of cumulative mobile ions (f). (From S. Mulay and L. Li. *Modelling and Simulation in Materials Science and Engineering*, DOI:10.1088/0965-0393/19/6/065009. With permission.)

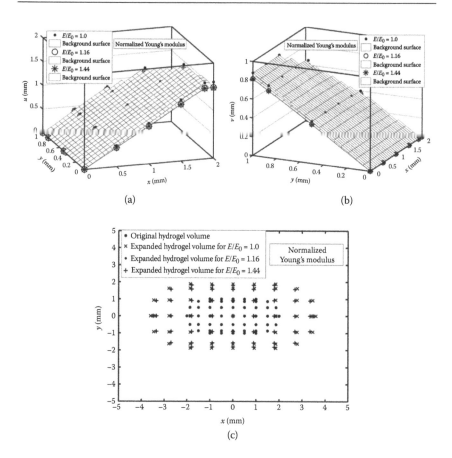

Figure 7.44 Displacements u (a) and v (b), and original and deformed shapes (c) for curved boundary hydrogel with pH = 7 and E = 25, 29, and 36. (From S. Mulay and L. Li. *Modelling and Simulation in Materials Science and Engineering*, DOI:10.1088/0965-0393/19/6/065009. With permission.)

Figure 7.50 shows the displacements and swelling at the field nodes of the hydrogel. The maximum values of the x and y coordinates in the original and deformed configurations from Figure 7.50c are obtained as 3.0 and 1.0, and 4.96 and 1.68 mm, respectively, and it is seen that their ratios are almost equal, namely (4.96/3) ≈ 1.66 and (1.68/1.0) = 1.68. This indicates that the hydrogel expanded almost equally in the x and y directions. This is expected as almost equal magnitudes of the osmotic pressure act on boundaries 3 and 4, leading to the uniform expansion of the hydrogel.

The hydrogel with a square cross-section as shown in Figure 7.38c is solved, and the profiles of field variables are plotted in Figure 7.51. Note from Figure 7.51f that the solution satisfies the electroneutrality condition and the distributions of all the field variables are smoothly captured across

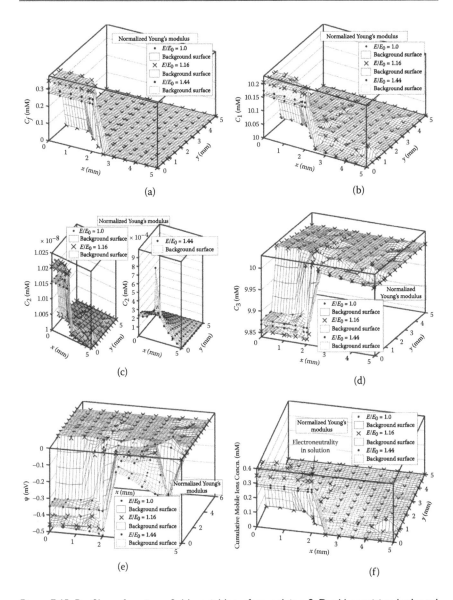

Figure 7.45 Profiles of various field variables after solving 2-D pH-sensitive hydrogel for pH = 11, and E = 25, 29, and 36, where different profiles are given as (a), (b), (c), (d), ψ (e), and distribution of cumulative mobile ions (f). (From S. Mulay and L. Li. *Modelling and Simulation in Materials Science and Engineering*, DOI:10.1088/0965-0393/19/6/065009. With permission.)

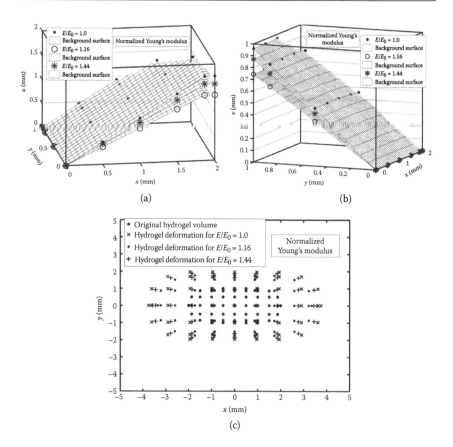

Figure 7.46 Displacements *u* (a) and *v* (b) and original and deformed shapes(c) for curved boundary hydrogel with pH = 11 and *E* = 25, 29, and 36. (From S. Mulay and L. Li. *Modelling and Simulation in Materials Science and Engineering,* DOI:10.1088/0965-0393/19/6/065009. With permission.)

the interface between the hydrogel and solution. Figure 7.52 shows the displacement values at the field nodes of the hydrogel and its expansion.

The analytical values of the displacements *u* and *v* along the boundaries *x* = 2 and *y* = 2, respectively, are computed by Equations (7.53) and (7.56), respectively, and plotted in Figure 7.52. The analytical and numerical values of the displacements match closely. As the analytical values by Equations (7.53) and (7.56) are computed via the numerical values of the osmotic pressure, there is a slight difference between the numerical and analytical displacements in Figure 7.52 away from the constrained boundaries. This shows that the analytical expressions given by Equations (7.53) and (7.56) offer a good judgment about the numerical values of the displacement at the hydrogel field nodes.

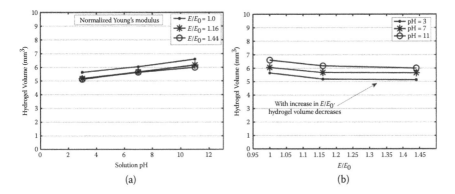

Figure 7.47 Influence of solution pH on hydrogel swelling at different values of normalized Young's modulus ε (a) and influence of ε on hydrogel swelling at different values of solution pH (b). (From S. Mulay and L. Li. *Modelling and Simulation in Materials Science and Engineering*, DOI:10.1088/0965-0393/19/6/065009. With permission.)

The maximum values of the x and y coordinates in the original and deformed configurations from Figure 7.52c are obtained as 2.0 and 2.0, and 3.1353 and 3.1353 mm, respectively, and it is seen that their ratios are exact, namely $(3.1353/2.0) \approx 1.57$. This indicates that the hydrogel is expanded equally in the xy plane.

The results of hydrogel deformation by different cross-sections indicate that when a domain is square ($r = 1:1$), equal magnitude of the osmotic pressure is imposed on all moving boundaries. As a result, the hydrogel

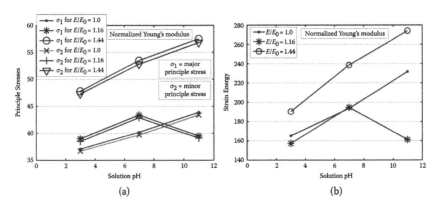

Figure 7.48 Maximum values of principal major and minor stresses in hydrogel at central fully constrained node (a) and total strain energy in hydrogel domain (b) for different values of solution pH. (From S. Mulay and L. Li. *Modelling and Simulation in Materials Science and Engineering*, DOI:10.1088/0965-0393/19/6/065009. With permission.)

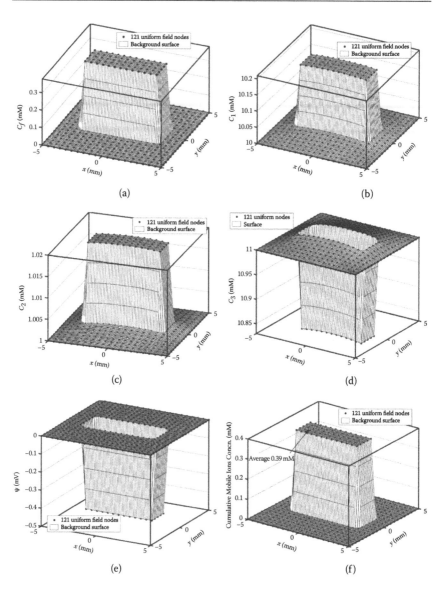

Figure 7.49 Profiles of various field variables after solving 2-D pH-sensitive hydrogel with rectangular shape for pH = 3 and E = 29, where different profiles are given as (a), (b), (c), (d), ψ (e), and distribution of cumulative mobile ions (f). (From S. Mulay and L. Li. *Modelling and Simulation in Materials Science and Engineering*, DOI:10.1088/0965-0393/19/6/065009. With permission.)

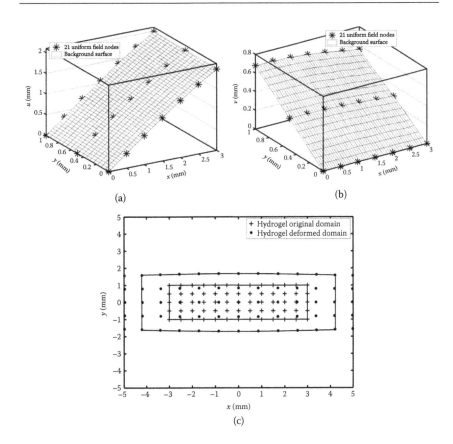

Figure 7.50 Displacements *u* (a) and *v* (b), and original and deformed shapes (c) for rect-angular hydrogel with pH = 3 and *E* = 29. (From S. Mulay and L. Li. *Modelling and Simulation in Materials Science and Engineering*, DOI:10.1088/0965-0393/19/6/065009. With permission.)

deforms equally in the *x* and *y* directions. Therefore, the normal stresses σ_{xx} and σ_{yy} for the square shape of the hydrogel are equal in magnitude as seen in Figure 7.53a. A lower shear stress generated in the hydrogel results in the lower value of the strain energy as seen in Figure 7.53b. For *r* = 1:3, however, the hydrogel deforms unequally in the *x* and *y* directions, lead-ing to unequal values of the stresses and higher strain energy, as observed in Figure 7.53a and b. When the hydrogel domain becomes nonuniform, the moving boundaries experience the unequal values of osmotic pressure, resulting in unequal deformations, (3.3408/2.0) = 1.67 and (1.74/1.0) = 1.74, in the *x* and *y* directions, respectively as shown in Figure 7.42c and high stresses and strain energy as shown in Figure 7.53.

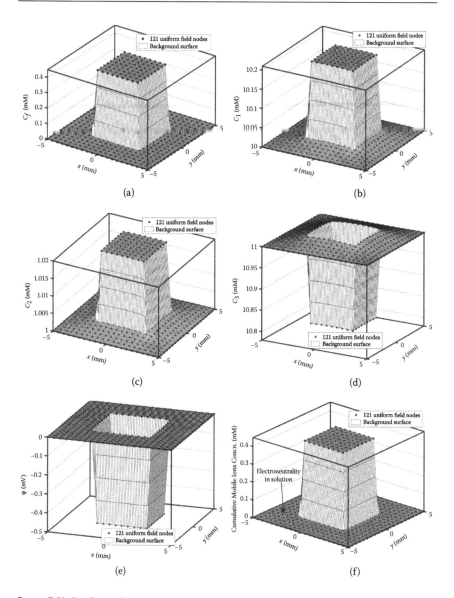

Figure 7.51 Profiles of various field variables after solving 2-D pH-sensitive hydrogel with square shape for pH = 3 and E = 29, where different profiles are given as (a), (b), (C), (d), ψ (e), and distribution of cumulative mobile ions (f). (From S. Mulay and L. Li. *Modelling and Simulation in Materials Science and Engineering*, DOI:10.1088/0965-0393/19/6/065009. With permission.)

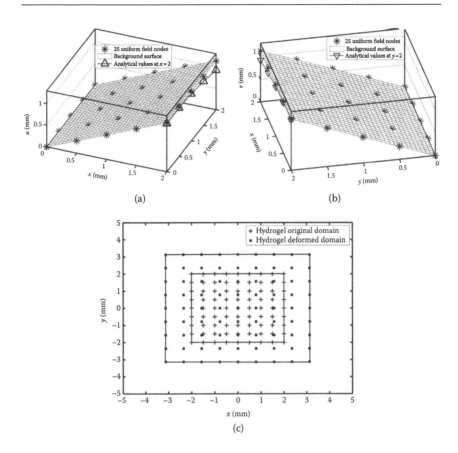

(a)

(b)

(c)

Figure 7.52 Displacements *u* (a) and *v* (b), and original and deformed shapes (c) for square hydrogel domain with pH = 3 and *E* = 29. (From S. Mulay and L. Li. *Modelling and Simulation in Materials Science and Engineering*, DOI:10.1088/0965-0393/19/6/065009. With permission.)

7.5 SUMMARY

Several meshless methods have been recently proposed and developed. Meshless methods are already hot topics in research in computational science and engineering. Because no mesh is used in meshless methods, these methods overcome the shortcomings of traditional methods such as FEM and FDM that depend on the mesh in the computation. Because of the attractive advantages, meshless methods have been widely used in the simulation and analysis of engineering and science problems.

Because of the rapid development of MEMS devices, numerical simulation already become an important tool for the analysis and design of practical

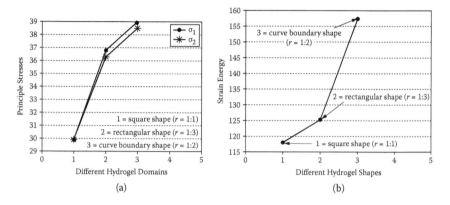

Figure 7.53 Comparison of principal stresses (a) and strain energy (b) for different geometrical shapes of hydrogel. (From S. Mulay and L. Li. *Modelling and Simulation in Materials Science and Engineering*, DOI:10.1088/0965-0393/19/6/065009. With permission.)

devices. Simulation of MEMS is usually very complex. The problems arise from the multiple coupled energy domains and media; dimensional scaling, and nonlinear issues. Therefore, the traditional analysis techniques (e.g., FEM) become difficult to achieve or ineffective. The meshless methods can avoid the disadvantages of the conventional numerical methods like FEM. Hence, they exhibit very good potential in numerical simulations of MEMS devices.

Several typical meshless methods are introduced and then used to simulate some MEMS devices. Numerical examples demonstrate that the meshless methods implemented are accurate, efficient, and convenient for simulation of MEMS devices. Meshless methods have already shown very good potential but further research is still required.

In the final application, the 2D simulation of pH-sensitive hydrogels is performed by the meshless RDQ method and the results are discussed. The effects of solution pH and initial fixed charge concentration on the swelling of hydrogels are studied with pH values of 3, 7, and 11 and initial fixed charge concentration values of 0.5, 1.05, and 1.5 mM. We can see from all the simulation results that the RDQ method is capable of capturing the moving boundary interfaces and the jump of the field variable distributions over the interface between multi-domain boundaries.

It is also seen that the RDQ method is capable of equally handling uniform and random distributions of the field nodes in the computational domains. It is observed from Figure 7.52 that the displacements u and v, computed by the analytical expressions given in Equations (7.53) and (7.56) respectively, closely match the corresponding numerical values for the field nodes located along boundaries 3 and 4. As a result, Equations (7.53) and (7.56) closely predict the displacements at the field nodes located along the Neumann boundaries 3 and 4 of hydrogels.

Further, the 2-D hydrogel model with an irregular domain is studied by changing the values of the Young's moduli as 25, 29, and 36 and the pH of solution as 3, 7, and 11. Based on simulation results, electroneutrality is always maintained in the solution for all the case studies and the jumps in the distributions of field variables are smoothly captured across the interface between hydrogel and solution. It is also noted from the results that the hydrogel swelling increases for a constant value of E with an increase in the value of solution pH. For a constant pH however, the hydrogel swelling reduces with an increase in the value of E.

This result is qualitatively in good agreement with other simulations and experiments. The different stresses across the hydrogel domain are computed. The principal stresses and planes are determined by Mohr's circle (Timoshenko and Goodier, 1970) by considering the state of stress corresponding to one field node at a time. The maximum values of the principal stresses and the strain energy in the hydrogel domain generally increase with an increase in the value of E.

Finally, three different geometrical shapes of hydrogel discs are considered to study their effects on deformation. The results indicate that the hydrogel expands equally in the x and y directions for the uniform geometrical shape, resulting in lower values of stress and strain energy. However, the hydrogel expands unequally for the nonuniform geometrical shape, leading to the higher values of stress and strain energy.

Appendix A: Derivation of characteristic polynomial $\phi(z)$

The following procedure is designed to derive the characteristic polynomial $\phi(z)$ from the discretised form of the governing partial differential equation (PDE). The procedure is explained by the central time and space with multi-step scheme and a similar procedure is followed for the rest of the schemes. From Equation (5.1)

$$\{a_{j,j-1} \quad a_{j,j} \quad a_{j,j+1}\} \left\{ \begin{array}{c} \phi(x_i, t_{j-1}) \\ \phi(x_i, t_j) \\ \phi(x_i, t_{j+1}) \end{array} \right\} = a \{a_{i,i-1} \quad a_{i,i} \quad a_{i,i+1}\} \left\{ \begin{array}{c} \phi(x_{i-1}, t_j) \\ \phi(x_i, t_j) \\ \phi(x_{i+1}, t_j) \end{array} \right\}$$

(A.1)

The weighting coefficients from Equation (A.1) are computed by Shu's general approach to get

$$\phi_m^{j+1} = \phi_m^{j-1} + r \left[-\phi_{m-1}^j + \phi_{m+1}^j \right]$$

(A.2)

where $m = i$ and $r = (a\,t/h)$, where t and h are the spacings of the time and space domains, respectively. Taking the Fourier inverse on both sides of Equation (A.2) results in

$$\frac{1}{\sqrt{2\pi}} \int_{-\infty}^{+\infty} e^{ihm\xi} \; \hat{\upsilon}^{n+1}(\xi) \; d\xi = \frac{1}{\sqrt{2\pi}} \int_{-\infty}^{+\infty} e^{ihm\xi} \; \hat{\upsilon}^{n-1}(\xi) \; d\xi$$

(A.3)

$$\frac{1}{\sqrt{2\pi}} \int_{-\infty}^{+\infty} r \left(-e^{ih(m-1)\xi} \; \hat{\upsilon}^n(\xi) \; d\xi + e^{ih(m+1)\xi} \; \hat{\upsilon}^n(\xi) \; d\xi \right)$$

where $n = j$. Rearranging Equation (A.3) results in

$$\frac{1}{\sqrt{2\pi}} \int_{-\infty}^{+\infty} e^{ihm\xi} \left\{ \hat{\upsilon}^{n+1}(\xi) - \hat{\upsilon}^{n-1}(\xi) + r\left[e^{-ih\xi} - e^{ih\xi}\right] \hat{\upsilon}^{n}(\xi)\right\} \, d\xi = 0 \qquad \text{(A.4)}$$

In order to satisfy the consistency

$$\hat{\upsilon}^{n+1}(\xi) - \hat{\upsilon}^{n-1}(\xi) + r\left[\cos(h\xi) - i\sin(h\xi) - \cos(h\xi) - i\sin(h\xi)\right] \hat{\upsilon}^{n}(\xi) = 0$$

$$\text{(A.5)}$$

where $e^{i\theta} = \cos(\theta) + i\,\sin(\theta)$ relation is used. Substitute $\hat{\upsilon}^{n}(\xi) = \phi^{n}(z)$ and $h\xi = \theta$ in Eq. (A.5) to get

$$\phi(z) = z^{2} - [2\,r\,i\,\sin(\theta)]\,z - 1, \text{ taking out } \phi^{n-1}(z) \text{ as common} \qquad \text{(A.6)}$$

Equation (A.6) is the same as Equation (5.2). A similar procedure is followed to derive all the characteristic polynomials cited in Chapter 5.

Appendix B: Definition of reduced polynomial $\phi_1(z)$

If $\phi(z)$ is any general nonzero polynomial of degree n, as given in Equation (5.3), and $\phi^*(z)$ is its complex conjugate, as given in Equation (5.5), the reduced polynomial $\phi_1(z)$ of at most the degree $n - 1$ can be defined as (Miller, 1971)

$$\phi_1(z) = \frac{\phi_0^*(0)\ \phi(z) - \phi(0)\ \phi_0^*(z)}{z} \tag{B.1}$$

The reduced polynomial $\phi_1(z)$ gives immediate criterion for the self inversiveness of the polynomial $\phi(z)$. As a result, $\phi(z)$ is self-inversive polynomial if and only if $\phi_1(z) = 0$. The proof can be seen in Miller (1971).

Appendix C: Derivation of discretisation equation by Taylor series

In this appendix, the derivation of the consistent equation from the discretised equation by Taylor series is explained for the scheme in Section 5.2.1.1 of Chapter 5. Identical procedure is followed to obtain the rest of the discretised equations by Taylor series (Mulay et al., 2009). From Equation (5.1)

$$\Rightarrow \{a_{j,j-1} \quad a_{j,j} \quad a_{j,j+1}\} \left\{ \begin{array}{c} \phi(x_i, t_{j-1}) \\ \phi(x_i, t_j) \\ \phi(x_i, t_{j+1}) \end{array} \right\} = a \{a_{i,i-1} \quad a_{i,i} \quad a_{i,i+1}\} \left\{ \begin{array}{c} \phi(x_{i-1}, t_j) \\ \phi(x_i, t_j) \\ \phi(x_{i+1}, t_j) \end{array} \right\}$$

$$(C.1)$$

The weighting coefficients $a_{i,j}$ are computed by Shu's general approach (Shu, 2000) to get

$$a_{m,\,m-1} = \frac{-1}{2h} \quad a_{m,\,m+1} = \frac{1}{2h}, \quad a_{m,\,m} = \frac{1}{2h} \quad \text{and} \tag{C.2}$$

$$a_{n,n-1} = \frac{-1}{2t} \quad a_{n,n+1} = \frac{1}{2t} \quad a_{n,n} = 0 \tag{C.3}$$

where $m = j$ and $n = i$. Substituting Equations (C.2) and (C.3) into Equation (C.1) leads to

$$\phi_m^{n+1} = \phi_m^{n-1} + r\left[-\phi_{m-1}^n + \phi_{m+1}^n\right] \tag{C.4}$$

The terms ϕ from Equation (C.4) are replaced by Taylor series expansion as

$$\phi_{m-1}^n = \phi(x_m - h) = \phi - h\phi_{,x} + \frac{h^2}{2}\phi_{,xx} - O(h)^3 \tag{C.5}$$

$$\phi^n_{m+1} = \phi(x_m + h) = \phi + h\phi_{,x} + \frac{h^2}{2}\phi_{,xx} + O(h)^3 \tag{C.6}$$

$$\phi^{n-1}_m = \phi(t_n - t) = \phi - t\phi_{,t} + \frac{t^2}{2}\phi_{,tt} - O(t)^3 \tag{C.7}$$

$$\phi^{n+1}_m = \phi(t_n + t) = \phi + t\phi_{,t} + \frac{t^2}{2}\phi_{,tt} + O(t)^3 \tag{C.8}$$

where $\phi = \phi^n_m$. Substituting Equations (C.5) to (C.8) into Equation (C.4) and simplifying it, we get

$$\phi_{,t} - a\,\phi_{,x} = \phi_{,xxx}\left[\left(\frac{a\,h^2}{6}\right) - \left(\frac{t^2\,a^3}{6}\right)\right] \tag{C.9}$$

where $\phi_{,ttt} = a^3\,\phi_{,xxx}$. Therefore, the consistent equation in (C.9) is the same as Equation (5.34).

Appendix D: Derivation of ratio of successive amplitude reduction values for fixed–fixed beam using explicit and implicit approaches

In this appendix, the derivation steps of the successive amplitude reduction ratio are given for the fixed–fixed beam with the explicit and implicit approaches. The ODE $\psi(t)$ obtained by the explicit approach is given in Equation (5.71) as

$$\frac{1}{\psi}\left[k\,\frac{\partial^3 \psi}{\partial t^3} + \frac{\partial^2 \psi}{\partial t^2} \right] = -\omega_n^2 \tag{D.1}$$

The roots of Equation (D.1) are computed by substituting $\psi = e^{st}$, and solving $ks^3 + s^2 + \omega_n^2 = 0$ in MATLAB® to get the general solution as

$$\psi(t) = C_1\,e^{\psi_1 t} + C_2\,e^{\psi_2 t} + C_3\,e^{\psi_3 t} \tag{D.2}$$

where

$$\psi_1 = \frac{1}{6\,k}\left[\beta + \frac{4}{\beta} - 2 \right] \quad \psi_2 = \frac{1}{12\,k}\left[-\beta - \frac{4}{\beta} - 4 \right] + \frac{i\,\sqrt{3}}{24\,k}\left[2\,\beta - \frac{8}{\beta} \right] \quad \text{and}$$

$$\psi_3 = \frac{1}{12\,k}\left[-\beta - \frac{4}{\beta} - 4 \right] - \frac{i\,\sqrt{3}}{24\,k}\left[2\,\beta - \frac{8}{\beta} \right] \tag{D.3}$$

where $\beta = \sqrt[3]{-108\,\omega_n^2\,k^2 - 8 + 12\sqrt{3}\,\left(\sqrt{27\omega_n^2\,k^2 + 4}\right)\,\omega_n\,k}$. The first root ψ_1 is neglected as it grows with time. The remaining roots, ψ_2 and ψ_3 are complex conjugates of each other. As a result, it is given from ψ_2 and ψ_3 that

$$v_1 = \frac{1}{12\,k}\left[\beta + \frac{4}{\beta} + 4 \right] \quad \text{and} \quad v_2 = \frac{\sqrt{3}}{24\,k}\left[2\,\beta - \frac{8}{\beta} \right] \tag{D.4}$$

Therefore, the solutions from Equation (D.2) are given as

$$sol_1 = e^{(-v_1 + iv_2)\, t} \quad \text{and} \quad sol_2 = e^{(-v_1 - iv_2)\, t} \tag{D.5}$$

such that

$$sol_1 = e^{-v_1 t}\left[C_1 \cos(v_2\, t) + C_3\, i\, \sin(v_2\, t)\right]$$

$$\text{and} \quad sol_2 = e^{-v_1 t}\left[C_4 \cos(v_2\, t) - i\, C_5 \sin(v_2\, t)\right] \tag{D.6}$$

The amplitudes from Equation (D.6) are obtained as

$$amp_1 = e^{-v_1 t}\, C_1\, \cos(v_2\, t) \quad \text{and} \quad amp_2 = e^{-v_1 t}\, C_4\, \cos(v_2\, t) \tag{D.7}$$

The total amplitude from the general solution of characteristic vibration is given as

$$A(t) = e^{-v_1 t}\left[C_1\, \cos(v_2\, t) + C_2\, \sin(v_2\, t)\right] \tag{D.8}$$

$A(0) = 0$ at $t = 0$, as a result $C_1 = 0$. Therefore, Equation (D.8) for times t_1 and $[t_1 + (2\pi / \omega_d)]$ is given as

$$A(t_1) = e^{-v_1\, t_1}\, C_2\, \sin(v_2\, t_1), \text{ and}$$

$$A\left(t_1 + \frac{2\pi}{\omega_d}\right) = e^{-v_1\left[t_1 + \frac{2\pi}{\omega_d}\right]}\, C_2\, \sin\left[v_2\left(t_1 + \frac{2\pi}{\omega_d}\right)\right] \tag{D.9}$$

Equation (D.9) is simplified as

$$A(t_1) = e^{-v_1\, t_1}\, C_2\, \sin(v_2\, t_1), \text{ and}$$

$$A(t_2) = A\left(t_1 + \frac{2\pi}{\omega_d}\right) = e^{-v_1\, t_1}\, e^{\left(-v_1 \frac{2\pi}{\omega_d}\right)}\, C_2\, \sin\left(v_2\, t_1\right) \tag{D.10}$$

As a result, the ratio $A(t_2) / A(t_1)$ is given as

$$\frac{A(t_2)}{A(t_1)} = e^{-v_1 \left[\frac{2\pi}{\omega_d}\right]} \qquad \because \omega_d = v_2 \tag{D.11}$$

It is verified that Equation (D.11) is the same as Equation (5.77). A similar procedure is adopted to compute the successive amplitude reduction ratios by the implicit approach.

Appendix E: Source code development

Several tips and sample object-oriented functions are provided in this appendix to assist readers in source code development for a specific meshless method. The functions provided here are written for the meshless RDQ method described in Chapter 3, but if readers keep in mind object-oriented programming by C++ language, the logic and concepts are fairly general and can be adopted conveniently in other languages for other meshless methods.

One of the best ways of programming is dividing a specific task into several subtasks and implementing the subtasks in small functions that can be called and executed by the main program as and when required. The advantage of this approach is that fairly general functions can be written, enabling easy debugging of the whole source code. This appendix is arranged as follows. At first, several small template functions are provided to perform frequently required operations. Second, several data structures are provided that store specific types of data. Finally, a flowchart is provided for the meshless RDQ method described in Chapter 3.

It is sometimes tedious to write fairly general template functions that may be called repeatedly while executing a main program. Thus, several functions are provided below that can be directly used while developing a source code for a meshless method. All the variables used in the following code are self-explanatory in their objective and the users of meshfree methods can interpret them.

TEMPLATE FUNCTION TO DELETE 1-D AND 2-D ARRAYS FROM MEMORY

A pointer to an array that has to be deleted is provided as an argument.

```
template <class T>
inline void Free_Array(T &x)
{
        if (x ! = NULL)
        {
```

```
                   delete []x; x = NULL;
        }
}
template <class T>
inline void Free_Array_2D(T &x, int &size)
{
        int i = 0;
        if (x ! = NULL)
        {
                for (i = 0; i<size; i++)
                {
                        if (x[i] ! = NULL)
                        {
                                delete x[i]; x[i] = NULL;
                        }
                }
                delete []x; x = NULL;
        }
}
```

SAMPLE TEMPLATE FOR THE MATRIX

```
template <class D> class Matrix
{
        public
        int m_irowsize;//maximum rows size
        int m_icolumnsize;//maximum column size
        D *data;
        Matrix ()
        {
                m_irowsize = 0; m_icolumnsize = 0;
                data = NULL;
        }
        void allocate()
        {
                delete []data;
                data = NULL;
                data = new D[m_irowsize*m_icolumnsize];
        }
        Matrix (Matrix &A)
        {
                this->m_icolumnsize = A.m_icolumnsize;
                this->m_irowsize = A.m_irowsize;
                this->data = A.data;
        }
```

```
        Matrix (int rowsize, int columnsize)
        {
                m_irowsize = rowsize; m_icolumnsize =
columnsize;
                data = new D[m_irowsize*m_icolumnsize];
        }
        int Getrowlength(){return (m_irowsize);}
        int Getcolumnlength(){return (m_icolumnsize);}
        BOOL Setvalue(int &row, int &column, D setvalue)
        {
                if ((row> = m_irowsize) || (column > = m_
icolumnsize)||(row < 0) || (column                    < 0))
                        return (0);
                data[row * m_icolumnsize + column] = setvalue;
                return (1);
        }
        void Getvalue(int &row, int &column, D &returnvalue,
BOOL bOK)
        {
                if ((row> = m_irowsize) || (column > = m_
icolumnsize)||(row < 0) || (column                    < 0))
                {
                        bOK = 0; return;
                }
                returnvalue = data[row * m_icolumnsize +
column];
                bOK = 1;
        }
```

Several matrix operations such as inverse, transpose, and product can be
written by the above template. Some operations of the matrix manipulation
are given as follows.

FUNCTION TO COPY MATRIX

```
void Copy_Matrix(Matrix<D> &source)
{
        m_irowsize = source. Getrowlength ();
        m_icolumnsize = source.Getcolumnlength();
        allocate();
        for (int i = 0; i < m_irowsize; i++)
        {
                for (int j = 0; j < m_icolumnsize; j++)
                {
                        D value;
                        BOOL success = 0;
```

```
                source. Getvalue (i, j, value, success);
                this->Setvalue(i, j, value);
            }
        }
    }
}
```

FUNCTION TO FILL MATRIX FROM ARRAY

```
void Fill_Matrix_From_Array(double *&SourceArray)
{
    int row = this-> Getrowlength (); int column = this-
>Getcolumnlength();
    int i = 0, j = 0, k = 0;
    double dtemp = 0.0;
    for (i = 0;i<row; i++)
    {
        for (j = 0; j<column; j++)
        {
            dtemp = 0.0;
            dtemp = SourceArray[k];
            this->Setvalue(i, j, dtemp);
            k++;
        }
    }
}
```

FUNCTION TO FILL ARRAY FROM MATRIX

```
void Fill_Array_From_Matrix(double **&DestinationArray)
{
    int row = this->Getrowlength();
    int column = this->Getcolumnlength();
    int i = 0, j = 0, bOK = 0;
    double dtemp = 0.0;
    for (i = 0; i<row; i++)
    {
        for (j = 0; j<column; j++)
        {
            this->Getvalue(i, j, dtemp, bOK);
            DestinationArray[i][j] = dtemp;
        }
    }
}
```

FUNCTIONS TO SORT ARRAY IN INCREMENTAL ORDER BASED ON STORED VALUES

Function to Sort 1-D Array Incrementally

```
void sort_Array_inAscending(int &quantity, double *&x)
{
      int i = 0, j = 0;BOOL bExchange = FALSE;
//- - - - - - - Initialization end- - - - - - - - - //
      for(i = 0;i<quantity;i++)
      {
            bExchange = FALSE;
            for (j = 0;j<(quantity -1);j++)
            {
                  if (x[j] > x[j+1])
                  {
                        bExchange = TRUE;
                        double dtemp = x[j];
                        x[j] = x[j+1];
                        x[j+1] = dtemp;
                  }
            }
            if (bExchange = = FALSE)
                  break;
      }
}
```

Function to Sort 2-D Array Incrementally

```
void sort_Array_inAscending(int &quantity, double *&x,
double *&y, double &highestx, double &highesty)
{
      int i = 0, j = 0;BOOL bExchange = FALSE;
      double dtemp = 0.0, dtemp1 = 0.0;
//- - - - - - - Initialization end- - - - - - - - -//
      for(i = 0; i<quantity; i++)
      {
            bExchange = FALSE;
            for (j = 0;j<(quantity -1);j++)
            {
                  if (x[j] > x[j+1])
                  {
                        bExchange = TRUE;
                        dtemp = x[j]; dtemp1 = y[j];
                        x[j] = x[j+1]; y[j] = y[j+1];
```

```
                                    x[j+1] = dtemp; y[j+1] = dtemp1;
                          }
                 }
                 if (bExchange = = FALSE)
                          break;
          }
          for (i = 0; i<quantity; i++)
          {
                 if (highestx = = x[i] && highesty = = y[i])
                 {
                          dtemp = x[quantity-1]; dtemp1 =
y[quantity-1];
                          x[quantity-1] = highestx; y[quantity-1]
= highesty;

                          x[i] = dtemp; y[i] = dtemp1;
                          break;
                 }
          }
}
```

Function to Sort 1-D Array in Ascending Order and Store Respective Node Numbers

```
void sort_Array_inAscending(int &itotalrealnodes, double
*&Distance, int *&NodeID)
{
      int i = 0, j = 0;BOOL bExchange = FALSE;
      double dtemp = 0.0;
      int dtemp2 = 0;
//- - - - - -Initialization end- - - - - - - - //
      for(i = 0;i<itotalrealnodes;i++)
      {
             bExchange = FALSE;
             for (j = 0;j<(itotalrealnodes -1);j++)
             {
                    if (Distance[j] > Distance[j+1])
                    {
                           bExchange = TRUE;
                           dtemp = Distance[j]; dtemp2 =
NodeID[j];
                           Distance[j] = Distance[j+1];
NodeID[j] = NodeID[j+1];
                           Distance[j+1] = dtemp;
NodeID[j+1] = dtemp2;
                    }
             }
```

```
            if (bExchange = = FALSE)
                break;
        }
}
```

Function to Search Specific Index in Array

```
int Search_Index(int &imaxrealnodes, int &nodeID, int
*&nodeidIndex)
{
        int i = 0;
        for (i = 0;i<imaxrealnodes; i++)
        {
                if (nodeidIndex[i] = = nodeID)
                        return (i);
        }
        return (-1);
}
```

SAMPLE FUNCTIONS TO GENERATE UNIFORM AND RANDOM NODES IN COMPUTATIONAL DOMAIN

Function to Generate Uniform Nodes in 2-D Domain

```
void Generate_Uniform_numbers(long int &totalnodesx, long
int &totalnodesy, double &lowestx, double &lowesty, double
&highestx, double &highesty, double *&x, double *&y, int
*&inodeID, int &nodecount)
{
        double deltax = 0.0, deltay = 0.0;
        double dtempx = 0.0, dtempy = 0.0;
        double dtempxold = 0.0, dtempyold = 0.0;
        int i = 0, j = 0, k = 0, nodeID = 0;
//- - - - - - - -Initialization end- - - - - - - -//
        deltax = (highestx - lowestx)/(totalnodesx-1);
        deltay = (highesty - lowesty)/(totalnodesy-1);
        for (j = 0; j<totalnodesy; j++)
        {
                for (i = 0; i<totalnodesx; i++)
                {
                        dtempx = lowestx + i* deltax;
                        x[k] = dtempx;
                        dtempy = lowesty + j* deltay;
```

```
                        y[k] = dtempy;
                        nodeID++; inodeID[k++] = nodeID; dtempx
= 0.0; dtempy = 0.0;
                    }
            }
}
```

Function to Generate Random Nodes in 2-D Domain

```
void Generate_Random_numbers_2D(long int &quantityx, long
int &totalnodesy, double &highestx, double &highesty,
double &lowestx, double &lowesty, double *&x, double *&y,
int *&inodeID)
{
        double range_x = 0.0, range_y = 0.0;
        BOOL bOK = FALSE;
        double dtemp = 0.0, dtemp1 = 0.0;
        int i = 0, count = 0, index = 0, totalquantity = 0;
//- - - - - - - - Initialization end- - - - - - - - -//
        range_x = highestx - lowestx; range_y = highesty
- lowesty;
        totalquantity = quantityx * totalnodesy;
        //srand(time(NULL));
        x[0] = lowestx; y[0] = lowesty; inodeID[count] =
count+1; count++;
        x[1] = lowestx; y[1] = highesty; inodeID[count] =
count+1; count++;
        x[2] = highestx; y[2] = lowesty; inodeID[count] =
count+1; count++;
        x[3] = highestx; y[3] = highesty; inodeID[count] =
count+1; count++;
        do
        {
                dtemp = lowestx + (((double)rand()/(double)
RAND_MAX)*range_x);
                dtemp1 = lowesty + (((double)rand()/(double)
RAND_MAX)*range_y);
                bOK = check_value(count,dtemp, dtemp1,x, y);
                inodeID[index] = i++;
                index++; count++;
                dtemp = 0.0; dtemp1 = 0.0;
        }while(index ! = totalquantity);
        sort_Array_inAscending(totalquantity, x, y,
highestx, highesty);
}
```

Several sample data structures are provided below, keeping in mind the meshless RDQ method, but they can be adopted well in other meshless methods as well.

DATA STRUCTURE TO STORE EACH FIELD NODE WITH ITS NUMBER AND *X* AND *Y* COORDINATES

```cpp
typedef struct point
{
        int id;
        double x; double y;
        char dbflg[1];
        point *next;
        point()
        {
                id = 0; x = 0.0; y = 0.0; dbflg[0] = '\0';
        }
        point *& operator = (point *right)
        {
                point *temp = new (point);
                temp->x = right->x; temp->y = right->y; temp-
>id = right->id;
                temp->next = right->next;
                return (temp);
        }
        point(const point *a)
        {
                x = a->x; y = a->y; id = a->id; next = a->next;
                dbflg[0] = a->dbflg[0];
        }
        ~point(){}
};
```

LINK LIST OF ALL FIELD NODES IN COMPUTATIONAL DOMAIN

```cpp
typedef struct point_Link
{
        point *head;
        point *tail;
        int totalpoints, totalpointsx, totalpointsy;
        point_Link()
        {
```

```
                head = NULL; tail = NULL;
                totalpoints = 0; totalpointsy = 0;
totalpointsx = 0;
        }
        void push (double &x, double &y, int &id) //function
to push nodes
        {
                point *temp = (point*)malloc(sizeof(point));
                temp->x = x; temp->y = y; temp->id = id;
                temp->next = NULL;
                if (this->head = = NULL)
                {
                        head = temp; this->head = temp; this-
>tail = temp;
                }
                else if (this->tail ! = NULL)
                {
                        this->tail->next = temp; this->tail =
temp;
                }
                temp = NULL;
        }
        point_Link(const point_Link *x) //copy constructor
        {
                this->head = x->head; this->tail = x->tail;
                this->totalpoints = x->totalpoints;
                this->totalpointsx = x->totalpointsx; this-
>totalpointsy = x->totalpointsy;
        }
        ~point_Link(){}
};
```

Data Structure to Store Field Node Located on Dirichlet Boundary (DB)

```
typedef struct DBPoint
{
        int id;
        double ux;//field variable value in x direction
        double uy;//field variable value in y direction
        DBPoint *next;
        DBPoint()
        {
                id = 0; ux = 0.0; uy = 0.0;
                next = NULL;
        }
```

```
        DBPoint(const DBPoint *x)
        {
                id = x->id; ux = x->ux; uy = x->uy; next =
x->next ;
        }
        ~DBPoint(){}
};
```

Link List of All Field Nodes Located on Dirichlet Boundary

```
typedef struct DBNodeList
{
        DBPoint *head;
        DBPoint *tail;
        int itoalDBnodes, itoalDBnodesx, itoalDBnodesy;
        int totalDBCFaces;//total faces which are on DBC
        int xl, xr, yb, yt;//xl:-DB nodes along x = 0, xr:-DB
nodes along x = L, yb:-DB
nodes along y = 0, yt:-DB nodes along y = L
        BOOL x00, x01, y10, y11;//flags to indicate, which
boundary is in DBC.x00:-
        x = 0, x01:-x = L, x10:-y = 0, x11:-y = L
        double value00, value01, value10, value11;//DBC
values along the faces x = 0,
                                      x = L, y = 0, y = L
        int *nodeID00, *nodeID01, *nodeID10, *nodeID11;//
array of real node ID's,                               which
are located along the 4 faces explained as above.
        DBNodeList()
        {
                head = NULL; tail = NULL;
                itoalDBnodes = 0; itoalDBnodesx = 0;
itoalDBnodesy = 0;
                totalDBCFaces = 0;
                x00 = FALSE; x01 = FALSE, y10 = FALSE, y11 =
FALSE;
                value00 = 0.0, value01 = 0.0, value10 = 0.0,
value11 = 0.0;
                nodeID00 = NULL; nodeID01 = NULL; nodeID10 =
NULL;
                nodeID11 = NULL;
                xl = 0; xr = 0; yb = 0; yt = 0;
        }
        void push (double &x, double &y, int &id);
        {
```

```
            DBPoint *temp = (DBPoint*)
malloc(sizeof(DBPoint));
            temp->ux = x; temp->uy = y; temp->id = id;
temp->next = NULL;
            if (this->head = = NULL)
            {
                    this->head = temp; this->tail = temp;
            }
            else if (this->tail ! = NULL)
            {
                    this->tail->next = temp; this->tail =
temp;
            }
            temp = NULL;
        }
        DBNodeList(const DBNodeList *x)
        {
            head = x->head; tail = x->tail; itoalDBnodes
= x->itoalDBnodes;
        }
        void Free_All_Nodes()//function to empty the list
        {
            if (this->nodeID00 ! = NULL)
            {
                    delete []this->nodeID00; this->nodeID00
= NULL;
            }
            if (this->nodeID01! = NULL)
            {
                    delete []this->nodeID01; this->nodeID01
= NULL;
            }
            if (this->nodeID10 ! = NULL)
            {
                    delete []this->nodeID10; this->nodeID10
= NULL;
            }
            if (this->nodeID11 ! = NULL)
            {
                    delete []this->nodeID11; this->nodeID11
= NULL;
            }
        }
        ~DBNodeList(){}
};
```

Similar data structures and link lists can be created for field nodes located on the Neumann boundary. The data structures given above are as general as possible. They can be extended for a specific meshless method. Finally, the complete implementation of the RDQ method is broadly summarized in Figure E.1 flowchart.

Figure E.1

Complete implementation flowchart of RDQ method.

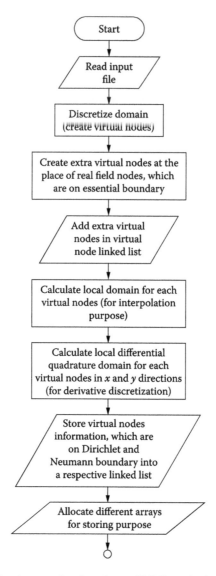

Figure E.1 Complete implementation flowchart of RDQ method.

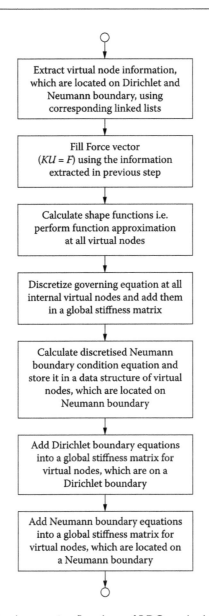

Figure E.1 Complete implementation flowchart of RDQ method.

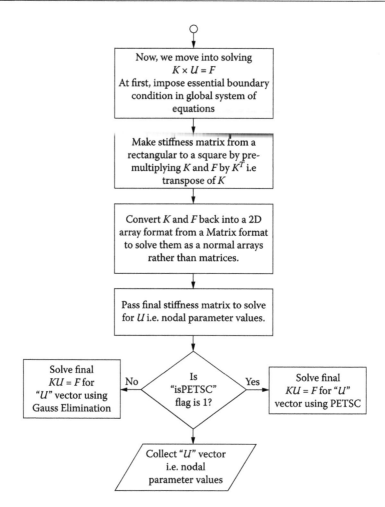

Figure E.1 Complete implementation flowchart of RDQ method.

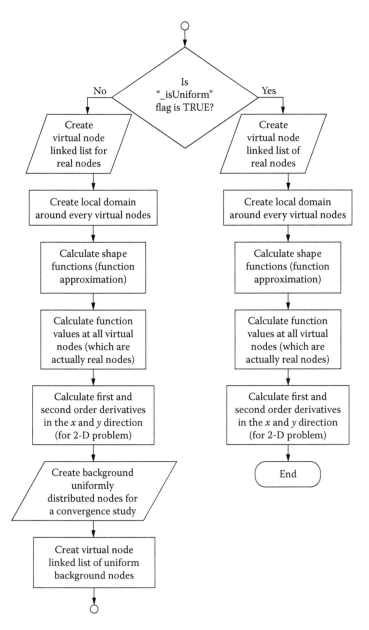

Figure E.1 Complete implementation flowchart of RDQ method.

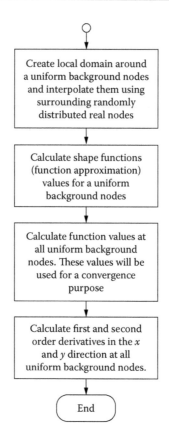

Figure E.1 Complete implementation flowchart of RDQ method.

References

L. Aceto, D. Trigiante. (2007). The stability problem for linear multistep methods Old and new results. *Journal of Computational and Applied Mathematics*, 210, 2-12

M. Ainsworth, J.T. Oden. (2000). *A posteriori error estimation in finite element analysis*, John Wiley & Sons, Inc

N.R. Aluru. (2000). A point collocation method based on reproducing kernel approximations. *International Journal for Numerical Methods in Engineering*, 47, 1083-1121

N.R. Aluru, G. Li. (2001). Finite cloud method: A true meshless technique based on fixed reproducing kernel approximation. *International Journal for Numerical Methods in Engineering*, 50, 2373-2410

G.K. Ananthasuresh, R.K. Gupta, S.D. Senturia. (1996). An approach to macromodeling of MEMS for nonlinear dynamic simulation. *Microelectromechanical Systems (MEMS), ASME Dynamic Systems and Control (DSC) series*, 59, 401-407

A. Arefmanesh, M. Najafi, H. Abdi. (2008). Meshless local Petrov–Galerkin method with unity test function for non-isothermal fluid flow. *CMES — Computer Modeling in Engineering and Sciences*, 25, 9-22

R. Ata, A. Soulaïmani. (2005). A stabilized SPH method for inviscid shallow water flows. *International Journal for Numerical Methods in Fluids*, 47, 139-159

S.N. Atluri. (2004). *The meshless method (MLPG) for domain & BIE discretisations*, Tech Science Press, Forsyth

S.N. Atluri. (2005). *Methods of computer modeling in engineering & the sciences*, Tech Science Press, Forsyth

S.N. Atluri, J.Y. Cho, H.G. Kim. (1999a). Analysis of thin beams, using the meshless local Petrov–Galerkin method, with generalized moving least squares interpolations. *Computational Mechanics*, 24, 334-347

S.N. Atluri, Z.D. Han, A.M. Rajendran. (2004). A new implementation of the meshless finite volume method, through the MLPG "mixed" approach. *CMES - Computer Modeling in Engineering and Sciences*, 6, 491-513

S.N. Atluri, H.G. Kim, J.Y. Cho. (1999b). A critical assessment of the truly meshless local Petrov–Galerkin (MLPG), and Local Boundary Integral Equation (LBIE) methods. *Computational Mechanics*, 24, 348-372

S.N. Atluri, H.T. Liu, Z.D. Han. (2006a). Meshless Local Petrov–Galerkin (MLPG) mixed collocation method for elasticity problems. *CMES - Computer Modeling in Engineering and Sciences*, 14, 141-152

S.N. Atluri, H.T. Liu, Z.D. Han. (2006b). Meshless Local Petrov–Galerkin (MLPG) mixed finite difference method for solid mechanics. *CMES - Computer Modeling in Engineering and Sciences*, 15, 1-16

S.N. Atluri, S. Shen. (2002). *The meshless local Petrov–Galerkin method*, Tech Science Press, Encino

S.N. Atluri, J. Sladek, V. Sladek, T. Zhu. (2000). Local boundary integral equation (LBIE) and it's meshless implementation for linear elasticity. *Computational Mechanics*, 25, 180-198

S.N. Atluri, T. Zhu. (1998). A new Meshless Local Petrov–Galerkin (MLPG) approach in computational mechanics. *Computational Mechanics*, 22, 117-127

I. Babuska. (1986). *Accuracy estimates and adaptive refinements in finite element computations*, John Wiley & Sons, Inc

I. Babuška, J.M. Melenk. (1997). The partition of unity method. *International Journal for Numerical Methods in Engineering*, 40, 727-758

I. Babuška, A. Miller. (1984a). Post-processing approach in the finite element method — Part 1: Calculation of displacements, stresses and other higher derivatives of the displacements. *International Journal for Numerical Methods in Engineering*, 20, 1085-1109

I. Babuska, A. Miller. (1984b). Post-processing approach in the finite element method — Part 2: The calculation of stress intensity factors. *International Journal for Numerical Methods in Engineering*, 20, 1111-1129

I. Babuska, A. Miller. (1984c). Post-processing approach in the finite element method — Part 3: A posteriori error estimates and adaptive mesh selection. *International Journal for Numerical Methods in Engineering*, 20, 2311-2324

I. Babuska, W.C. Rheinboldt. (1978). *A-posteriori* error estimates for the finite element method. *International Journal for Numerical Methods in Engineering*, 12, 1597-1615

K. Balakrishnan, P.A. Ramachandran. (2001). Osculatory interpolation in the method of fundamental solution for nonlinear Poisson problems. *Journal of computational physics*, 172, 1-18

D. Ballhause, T. Wallmersperger. (2008). Coupled chemo-electro-mechanical finite element simulation of hydrogels: I. Chemical stimulation. *Smart Materials and Structures*, 17

D.S. Balsara. (1995). Von Neumann stability analysis of smoothed particle hydrodynamics — Suggestions for optimal algorithms. *Journal of Computational Physics*, 121, 357-372

R.E. Bank, A. Weiser. (1985). Some a posteriori error estimators for elliptic partial differential equations. *Mathematics of Computation*, 44, 283-301

R. Bellman, B.G. Kashef, J. Casti. (1972). Differential quadrature: A technique for the rapid solution of nonlinear partial differential equations. *Journal of Computational Physics*, 10, 40-52

T. Belytschko, L. Gu, Y.Y. Lu. (1994a). Fracture and crack growth by element free Galerkin methods. *Modelling and Simulation in Materials Science and Engineering*, 2, 519-534

T. Belytschko, Y. Guo, W.K. Liu, S.P. Xiao. (2000). A unified stability analysis of meshless particle methods. *International Journal for Numerical Methods in Engineering*, 48, 1359-1400

T. Belytschko, Y. Krongauz, D. Organ, M. Fleming, P. Krysl. (1996). Meshless methods: An overview and recent developments. *Computer Methods in Applied Mechanics and Engineering*, 139, 3-47

T. Belytschko, Y.Y. Lu, L. Gu. (1994b). Element-free Galerkin methods. *International Journal for Numerical Methods in Engineering*, 37, 229 - 256

T. Belytschko, Y.Y. Lu, L. Gu. (1995a). Crack propagation by element-free Galerkin methods. *Engineering Fracture Mechanics*, 51, 295-315

T. Belytschko, Y.Y. Lu, L. Gu, M. Tabbara. (1995b). Element-free Galerkin methods for static and dynamic fracture. *International Journal of Solids and Structures*, 32, 2547-2570

E. Birgersson, H. Li, S. Wu. (2008). Transient analysis of temperature-sensitive neutral hydrogels. *Journal of the Mechanics and Physics of Solids*, 56, 444-466

H. Blum, A. Rademacher, A. Schröder. (2009). Space adaptive finite element methods for dynamic Signorini problems. *Computational Mechanics*, 44, 481-491

J. Bonet, T.S.L. Lok. (1999). Variational and momentum preservation aspects of Smooth Particle Hydrodynamic formulations. *Computer Methods in Applied Mechanics and Engineering*, 180, 97-115

S. Børve, M. Omang, J. Trulsen. (2004). Two-dimensional MHD smoothed particle hydrodynamics stability analysis. *Astrophysical Journal, Supplement Series*, 153, 447-462

J. Braun, M. Sambridge. (1995). A numerical method for solving partial differential equations on highly irregular evolving grids. *Nature*, 376, 655-660

D. Brock, W. Lee. (1994). Dynamic model of a linear actuator based on polymer hydrogel. *Journal of Intelligent Material Systems and Structures*, 5, 764-771

Y.C. Cai, H.H. Zhu. (2008). A local meshless Shepard and least square interpolation method based on local weak form. *CMES - Computer Modeling in Engineering and Sciences*, 34, 179-204

C. Carstensen. (2009). Convergence of adaptive finite element methods in computational mechanics. *Applied Numerical Mathematics*, 59, 2119-2130

L. Champaney, P.A. Boucard, S. Guinard. (2008). Adaptive multi-analysis strategy for contact problems with friction: Application to aerospace bolted joints. *Computational Mechanics*, 42, 305-315

D.D.J. Chandar, M. Damodaran. (2009). Computational fluid-structure interaction of a flapping wing in free flight using overlapping grids, *27th AIAA Applied Aerodynamics Conference*, San Antonio, TX

A.N. Chatterjee, Q. Yu, J.S. Moore, N.R. Aluru. (2003). Mathematical modeling and simulation of dissolvable hydrogels. *Journal of Aerospace Engineering*, 16, 55-64

J. Chen, H. Li, K.Y. Lam. (2005). Transient simulation for kinetic responsive behaviors of electric-sensitive hydrogels subject to applied electric field. *Materials Science and Engineering C*, 25, 710-712

S.S. Chen, Y.H. Liu, Z.Z. Cen. (2008). A combined approach of the MLPG method and nonlinear programming for lower-bound limit analysis. *CMES - Computer Modeling in Engineering and Sciences*, 28, 39-55

J.Y. Cho, H.G. Kim, S.N. Atluri. (2000). Analysis of shear flexible beams, using the meshless local Petrov–Galerkin method based on locking-free formulation. *Proceedings of Advances in Computational Engineering and Science*, Los Angeles, 1404-1409

U. Christoph. (1997). *Numerical computation: Methods, software, and analysis*, Springer-Verlag, Germany

G. Dahlquist, A. Björck (2003). 4.3.4. Equidistant Interpolation and the Runge Phenomenon In: Anderson N (ed.) *Numerical methods*, Dover Publications, New York, pp 101-103

K.Y. Dai, G.R. Liu, K.M. Lim, Y.T. Gu. (2003). Comparison between the radial point interpolation and the Kriging interpolation used in meshfree methods. *Computational Mechanics*, 32, 60-70

T.D. Dane, B.V. Sankar. (2008). Meshless local Petrov–Galerkin micromechanical analysis of periodic composites including shear loadings. *CMES - Computer Modeling in Engineering and Sciences*, 26, 169-187

S.K. De, N.R. Aluru. (2004). A chemo-electro-mechanical mathematical model for simulation of pH sensitive hydrogels. *Mechanics of Materials*, 36, 395-410

K.S. De, N.R. Aluru, B. Johnson, W.C. Crone, D.J. Beebe, J. Moore. (2002). Equilibrium swelling and kinetics of pH-responsive hydrogels: Models, experiments, and simulations. *Journal of Microelectromechanical Systems*, 11, 544-555

S.K. De, R.R. Ohs, N.R. Aluru (2001). Modeling of hydrogel swelling in buffered solutions In: Bar-Cohen Y. (ed.) (*Electroactive polymer, actuators and devices-smart structures and materials 2001-*), Newport Beach, CA, pp 285-291

L. Demkowicz, J.T. Oden, T. Strouboulis. (1984). Adaptive finite elements for flow problems with moving boundaries. Part I: Variational principles and a posteriori estimates. *Computer Methods in Applied Mechanics and Engineering*, 46, 217-251

R. Di Lisio, E. Grenier, M. Pulvirenti. (1998). The convergence of the SPH method. *Computers and Mathematics with Applications*, 35, 95-102

L. Diening, C. Kreuzer. (2008). Linear convergence of an adaptive finite element method for the p-Laplacian equation. *SIAM Journal on Numerical Analysis*, 46, 614-638

H. Ding, C. Shu, K.S. Yeo, D. Xu. (2006). Numerical computation of three-dimensional incompressible viscous flows in the primitive variable form by local multiquadric differential quadrature method. *Computer Methods in Applied Mechanics and Engineering*, 195, 516-533

C.A. Duarte, J.T. Oden. (1996). An h-p adaptive method using clouds. *Computer Methods in Applied Mechanics and Engineering*, 139, 237-262

P.J. Flory, J. Rehner Jr. (1943a). Statistical mechanics of cross-linked polymer networks I. Rubberlike elasticity. *The Journal of Chemical Physics*, 11, 512-520

P.J. Flory, J. Rehner Jr. (1943b). Statistical mechanics of cross-linked polymer networks II. Swelling. *The Journal of Chemical Physics*, 11, 521-526

T.C. Fung. (2002). Stability and accuracy of differential quadrature method in solving dynamic problems. *Computer Methods in Applied Mechanics and Engineering*, 191, 1311-1331

L. Gavete, S. Falcón, A. Ruiz. (2001). An error indicator for the element free Galerkin method. *European Journal of Mechanics, A/Solids*, 20, 327-341

D.F. Gilhooley, J.R. Xiao, R.C. Batra, M.A. McCarthy, J.W. Gillespie Jr. (2008). Two-dimensional stress analysis of functionally graded solids using the MLPG method with radial basis functions. *Computational Materials Science*, 41, 467-481

R.A. Gingold, J.J. Monaghan. (1977). Smoothed particle hydrodynamics: Theory and application to non-spherical stars. *Monthly Notices of the Royal Astronomical Society* 181, 375-389

M.A. Golberg, C.S. Chen, H. Bowman. (1999). Some recent results and proposals for the use of radial basis function in the BEM. *Engineering Analysis with Boundary Elements*, 23, 285-296

L. Gu. (2003). Moving kriging interpolation and element-free Galerkin method. *International Journal for Numerical Methods in Engineering*, 56, 1–11

Y.T. Gu, G.R. Liu. (2001). A coupled element free Galerkin/Boundary element method for stress analysis of two-dimensional solids. *Computer Methods in Applied Mechanics and Engineering*, 190, 4405-4419

Y.T. Gu, G.R. Liu. (2003). A boundary radial point interpolation method (BRPIM) for 2-D structural analyses. *Structural Engineering and Mechanics*, 15(5), 535-550

R.W. Guelch, J. Holdenried, A. Weible, T. Wallmersperger, B. Kroeplin. (2000). Polyelectrolyte gels in electric fields: A theoretical and experimental approach. *Proceedings of SPIE – The International Society for Optical Engineering*, 3987, 193-202

W. Han, X. Meng. (2001). Error analysis of the reproducing kernel particle method. *Computer Methods in Applied Mechanics and Engineering*, 190, 6157-6181

Z.D. Han, H.T. Liu, A.M. Rajendran, S.N. Atluri. (2006). The applications of Meshless Local Petrov–Galerkin (MLPG) approaches in high-speed impact, penetration and perforation problems. *CMES - Computer Modeling in Engineering and Sciences*, 14, 119-128

Z.D. Han, A.M. Rajendran, S.N. Atluri. (2005). Meshless Local Petrov–Galerkin (MLPG) approaches for solving nonlinear problems with large deformations and rotations. *CMES: Computer Modeling in Engineering and Sciences*, 10, 1-12

D. Hegen. (1996). Element-free Galerkin methods in combination with finite element approaches. *Computer Methods in Applied Mechanics and Engineering*, 135, 143-166.

Y.C. Hon, M.W. Lu, W.M. Xue, X. Zhou. (1999). New formulation and computation of the triphasic model for mechano-electrochemical mixtures. *Computational Mechanics*, 24, 155-165

W. Hong, X. Zhao, J. Zhou, Z. Suo. (2008). A theory of coupled diffusion and large deformation in polymeric gels. *Journal of the Mechanics and Physics of Solids*, 56, 1779-1793

E.S. Hung, S.D. Senturia. (1999). Generating efficient dynamical models for micro-electromechanical systems from a few finite-element simulation runs. *IEEE Journal of Microelectromechanical Systems*, 8(3), 280-289.

S.R. Idelsohn, E. Onate, N. Calvo, F. Del Pin. (2003). The meshless finite element method. *International Journal for Numerical Methods in Engineering*, 58, 893-912

A.L. Iordanskii, M.M. Feldstein, V.S. Markin, J. Hadgraft, N.A. Plate. (2000). Modeling of the drug delivery from a hydrophilic transdermal therapeutic system across polymer membrane. *European Journal of Pharmaceutics and Biopharmaceutics*, 49, 287-293

T. Jarak, J. Sorić, J. Hoster. (2007). Analysis of shell deformation responses by the Meshless Local Petrov–Galerkin (MLPG) approach. *CMES - Computer Modeling in Engineering and Sciences*, 18, 235-246

E.T. Jaynes (1957). Information theory and statistical mechanics. *Physical Review*, 106 (4), 620-630

X. Jin, G. Li, N.R. Aluru. (2001). On the equivalence between least-squares and kernel approximations in meshless methods. *CMES - Computer Modeling in Engineering and Sciences*, 2, 447-462

B. Kang, Y.d. Dai, X.h. Shen, D. Chen. (2008). Dynamical modeling and experimental evidence on the swelling/deswelling behaviors of pH sensitive hydrogels. *Materials Letters*, 62, 3444-3446

B.B.T. Kee, G.R. Liu, C. Lu. (2007). A regularized least-squares radial point collocation method (RLS-RPCM) for adaptive analysis. *Computational Mechanics*, 40, 837-853

Y. Kondratyuk, R. Stevenson. (2008). An optimal adaptive finite element method for the stokes problem. *SIAM Journal on Numerical Analysis*, 46, 747-775

D.G. Krige (1975) A review of the development of geostatistics in South Africa In: Guarascio M., David M.,Huijbregts C. (eds.) *Advanced geostatistcs in the mining industry*, D. Reidel Publishing Company, Holland

Y. Krongauz, T. Belytschko. (1996). Enforcement of essential boundary conditions in meshless approximations using finite elements. *Computer Methods in Applied Mechanics and Engineering*, 131, 133-145

M. Kvarnström, A. Westergård, N. Lorén, M. Nydén. (2009). Brownian dynamics simulations in hydrogels using an adaptive time-stepping algorithm. *Physical Review E - Statistical, Nonlinear, and Soft Matter Physics*, 79

W.M. Lai, J.S. Hou, V.C. Mow. (1991). A triphasic theory for the swelling and deformation behaviors of articular cartilage. *Journal of Biomechanical Engineering*, 113, 245-258

P. Lancaster, K. Salkauskas. (1986). *Curve and surface fitting: An introduction.* Academic Press

K.Y. Lam, H. Li, T.Y. Ng, R. Luo. (2006). Modeling and simulation of the deformation of multi-state hydrogels subjected to electrical stimuli. *Engineering Analysis with Boundary Elements*, 30, 1011-1017

K.Y. Lam, Q.X. Wang, H. Li. (2004) A novel meshless approach: Local Kriging (LoKriging) method with two-dimensional structural analysis. *Computational Mechanics*, 33, 235-244

K.Y. Lam, Q.X. Wang, Z. Zong. (2002). A nonlinear fluid-structure interaction analysis of a near-bed submarine pipeline in a current. *Journal of Fluids and Structures*, 16, 1177-1191

K. Lebedev, S. Mafé, P. Stroeve. (2006). Convection, diffusion and reaction in a surface-based biosensor: Modeling of cooperativity and binding site competition on the surface and in the hydrogel. *Journal of Colloid and Interface Science*, 296, 527-537

C.K. Lee, Y.Y. Shuai. (2007a). An automatic adaptive refinement procedure for the reproducing kernel particle method. Part I: Stress recovery and a posteriori error estimation. *Computational Mechanics*, 40, 399-413

C.K. Lee, Y.Y. Shuai. (2007b). An automatic adaptive refinement procedure for the reproducing kernel particle method. Part II: Adaptive refinement. *Computational Mechanics*, 40, 415-427

C.K. Lee, Y.Y. Shuai. (2007c). On adaptive refinement analysis for the coupled boundary element method: Reproducing kernel particle method. *International Journal of Computational Methods in Engineering Science and Mechanics*, 8, 263-272

H. Li. (2009). Kinetics of smart hydrogels responding to electric field: A transient deformation analysis. *International Journal of Solids and Structures*, 46, 1326-1333

H. Li, J. Chen, K.Y. Lam. (2004a). Multiphysical modeling and meshless simulation of electric-sensitive hydrogels. *Journal of Polymer Science, Part B: Polymer Physics*, 42, 1514-1531

H. Li, J.Q. Cheng, T.Y. Ng, J. Chen, K.Y. Lam. (2004b). A meshless Hermite-cloud method for nonlinear fluid-structure analysis of near-bed submarine pipelines under current. *Engineering Structures*, 26, 531-542

H. Li, R. Luo. (2009). Modeling and characterization of glucose-sensitive hydrogel: Effect of Young's modulus. *Biosensors and Bioelectronics*, 24, 3630-3636

H. Li, R. Luo, E. Birgersson, K.Y. Lam. (2009a). A chemo-electro-mechanical model for simulation of responsive deformation of glucose-sensitive hydrogels with the effect of enzyme catalysis. *Journal of the Mechanics and Physics of Solids*, 57, 369-382

H. Li, R. Luo, K.Y. Lam. (2007a). Modeling of ionic transport in electric-stimulus-responsive hydrogels. *Journal of Membrane Science*, 289, 284-296

H. Li, S.S. Mulay. (2011) 2D simulation of the deformation of pH-sensitive hydrogel by novel strong-form meshless random differential quadrature method, *Computational Mechanics*, 48, 729-753

H. Li, S.S. Mulay, S. See. (2009a). On the convergence of random differential quadrature (RDQ) method and its application in solving nonlinear differential equations in mechanics. *CMES - Computer Modeling in Engineering and Sciences*, 48, 43-82

H. Li, S.S. Mulay, S. See. (2009b). On the location of zeroes of polynomials from the stability analysis of novel strong-form meshless random differential quadrature method. *CMES - Computer Modeling in Engineering and Sciences*, 54, 147-199

H. Li, T.Y. Ng, J.Q. Cheng, K.Y. Lam. (2003). Hermite-Cloud: A novel true meshless method. *Computational Mechanics*, 33, 30-41

H. Li, T.Y. Ng, Y.K. Yew, K.Y. Lam. (2005a). Modeling and simulation of the swelling behavior of pH-stimulus-responsive hydrogels. *Biomacromolecules*, 6, 109-120

H. Li, T.Y. Ng, Y.K. Yew, K.Y. Lam. (2007b). Meshless modeling of pH-sensitive hydrogels subjected to coupled pH and electric field stimuli: Young modulus effects and case studies. *Macromolecular Chemistry and Physics*, 208, 1137–1146

H. Li, Q.X. Wang, K.Y. Lam. (2004c). Development of a novel meshless Local Kriging (LoKriging) method for structural dynamic analysis. *Computer Methods in Applied Mechanics and Engineering*, 193, 2599-2619

H. Li, Q.X. Wang, K.Y. Lam. (2004d). A variation of local point interpolation method (vLPIM) for analysis of microelectromechanical systems (MEMS) device. *Engineering Analysis with Boundary Elements*, 28, 1261-1270

H. Li, X. Wang, G. Yan, K.Y. Lam, S. Cheng, T. Zou, R. Zhuo. (2005b). A novel multiphysic model for simulation of swelling equilibrium of ionized thermal-stimulus responsive hydrogels. *Chemical Physics*, 309, 201-208

H. Li, Y.K. Yew. (2009). Simulation of soft smart hydrogels responsive to pH stimulus: Ionic strength effect and case studies. *Materials Science and Engineering C*, 29, 2261-2269

H. Li, Y.K. Yew, K.Y. Lam, T.Y. Ng. (2004e). Numerical simulation of pH-stimuli responsive hydrogel in buffer solutions. *Colloids and Surfaces A: Physicochemical and Engineering Aspects*, 249, 149-154

H. Li, Y.K. Yew, T.Y. Ng, K.Y. Lam. (2005c). Meshless steady-state analysis of chemo-electro-mechanical coupling behavior of pH-sensitive hydrogel in buffered solution. *Journal of Electroanalytical Chemistry*, 580, 161-172

H. Li, Z. Yuan, K.Y. Lam, H.P. Lee, J. Chen, J. Hanes, J. Fu. (2004f). Model development and numerical simulation of electric-stimulus-responsive hydrogels subject to an externally applied electric field. *Biosensors and Bioelectronics*, 19, 1097-1107

L. Li, S. Liu, H. Wang. (2008). Meshless analysis of ductile failure. *CMES - Computer Modeling in Engineering and Sciences*, 36, 173-191

S. Li, S.N. Atluri. (2008). Topology-optimization of structures based on the MLPG mixed collocation method. *CMES - Computer Modeling in Engineering and Sciences*, 26, 61-74

S.F. Li and W.K. Liu. (2002). Meshfree and particle methods and their application. *Applied Mechanics Reviews*, 55(1), 1-34

K.M. Liew, Y.Q. Huang, J.N. Reddy. (2002). A hybrid Moving Least Squares and Differential Quadrature (MLSDQ) meshfree method. *International Journal of Computational Engineering Science*, 3, 1-12

K.M. Liew, Y.Q. Huang, J.N. Reddy. (2003a). Moving least squares differential quadrature method and its application to the analysis of shear deformable plates. *International Journal for Numerical Methods in Engineering*, 56, 2331-2351

K.M. Liew, Y.Q. Huang, J.N. Reddy. (2004). Analysis of general shaped thin plates by the moving least-squares differential quadrature method. *Finite Elements in Analysis and Design*, 40, 1453-1474

K.M. Liew, J.Z. Zhang, T.Y. Ng, S.A. Meguid. (2003b). Three-dimensional modelling of elastic bonding in composite laminates using layerwise differential quadrature. *International Journal of Solids and Structures*, 40, 1745-1764

T. Liszka. (1984). An interpolation method for an irregular net of nodes. *International Journal for Numerical Methods in Engineering*, 20, 1599-1612

T. Liszka, J. Orkisz. (1980). The finite difference method at arbitrary irregular grids and its application in applied mechanics. *Computers and Structures*, 11, 83-95

T.J. Liszka, C.A.M. Duarte, W.W. Tworzydlo. (1996). hp-Meshless cloud method. *Computer Methods in Applied Mechanics and Engineering*, 139, 263-288

G.R. Liu. (2003). *Mesh free methods: Moving beyond the finite element method*, CRC Press

G.R. Liu, Y.T. Gu. (2001). A point interpolation method for two-dimensional solids. *International Journal for Numerical Methods in Engineering*, 50, 937-951

W.K. Liu, Y. Chen, R.A. Uras, C.T. Chang. (1996). Generalized multiple scale reproducing kernel particle methods. *Computer Methods in Applied Mechanics and Engineering*, 139, 91-157

W.K. Liu, S. Jun. (1998). Multiple-scale reproducing kernel particle methods for large deformation problems. *International Journal for Numerical Methods in Engineering*, 41, 1339-1362

W.K. Liu, S. Jun, Y.F. Zhang. (1995). Reproducing kernel particle methods. *International Journal for Numerical Methods in Fluids*, 20, 1081-1106

G.R. Liu, S.S. Quek. (2003). *The finite element method: A practical course*, Butterworth-Heinemann, Oxford

G.R. Liu, J. Zhang, K.Y. Lam, H. Li, G. Xu, Z.H. Zhong, G.Y. Li, X. Han. (2008b). A gradient smoothing method (GSM) with directional correction for solid mechanics problems. *Computational Mechanics*, 41, 457-472

G.R. Liu, J. Zhang, H. Li, K.Y. Lam, B.B.T. Kee. (2006). Radial point interpolation based finite difference method for mechanics problems. *International Journal for Numerical Methods in Engineering*, 68, 728-754

Y.H. Liu, S.S. Chen, J. Li, Z.Z. Cen. (2008a). A meshless local natural neighbour interpolation method applied to structural dynamic analysis. *CMES - Computer Modeling in Engineering and Sciences*, 31, 145-156

S.Y. Long, K.Y. Liu, G.Y. Li. (2008). An analysis for the elasto-plastic fracture problem by the meshless local Petrov–Galerkin method. *CMES - Computer Modeling in Engineering and Sciences*, 28, 203-216

L.B. Lucy. (1977). A numerical approach to testing the fission hypothesis. *Astronomical Journal*, 82, 1013-1024

R. Luo, H. Li. (2009). Simulation analysis of effect of ionic strength on physiochemical and mechanical characteristics of glucose-sensitive hydrogels. *Journal of Electroanalytical Chemistry*, 635, 83-92

R. Luo, H. Li, K.Y. Lam. (2007). Modeling and simulation of chemo-electro-mechanical behavior of pH-electric-sensitive hydrogel. *Analytical and Bioanalytical Chemistry*, 389, 863-873

R. Luo, H. Li, E. Birgersson, Y.L. Khin. (2008). Modeling of electric-stimulus-responsive hydrogels immersed in different bathing solutions. *Journal of Biomedical Materials Research - Part A*, 85, 248-257

S.E. Lyshevski. (2002). *MEMS and NEMS systems, devices, and structures*. CRC Press, Boca Raton, Florida

N.C. MacDonald, L.Y. Chen, J.J. Yao, J.A. Zhang, J.A. Mcmillan, D.C. Thomas. (1989). Selective chemical vapor deposition of tungsten for microelectromechanical structures. *Sensors and Actuators*, 20, 123-133

B.A. Mann, K. Kremer, C. Holm. (2006). The swelling behavior of charged hydrogels. *Macromolecular Symposia*, 237, 90-107

J.M. Melenk, I. Babuška. (1996). The partition of unity finite element method: Basic theory and applications. *Computer Methods in Applied Mechanics and Engineering*, 139, 289-314

J.J.H. Miller. (1971). On the location of zeros of certain classes of polynomials with applications to numerical analysis. *IMA J Appl Math*, 8, 397-406

Y.X. Mukherjee, S. Mukherjee. (1997). On boundary conditions in the element-free Galerkin method. *Computational Mechanics*, 19, 264-270

S.S. Mulay, H. Li. (2009). Analysis of microelectromechanical systems using the meshless random differential quadrature method, *International Conference on Materials for Advanced Technologies, ICMAT 2009*, Singpore

S.S. Mulay, H. Li. (2011). Influence of Young's modulus and geometrical shapes on the 2-D simulation of pH-sensitive hydrogels by the meshless random differential quadrature method, *Modelling and Simulation in Materials Science and Engineering*, 19, 065009

S.S. Mulay, H. Li, S. See, (2009). On the random differential quadrature (RDQ) method: Consistency analysis and application in elasticity problems. *Computational Mechanics*, 44, 563-590

S.S. Mulay, H. Li, S. See. (2010). On the development of adaptive random differential quadrature method with an error recovery technique and its application in the locally high gradient problems. *Computational Mechanics*, 45, 467-493

G. Naadimuthu, R. Bellman, K.M. Wang, E.S. Lee. (1984). Differential quadrature and partial differential equations: Some numerical results. *Journal of Mathematical Analysis and Applications*, 98, 220-235

B. Nayroles, G. Touzot, P. Villon. (1992). Generalizing the finite element method: Diffuse approximation and diffuse elements. *Computational Mechanics*, 10, 307-318

T.Y. Ng, H. Li, J.Q. Cheng, K.Y. Lam. (2003). A new hybrid meshless-differential order reduction (hM-DOR) method with applications to shape control of smart structures via distributed sensors/actuators. *Engineering Structures*, 25, 141-154

Y.F. Nie, S.N. Atluri, C.W. Zuo. (2006). The optimal radius of the support of radial weights used in moving least squares approximation. *CMES - Computer Modeling in Engineering and Sciences*, 12, 137-147

R.A. Olea. (1999). *Geostatistics for engineers and earth scientists*. Kluwer Academic Publishers, Boston

E. Oñate, S. Idelsohn, O.C. Zienkiewicz, R.L. Taylor. (1996a). A finite point method in computational mechanics. Applications to convective transport and fluid flow. *International Journal for Numerical Methods in Engineering*, 39, 3839-3866

E. Oñate, S. Idelsohn, O.C. Zienkiewicz, R.L. Taylor, C. Sacco. (1996b). A stabilized finite point method for analysis of fluid mechanics problems. *Computer Methods in Applied Mechanics and Engineering*, 139, 315-346

A. Ortiz, M.A. Puso, N. Sukumar (2010). Maximum-entropy meshfree method for compressible and near-incompressible elasticity. *Computer Methods in Applied Mechanics and Engineering*, 199, 1859-1871

A. Ortiz, M.A. Puso, N. Sukumar (2011). Maximum-entropy meshfree method for incompressible media problems. *Finite Elements in Analysis and Design*, 47, 572-585

P.M. Osterberg, S.D. Senturia. (1997). M-TEST: A test chip for MEMS material property measurement using electrostatically actuated test structures. *Journal of Microelectromechanical Systems*, 6(2), 107-118

J. Ostroha, M. Pong, A. Lowman, N. Dan. (2004). Controlling the collapse/swelling transition in charged hydrogels. *Biomaterials*, 25, 4345-4353

C.K. Park (2009). The development of a generalized meshfree approximation for solid and fracture analysis. Dissertation, George Washington University, Washington, DC

C.K. Park, C.T. Wu, C.D. Kan (2011). On the analysis of dispersion property and stable time step in meshfree method using the generalized meshfree approximation. *Finite Elements in Analysis and Design*, 47, 683-697

A. Phan, F. Trochu (1998). Application of dual kriging to structural shape optimization based on the boundary contour method. *Archive of Applied Mechanics*, 68, 539–551

G. Pini, A. Mazzia, F. Sartoretto. (2008). Accurate MLPG solution of 3D potential problems. *CMES - Computer Modeling in Engineering and Sciences*, 36, 43-63

E.P. Popov. (1990). *Engineering mechanics of solids*, Prentice Hall, Englewood Cliffs, NJ.

J.R. Quan, C.T. Chang. (1989a). New insights in solving distributed system equations by the quadrature method-I. Analysis. *Computers and Chemical Engineering*, 13, 779-788

J.R. Quan, C.T. Chang. (1989b). New insights in solving distributed system equations by the quadrature method-II. Numerical experiments. *Computers and Chemical Engineering*, 13, 1017-1024

J.N. Reddy. (1993). *An introduction to the finite element method*, McGraw-Hill Inc, New York

N. Roquet, P. Saramito. (2008). An adaptive finite element method for viscoplastic flows in a square pipe with stick-slip at the wall. *Journal of Non-Newtonian Fluid Mechanics*, 155, 101-115

M. Rüter, R. Stenberg. (2008). Error-controlled adaptive mixed finite element methods for second-order elliptic equations. *Computational Mechanics*, 42, 447-456

E. Samson, J. Marchand, J.L. Robert, J.P. Bournazel. (1999). Modelling ion diffusion mechanisms in porous media. *International Journal for Numerical Methods in Engineering*, 46, 2043-2060

S.A. Sarra. (2006). Chebyshev interpolation: An interactive tour. *Journal of Online Mathematics and Its Applications*, 6, 1-13

R. Schaback, H. Wendland. (2000). Characterization and construction of radial basis functions. In: *Multivariate approximation and applications*, Dyn N, Leviatan D, Levin D, Pinkus A(eds.), Cambridge University Press, Cambridge, UK

E. Scheinerman. (2006). *C++ for mathematicians: An introduction for students and professionals*, CRC Press, Boca Raton, FL

S.D. Senturia. (1998). CAD challenges for microsensors, microactuators, and microsystems. *Proc. IEEE*, 86, 1611-1626.

Y.Y. Shan, C. Shu, Z.L. Lu. (2008). Application of local MQ-DQ method to solve 3D incompressible viscous flows with curved boundary. *CMES - Computer Modeling in Engineering and Sciences*, 25, 99-113

C.E. Shannon (1948). A mathematical theory of communication. *The Bell Systems Technical Journal*, 27, 379-423

F. Shi, P. Ramesh, S. Mukherjee. (1995), Simulation methods for micro-electromechanical structures (MEMS) with application to a microtweezer. *Computers and Structures*, 56 (5), 769-783

C. Shu. (2000). *Differential quadrature and its application in engineering*, Springer-Verlag, London

C. Shu, B.C. Khoo, K.S. Yeo. (1994). Numerical solutions of incompressible Navier-Stokes equations by generalized differential quadrature. *Finite Elements in Analysis and Design*, 18, 83-97

L.D.G. Sigalotti, H. López. (2008). Adaptive kernel estimation and SPH tensile instability. *Computers and Mathematics with Applications*, 55, 23-50

J. Sladek, V. Sladek, C. Zhang, P. Solek. (2007a). Application of the MLPG to thermo-piezoelectricity. *CMES - Computer Modeling in Engineering and Sciences*, 22, 217-233

J. Sladek, V. Sladek, C. Zhang, P. Solek, L. Starek. (2007b). Fracture analyses in continuously nonhomogeneous piezoelectric solids by the MLPG. *CMES - Computer Modeling in Engineering and Sciences*, 19, 247-262

J. Sladek, V. Sladek, P. Solek, S.N. Atluri. (2008a). Modeling of intelligent material systems by the MLPG. *CMES - Computer Modeling in Engineering and Sciences*, 34, 273-300

J. Sladek, V. Sladek, P. Solek, P.H. Wen. (2008b). Thermal bending of Reissner-Mindlin plates by the MLPG. *CMES - Computer Modeling in Engineering and Sciences*, 28, 57-76

J. Sladek, V. Sladek, C.L. Tan, S.N. Atluri. (2008c). Analysis of transient heat conduction in 3D anisotropic functionally graded solids, by the MLPG method. *CMES - Computer Modeling in Engineering and Sciences*, 32, 161-174

D. Šnita, M. Pačes, J. Lindner, J. Kosek, M. Marek. (2001). Nonlinear behaviour of simple ionic systems in hydrogel in an electric field. *Faraday Discussions*, 120, 53-66

E. Stein. (2003). *Error-controlled adaptive finite elements in solid mechanics*, John Wiley & Sons Ltd, England

M.L. Stein. (1999). *Interpolation of spatial data: Some theory for Kriging*. Springer, New York

J. Strikwerda. (1989). *Finite difference schemes and partial differential equations*, Wadsworth and Brooks, California

N. Sukumar (2004). Construction of polygonal interpolants: A maximum entropy approach. *International Journal for Numerical Methods in Engineering*, 61, 2159-2181

N. Sukumar, B. Moran, T. Belytschko. (1998). The natural element method in solid mechanics. *International Journal for Numerical Methods in Engineering*, 43, 839-887

N. Sukumar, B. Moran, A.Y. Semenov, V.V. Belikov. (2001). Natural neighbour Galerkin methods. *International Journal for Numerical Methods in Engineering*, 50, 1-27

K.J. Suthar, M.K. Ghantasala, D.C. Mancini (2008). Simulation of hydrogel micro-actuation (*Microelectronics: Design, technology, and packaging III*), Canberra, ACT

J.W. Swegle, D.L. Hicks, S.W. Attaway. (1995). Smoothed particle hydrodynamics stability analysis. *Journal of Computational Physics*, 116, 123-134

S.P. Timoshenko, J.N. Goodier. (1970). *Theory of elasticity*, McGraw-Hill, New York

S. Tomasiello. (2003). Stability and accuracy of the iterative differential quadrature method. *International Journal for Numerical Methods in Engineering*, 58, 1277-1296

T. Traitel, J. Kost, S.A. Lapidot. (2003). Modeling ionic hydrogels swelling: Characterization of the non-steady state. *Biotechnology and Bioengineering*, 84, 20-28

F. Trochu. (1993). A contouring program based on dual Kriging interpolation. *Engineering with Computers*, 9, 160-177

T. Wallmersperger, D. Ballhause. (2008). Coupled chemo-electro-mechanical finite element simulation of hydrogels: II. Electrical stimulation. *Smart Materials and Structures*, 17, 045012

T. Wallmersperger, D. Ballhause, B. Kröplin, M. Günther, Z. Shi, G. Gerlach (2008). Coupled chemo-electro-mechanical simulation of polyelectrolyte gels as actuators and sensors (Electroactive Polymer Actuators and Devices (EAPAD) 2008), San Diego, CA

T. Wallmersperger, D. Ballhause, B. Kraplin, M. Nther, G. Gerlach. (2009). Coupled multi-field formulation in space and time for the simulation of intelligent hydrogels. *Journal of Intelligent Material Systems and Structures*, 20, 1483-1492

T. Wallmersperger, B. Kröplin, J. Holdenried, R.W. Gülch (2001). A coupled multi-field-formulation for ionic polymer gels in electric fields In: Bar-Cohen Y. (ed.) (Electroactive polymer, actuators and devices: Smart structures and materials 2001-), Newport Beach, CA, pp 264-275

Q.X. Wang, H. Li, K.Y. Lam. (2005). Development of a new meshless-point weighted least-squares (PWLS) method for computational mechanics. *Computational mechanics*, 35, 170-181

Q.X. Wang, H. Li, K.Y. Lam. (2006). Meshless simulation of equilibrium swelling/deswelling of PH-sensitive hydrogels. *Journal of Polymer Science, Part B: Polymer Physics*, 44, 326-337

Q.X. Wang, H. Li, K.Y. Lam. (2007). Analysis of microelectromechanical systems (MEMS) devices by the meshless point weighted least-squares method. *Computational Mechanics*, 40, 1-11

S. Wong, Y. Shie. (2008). Large deformation analysis with Galerkin-based smoothed particle hydrodynamics. *CMES - Computer Modeling in Engineering and Sciences*, 36, 97-117

C.P. Wu, K.H. Chiu, Y.M. Wang. (2008). A differential reproducing kernel particle method for the analysis of multilayered elastic and piezoelectric plates. *CMES - Computer Modeling in Engineering and Sciences*, 27, 163-186

C.T. Wu, C.K. Park, J.S. Chen (2011). A generalized meshfree approximation for the meshfree analysis of solids. *International Journal for Numerical Methods in Engineering*, 85, 693–722

S. Wu, H. Li, J.P. Chen, K.Y. Lam. (2004). Modeling investigation of hydrogel volume transition. *Macromolecular Theory and Simulations*, 13, 13-29

X. Wu, S. Shen, W. Tao. (2007). Meshless local Petrov–Galerkin collocation method for two-dimensional heat conduction problems. *CMES - Computer Modeling in Engineering and Sciences*, 22, 65-76

Y. Wu, S. Joseph, N.R. Aluru. (2009). Effect of cross-linking on the diffusion of water, ions, and small molecules in hydrogels. *Journal of Physical Chemistry B*, 113, 3512-3520

Z. Xu. (1992). *Applied elasticity* (Eng), New Age International (p) Ltd, India

Y.K. Yew, T.Y. Ng, H. Li, K.Y. Lam. (2007). Analysis of pH and electrically controlled swelling of hydrogel-based micro-sensors/actuators. *Biomedical microdevices*, 9, 487-499

Y.Q. Yu, Z.Z. Li, H.J. Tian, S.S. Zhang, P.K. Ouyang. (2007). Synthesis and characterization of thermoresponsive hydrogels cross-linked with acryloyloxyethylaminopolysuccinimide. *Colloid and Polymer Science*, 285, 1553-1560

Y.G. Yu, Y. Xu, H. Ning, S.S. Zhang. (2008). Swelling behaviors of thermoresponsive hydrogels cross-linked with acryloyloxyethylaminopolysuccinimide. *Colloid and Polymer Science*, 286, 1165-1171

Z. Yuan, L. Yin, H. Jiang (2007). Numerical simulation of transient nonlinear behaviors of electric-sensitive hydrogel membrane under an external electric field (Microfluidics, BioMEMS, and Medical Microsystems V), San Jose, CA

J. Zhang, G.R. Liu, K.Y. Lam, H. Li, G. Xu. (2008). A gradient smoothing method (GSM) based on strong form governing equation for adaptive analysis of solid mechanics problems. *Finite Elem Anal Des*, 44, 889-909

J. Zhang, X. Zhao, Z. Suo, H. Jiang. (2009). A finite element method for transient analysis of concurrent large deformation and mass transport in gels. *Journal of Applied Physics*, 105

X. Zhang, X.H. Liu, K.Z. Song and M.W. Lu. (2001). Least-squares collocation meshless method. *International Journal for Numerical Methods in Engineering*, 51, 1089-1100.

X. Zhao, Z. Suo. (2009). Electromechanical instability in semicrystalline polymers. *Applied Physics Letters*, 95

J. Zheng, S. Long, Y. Xiong, G. Li. (2009). A finite volume meshless local petrov–galerkin method for topology optimization design of the continuum structures. *CMES - Computer Modeling in Engineering and Sciences*, 42, 19-34

J. Zhou, W. Hong, X. Zhao, Z. Zhang, Z. Suo. (2008). Propagation of instability in dielectric elastomers. *International Journal of Solids and Structures*, 45, 3739-3750

X. Zhou, Y.C. Hon, S. Sun, A.F.T. Mak. (2002). Numerical simulation of the steady-state deformation of a smart hydrogel under an external electric field. *Smart Materials and Structures*, 11, 459-467

T. Zhu, J.D. Zhang, S.N. Atluri. (1998). A local boundary integral equation (LBIE) method in computational mechanics and a meshless discretisation approach *Computational Mechanics*, 21, 223–235

O.C. Zienkiewicz, J.Z. Zhu. (1987). Simple error estimator and adaptive procedure for practical engineering analysis. *International Journal for Numerical Methods in Engineering*, 24, 337-357

O.C. Zienkiewicz, R.L. Taylor, J.Z. Zhu. (2005). *The finite element method: Its basis and fundamentals*, Elsevier Butterworth-Heinemann

O.C. Zienkiewicz, J.Z. Zhu. (1992a). Superconvergent patch recovery and a posteriori error estimates. Part 1: The recovery technique. *International Journal for Numerical Methods in Engineering*, 33, 1331-1364

O.C. Zienkiewicz, J.Z. Zhu. (1992b). Superconvergent patch recovery and a posteriori error estimates. Part 2: Error estimates and adaptivity. *International Journal for Numerical Methods in Engineering*, 33, 1365-1382

Index

Milton Keynes UK
Ingram Content Group UK Ltd.
UKHW031139141024
449569UK00024B/1220